Beef Cattle Production:
An Integrated Approach

Beef Cattle Production:

An Integrated Approach

Verl M. Thomas, Ph.D.

Animal and Range Science Division
Montana State University
Bozeman, Montana

Lea & Febiger   *Philadelphia   1986*

Lea & Febiger
600 Washington Square
Philadelphia, PA 19106-4198
U.S.A.
(215) 922-1330

Library of Congress Cataloging-in-Publication Data

Thomas, Verl M.
    Beef cattle production.

    Bibliography: p.
    Includes index.
    1. Beef cattle.   2. Beef cattle—United States.
I. Title.
SF207.T48   1986   636.2'13   86-144
ISBN 0-8121-1024-2

PRINTED IN THE UNITED STATES OF AMERICA

Print No.   4 3 2 1

This book is dedicated to my parents, Earl and Vera Thomas, who instilled in me, at a very young age, the importance of education, and to my wife, Christine, for her love, support, and understanding during the writing of this book.

Preface

Beef Cattle Production: An Integrated Approach was written as a text to be used in teaching undergraduate courses in beef cattle production. Only information pertinent to such a course has been included. Whether the text is used in an introductory or advanced course depends on the university, the instructor, course objectives, and the background of the students. It became apparent to me, while teaching beef production, that the majority of students did not have extensive beef cattle experience prior to entering college, and that some, even though they may have had the necessary prerequisite courses, needed a refresher in some disciplines related to beef production. Therefore, I have included some general information on nutrition, reproduction, and genetics.

This book is organized so that an instructor can design a course around the biologic cycle of the cow. The first part of the text deals primarily with the management of the cow herd, followed by management of the pre-weaned and post-weaned calf. A chapter on business management has been included. The business of raising cattle is becoming more complex, and today's students of beef cattle production must understand and appreciate the importance of a good business background if they wish to be successful in the beef cattle industry.

The author is indebted to the Cooperative Extension Service for compiling excellent information on beef cattle production over the years, which was used extensively in the preparation of the text.

I would also like to acknowledge those who reviewed and offered constructive criticism of the following chapters.

Chapter 1—
D. Craig Anderson, Livestock Management and Consulting Service, Salem, OR.
M.J. McInerney, Animal and Range Sciences, Montana State University.
Chapter 3—
D.D. Kress, Animal and Range Sciences, Montana State University.
Chapter 5—
A.C. Linton, Animal and Range Sciences, Montana State University.
R.R. Loucks, Cooperative Extension Service, University of Idaho, Salmon, ID.
Chapter 6—
R.G. Sasser, Animal Sciences Department, University of Idaho, Salmon, ID.
D.G. Falk, Animal Sciences Department, University of Idaho, Salmon, ID.
Chapter 8—
R.C. Bull, Animal Sciences Department, University of Idaho, Salmon, ID.
Chapter 9—
Norm Gates, College of Veterinary Medicine, Washington State University.
Peter J. South, Veterinary Medicine Department, University of Idaho.
Chapter 11—
D.A. Yates, Elanco Products Company, Indianapolis, Indiana.
R.M. Brownson, Cooperative Extension Service, Montana State University.
Chapter 12—
C.B. Marlow, Animal and Range Sciences, Montana State University.
Chapter 13—
K.S. Hendrix, Department of Animal Science, Purdue University.
Chapter 14—
E.T. Evans, Livestock Producer, Boise, ID.
Gilbert Ball, Jr., Livestock Producer, Weiser, ID.

I am indebted to Mrs. Beverly Hawk and Rolinda Coffey for typing the rough and final drafts of the manuscript. Their patience and expertise were very much appreciated.

Finally, a special thanks to Christian C.F. Spahr, Jr., and others at Lea & Febiger for their cooperation and support.

Bozeman, Montana   Verl M. Thomas

Contents

Chapter One

Beef Production—U.S.A.

THE INDUSTRY

The beef cattle industry is the largest segment of agriculture in the United States and accounts for 23% of all farm marketings (NCA, 1982b). Over the last 25 years, cattle numbers have increased in a cyclical manner by over one million per year (Table 1–1). Per capita consumption of meat has grown at an annual rate of 1.1%. Growth in average per capita consumption of protein food appears to be over, however (Table 1–2). Consequently, choices among animal products, rather than additional consumption, become the primary consideration of the consumer. For beef, future industry growth or maintenance of market share will depend on:

1. Domestic population growth (estimated at 0.8% per year to 1990)
2. Increased exports
3. New forms and kinds of beef products
4. Improvement of beef's competitive position in terms of relative prices

Beef cattle are harvesters of forage, with approximately 83% of all cattle feed provided by forage. Only 15 to 20% of all feed for beef cattle is from concentrate feed resources (Byerly, 1975). No other factory, with the exception of other ruminants, can produce so much from so little. Much of the land used by cattle is not suitable for other use. Cattle can harvest the forage on these lands, which is inedible, indigestible and undesirable to man, and convert it into a high-quality protein food.

The beef cattle industry must take advantage of the cow's unique ability to utilize forage crop residues, feed by-products, and non-protein nitrogen to improve its production and distribution efficiency and to provide the consumer with a variety of conveniently packaged and competitively priced products (Koch and Algeo, 1983). Beef cattle producers must adopt new technology to improve production, as well as economic efficiency, at a much faster rate than in the past.

Beef is a valuable source of balanced protein and other essential nutrients. A 3-oz serving of lean, roasted beef supplies only 8% of one's daily caloric needs (in a 2000-calorie diet), but 57% of the recommended daily allowance of protein, 34% of vitamin B_{12}, 32% of zinc, 18% of niacin, and 12% of iron.

Beef does not contain appreciably more cholesterol than other animal proteins (NCA, 1982a). Contrary to the views of many in the medical profession, comparative data from other food sources, shown in Table 1–3, illustrate that beef consumption and high cholesterol levels do not go hand in hand. The cholesterol scare and its relationship to heart disease has caused individuals to reduce their consumption of animal protein. In most persons, however, dietary cholesterol and blood cholesterol are not closely related. In fact, in the majority of people, heredity, weight, and the amount of fatty tissue in the body are more important than diet, including dietary cho-

1

TABLE 1-1. Trends in Cattle Inventory, Beef Production and Per Capita Consumption

Year	Cattle Numbers (million head)		Beef Production (billion lb)	Per Capita Consumption, Retail Wt (lb)
	Total Cattle	Cows		
1955	96.6	49.1	12.12	63.8
6	95.9	48.3	14.08	66.0
7	92.9	46.9	13.86	64.9
8	91.2	45.4	12.98	61.4
9	93.3	45.2	13.18	61.6
1960	96.2	45.9	14.37	64.0
1	97.5	46.6	14.87	65.1
2	100.0	47.7	14.87	66.0
3	103.8	49.0	15.97	69.7
4	107.9	50.4	17.95	73.7
1965	109.0	48.3	18.26	74.8
6	108.9	48.0	19.47	77.0
7	108.8	47.5	19.95	78.5
8	109.4	47.7	20.66	80.9
9	110.0	48.0	20.97	81.8
1970	112.4	48.8	21.45	83.8
1	114.5	49.8	21.45	83.2
2	117.9	50.6	22.15	85.4
3	121.5	52.5	21.05	80.3
4	127.7	54.3	22.75	85.4
1975	131.8	56.7	23.65	87.8
6	128.0	55.0	25.65	94.2
7	122.8	52.4	24.95	91.5
8	116.4	49.6	23.96	87.1
9	110.9	47.9	21.25	77.9
1980	111.2	47.9	21.45	76.3
1	114.3	49.6	22.04	77.9
2	115.7	50.4	22.15	77.9
3	115.2	50.3	23.10	78.7
4	114.0	48.8	23.10	77.5

From USDA. 1983. Livestock and Meat Statistics, United States Dept. of Agriculture. Washington, DC.

lesterol, in determining blood cholesterol level.

Many important by-products, edible and inedible, are also produced from cattle (Fig. 1–1). These include pharmaceuticals (insulin), inedible by-products (leather), edible by-products (gelatin), and variety meats (liver). Often, consumers do not realize that these items are cattle by-products.

EFFICIENCY OF PRODUCTION

Production and distribution efficiency must be improved in the future for the beef cattle industry to remain competitive with industries of other meat animals. According to the National Cattleman's Associa-

tion (NCA, 1982b), the present supply of beef cannot be produced and sold profitably in competition with less expensive competing products. The retail price of chicken dropped from 80% of beef prices in the 1950s to 30% in the 1980s with a resulting shift in per capita consumption of poultry (Table 1–2).

The average beef cow produces 0.7 progeny per year, which in terms of slaughter weight is about 70% of her body weight. The average sow produces 14 progeny per year, and those pigs represent a total market weight that is eight times the sow's body weight. The average meat-type hen produces 150 progeny per year and 100 times her body weight. Beef cattle are not

TABLE 1-2. Per Capita Consumption of Animal Protein Foods (lb)

	1960	1965	1970	1975	1976	1980	1981[a]
Beef	64.0	73.5	83.8	87.8	94.2	76.3	77.9
Pork	60.3	54.6	61.8	50.6	53.7	68.2	64.9
Veal and lamb	9.5	7.5	5.3	5.3	4.8	2.9	3.1
Poultry	34.3	40.9	48.6	48.8	52.1	60.7	61.8
Total	168.1	176.5	199.5	192.5	204.8	208.1	207.7
Fish	13.6	14.1	15.8	16.3	17.2	16.7	16.9
Cheese	8.4	9.7	11.4	14.3	15.6	17.6	18.0
Eggs	43.8	40.9	40.3	36.1	35.2	35.4	35.0
Total	65.8	64.7	67.5	66.7	68.0	69.7	69.9
Fluid milk[b]	184.6	181.7	173.6	166.5	164.6	156.6	156.6
Total animal protein	418.5	422.9	440.6	425.7	437.4	434.4	434.2

[a] Estimated.
[b] Weight reduced 30% to render milk (about 87% water) comparable in moisture content to retail meat cuts (about 60% water).
From NCA. 1982b. The future for beef. A report by the Special Advisory Committee. National Cattlemen's Association. Beef Business Bulletin, Englewood, CO. March 5.

strong competitors with species that have a higher reproductive rate. If the national calf crop average of 70 to 75% were improved, however, beef could become more competitive and profitable.

Because of their larger size, beef cattle require more feed for maintenance of breeding stock and for meat production than do other meat animals. A mature cow weighing 1000 lb requires approximately 4 tons of forage (88% hay dry matter equivalent) per year, and 40% of all feed consumed by beef cattle is used for maintenance of the cow herd.

Beef can be produced with little or no grain, and that is its competitive advantage. Cattle require an average of 2.3 lb of feed grain and oilseed meal protein to produce 1 lb of animal protein in the form of beef. This compares with 2.5 lb for broilers, 4.0 lb for turkeys, and 5.5 lb for swine. Only dairy cattle, sheep, and laying hens are more efficient users of protein from grain and oilseed meals. The remainder of the nutrient requirements for beef cattle are met by forage. High grain feedlot rations result in faster gains, which reduces labor, interest, and depreciation costs per pound of grain, resulting in lower cost per pound of beef produced for the consumer.

Through improved forage management, the cattle industry can increase the competitive edge that ruminant animals have over non-ruminants, thus helping to make the industry more competitive and profitable. Future beef cattle management systems must be more flexible to adapt to changing feed grain supplies and feed more or less cattle for shorter or longer time periods, according to changing economic conditions.

TABLE 1-3. Cholesterol Intake from Different Food Sources

Food Source	Cholesterol Consumption (mg)
Beef, roasted lean portion (3 oz)	72
Chicken, light meat without skin (3 oz)	72
Flounder, fried (3 oz)	65
Shrimp (3 oz)	130
Cheddar cheese (3 oz)	90
Pork, roasted lean portion (3 oz)	80

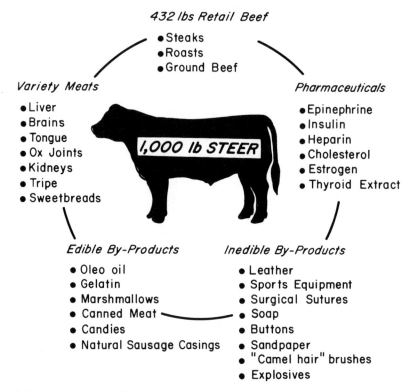

432 lbs Retail Beef
- Steaks
- Roasts
- Ground Beef

Variety Meats
- Liver
- Brains
- Tongue
- Ox Joints
- Kidneys
- Tripe
- Sweetbreads

1,000 lb STEER

Pharmaceuticals
- Epinephrine
- Insulin
- Heparin
- Cholesterol
- Estrogen
- Thyroid Extract

Edible By-Products
- Oleo oil
- Gelatin
- Marshmallows
- Canned Meat
- Candies
- Natural Sausage Casings

Inedible By-Products
- Leather
- Sports Equipment
- Surgical Sutures
- Soap
- Buttons
- Sandpaper
- "Camel hair" brushes
- Explosives

FIG. 1-1.　Good things from cattle. (Courtesy of the National Livestock and Meat Board, Chicago, IL.)

IMPROVING THE EFFICIENCY OF BEEF PRODUCTION

Beef cattle producers must adopt known technology to be competitive in the future. Management tools such as crossbreeding, artificial insemination, growth-promoting agents, selection of performance-tested bulls, computerized recordkeeping, and embryo transfer must be used to improve the efficiency of production.

Beef will be more competitive with other animal protein sources when it becomes unnecessary to trim as much as 30% of a carcass as fat. Leaner carcasses require less processing than those with more fat. Changes in quality grading may once again be needed.

Rather than using feeding management to achieve the desired tenderness required by the consumer, technologic advancements such as electrical stimulation, and mechanical and enzymatic tenderization, are being and will continue to be used to tenderize meat. These reduce carcass processing time, and the subsequent energy savings means cheaper cost per pound of retail beef produced.

The development of new intermediate-value beef meat products could make the beef industry more competitive and profitable (NCA, 1982b). For example, if hamburger sold for $1.59 per pound at the retail level, and the rib and loin cuts (middle of the carcass meats) sold for $4.98 per pound, today's meat technology could restructure the cuts used for hamburger into an intermediate-value cut ($2.69 per pound). This would lower the pressure on the "middle" meats, which could then be sold for $3.69 per pound. Also, if mechanically deboned beef could be used to extend ground beef, ground beef could be sold at a more competitive price.

Numerous experiment stations have demonstrated that bulls less than 14 to 15 months of age gain weight significantly faster and have improved feed efficiency and leaner carcasses compared with steers. Traditionally, bull meat has been regarded as less tender and flavorful than steer meat. Bulls are less docile and present more management problems than steers; however, the growing demand by the consumer for lean beef and the use of electrical stimulation to improve meat tenderness are two factors that have changed the attitude toward bull meat. Production and feeding of young bulls may not work for all producers or feeders, but they offer an opportunity for improving efficiency in lean beef production. Grading, marketing, and management problems must be overcome before feeding bulls becomes a common management practice.

Electronic marketing allows cattle to be bought and sold without being transported great distances. Cattle remain on the farm or ranch until the sale has been completed, giving the seller more access to potential buyers. Cooperative marketing and various joint venturing between producers, feeders, and packers are expected to become more prevalent. Cooperative ventures allow smaller producers and feeders the purchasing and marketing benefits of scale and reduced costs, and shrink and extra commission, associated with our current segmented industry (Koch and Algeo, 1983).

Most cattle are marketed at a minimum of 24 months of age and may be owned by four or five owners during their lifetime. Hogs and chickens seldom change hands more than once and are part of an integrated, closely coordinated business complex. If cattle were fed to slaughter weight by 12 to 14 months of age, the cost of producing beef would be greatly reduced. Cattlemen in the future may also produce specific types of cattle for specific retail and institutional markets.

Table 1–4 summarizes areas that deserve

TABLE 1-4. Improving Efficiency of Beef Production

Dilute maintenance costs: Sell more weight per cow.

Retain ownership of calves to slaughter if economically feasible.

Reduce time on feed to minimum needed for acceptable palatability.

Reduce emphasis on marbling; stress lean growth in feedlot cattle.

Fine-tune the tradeoffs between lean growth and more energy to maintain fertility; lean growth and dystocia.

Adopt new technology in processing and merchandising beef.

Implement known management tools currently available (e.g., crossbreeding, bull feeding, growth-promoting agents).

Adapted from Ritchie, H.D. 1981. Future and direction of the beef cattle industry in the United States: Cow-calf production. Michigan State University. Beef Cattle Report AC-LC-8204. East Lansing, MI.

attention to improve the efficiency of beef production in the future.

DEVELOPMENT OF THE BEEF CATTLE INDUSTRY

Beef is one of the oldest foods known to man. Cattle were probably first domesticated during the New Stone Age, 8000 to 9000 years ago. The origin of today's cattle breeds can be traced to either or both of two species: *Bos indica* and *Bos taurus*. *Bos indica* cattle are represented by the humped cattle (Zebu) of India and Africa. They are characterized by a hump of fleshy tissue over the withers, a large dewlap, and resistance to heat and to certain diseases and parasites that affect *Bos taurus* cattle. *Bos taurus* cattle include the ancestors of European cattle common to the more temperate climates of the world. A few examples of *Bos taurus* cattle are Hereford, Shorthorn, Angus, Simmental, Charolais, and Limousin breeds. Interestingly, no cattle existed in the Western Hemisphere until they were introduced by European explorers. The only member of the bovine family native to

North America is the American bison (*Bos bison*).

The first beef cattle came to the North American continent (West Indies) with Christopher Columbus on his second voyage in 1493. These cattle were intended as work oxen for the colonists. The Spanish explorer Cortez in 1519 brought cattle from Spain to Mexico. Beginning about 1600, Christian missions were established along the Rio Grande River, and across the mountains to the Pacific Coast. The missionaries brought cattle, which were used for meat and milk. The cattle reproduced rapidly because of the abundant forage that was available. It is estimated that by 1833, the missions owned more than 400,000 cattle. These Spanish cattle were the progenitors of the Longhorn breed known today.

The first cattle established in the British colonies were shipped from England to Jamestown, Virginia in 1609. When a supporting ship returned the next year, however, the settlers and cattle had vanished. In 1611, Sir Thomas Dole brought in more cattle, and these cattle reproduced. The first shipment of cattle to the Plymouth colony was brought by Governor Edward Winslow in 1623. A prosperous cattle industry developed, and surplus cattle were driven north. Colonists valued cattle for their work, milk, butter, and hides, attaching little value to their meat.

The beef cattle industry did not change dramatically from early colonial times to the late eighteenth century. Cattle were expected to survive on the available forage, and no harvested winter feed was provided. There was limited interest in improving cattle. Near the end of the eighteenth century, however, cattle breeders in the United States became interested in acquiring purebred British stock. The first purebred stock brought to the United States consisted of a few head of the English Shorthorn breed in 1783. Through the first half of the nineteenth century, frequent but small importations of the British

breeds occurred. The popular breed was the Shorthorn, followed by the Hereford and later the Angus. The British breeds were used to upgrade native Eastern cattle, Florida scrub cattle (survivors of the Spanish cattle that came to Florida by way of the West Indies), and the Longhorns of Texas.

After the Civil War, the cattle industry expanded westward as the population of the eastern United States increased. More purebred cattle were imported from Britain, and several purebred herds were established to supply high-quality range bulls to commercial cattlemen, and foundation heifers and herd bulls to other purebred producers.

As the railroads expanded westward, "cow towns" sprang up as shipping points. Cattle were driven to the railheads over such famous trails as the Chisholm Trail to be shipped eastward for further feeding (Fig. 1–2). By the end of the nineteenth century, cattle territory was in the Great Plains and moving farther westward. The Western range had plenty of water and unlimited forage and land for grazing. Westward expansion progressed well until the severe winters of 1882, 1889, and 1890. In some areas, losses reportedly ranged from 50 to 90%, lessening slightly in 1890. Prior to the 1880s, cattle herds roamed the West at will. The ambition of each rancher was to increase his herd size without restriction; little thought was given to the welfare of the range. A cattle industry based upon no herding, no conservation, and no deferred grazing could not exist on the bunch grass range of the West. Out of such disaster, however, ranchers learned to respect the range. They learned to avoid overexpansion and overgrazing, as well as the importance of providing an adequate winter feed supply for their cattle.

The cattle industry became more of a business as it developed. The Western range continued to supply cattle that were fed in the Midwest or East. The farmland between the Western range and the Corn Belt became an area where cattle were

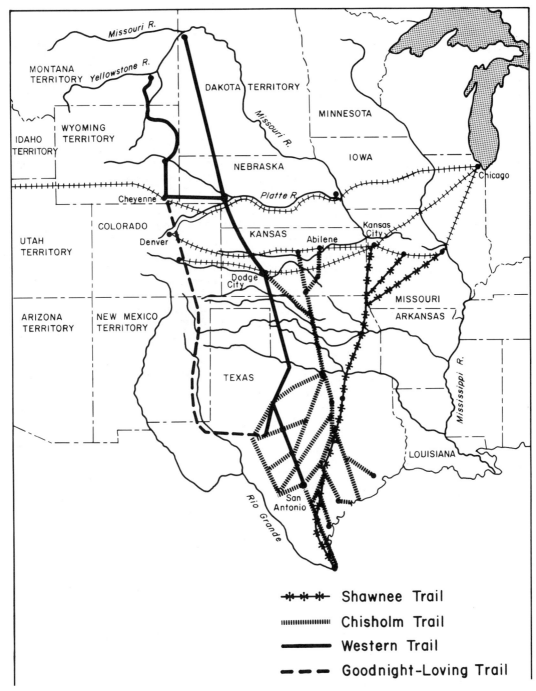

FIG. 1–2. Cattle drive trails of the United States. (Drawn by Kathy Dawes Graphics, Moscow, ID.)

grown out after leaving the range and before going to the feedlot. Large packing plants were built in cattle feeding areas.

The feedlot industry became more specialized. After World War II, Colorado, California, and Arizona became the predominant areas for cattle feeding. In the early 1960s, a geographic shift in the number of cattle on feed took place. The feeding of cattle shifted to eastern New Mexico and the Texas and Oklahoma panhandles. The nation's largest meat-packing companies built modern meat-packing establishments on the high plains, and cattle were slaughtered in the areas where they were fed. The movement of commercial cattle-feeding operations on the Great Plains now extends from Texas into Nebraska.

Historical Milestones

The beef industry is relatively young. The most important dates in the development of this industry in the United States are:

1493   Columbus brought cattle to the West Indies on his second trip.

1611   Sir Thomas Dole accomplished the first successful importation of cattle to the colonies.

1783   The English Shorthorn breed was imported to the United States from Britain.

1817   Henry Clay of Kentucky imported two Hereford bulls and two heifers from Britain.

1849   Zebu cattle were first imported to the United States.

1855   Michigan State University, the first college of agriculture, was established.

1860   Several cattle breed associations were established in the United States.

1862   Land-grant colleges were created by the Morrill Act.

1867   Abilene, Kansas was established as a rail shipping point by Mr. J.G. McCoy, a prominent Illinois stockman.

1868   Outbreak of cattle tick fever (Texas fever) occurred, causing strong prejudice against Texas cattle.

1873   George Grant of Victoria, Kansas imported three Aberdeen Angus bulls from Scotland.

1874   A patent for barbed wire was granted to Joseph F. Glidden, an Illinois farmer.

1884   Colonel Robert D. Hunter organized and promoted the National Convention of Cattlemen in St. Louis.

1887   Agriculture experiment stations were established by the Hatch Act.

1908   American Society of Animal Production was formed.

1914   Agricultural Extension Service was established by the Smith-Lever Act.

1918   Development of the Santa Gertrudis breed was begun by the King Ranch, Kingsville, Texas.

1936   Charolais cattle were imported from Mexico by the King Ranch, Kingsville, Texas.

1939   Artificial insemination of cattle became available commercially.

1940   First research on selection to improve cattle growth took place at the USDA-ARS Livestock and Range Research Station, Miles City, Montana.

1946   First bull test station was established in Texas.

1965   Yield grades were adopted for optional use with quality grades.

1968   Beef Improvement Federation was formed.

1973   First national sire evaluation report (beef) was established by American Simmental Association.

1978   First frozen-thawed embryo transplant calf was born in California.

BEEF PRODUCTION SYSTEMS

Three dominant cattle-raising systems in the United States are cow-calf, both commercial and purebred; stocker; and feedlot. The following sections describe each system.

Commercial Cow-Calf

In commercial cow-calf production, cows that are not registered in a breed association registry produce feeder calves. In this system, most calves are sold at weaning or shortly thereafter. This system was concentrated for years in areas that had an abundance of relatively inexpensive forage. The regions of the Southwest, the southern parts of the Great Plains, the West, and the Southeast also had advantages such as shorter and less severe winters and opportunities for yearlong grazing (Fig. 1–3).

Cow-calf systems have been modified as needed to improve the profit potential of herds. Although calves may be sold at weaning, many producers retain ownership of their calves and graze them on pas-

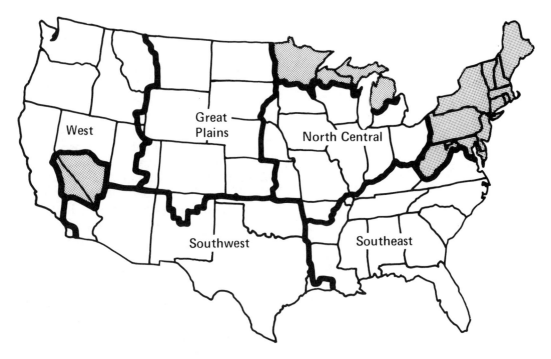

FIG. 1-3. Cattle-raising regions of the United States. (From Boykin, C.C., et al. 1980. Structural characteristics of beef cattle raising in the United States. USDA, Washington, DC. AER No. 450.)

tures and crop residues. The amount of available winter feed, the price of calves at weaning, and the projected price for feeder calves in the spring are the most important factors dictating whether a producer will sell his calves or grow them out over the winter.

Approximately 89% of the farms and ranches have less than 100 cows, but only 52% of the total number of cows are in these smaller herds (Boykin et al., 1980) (Table 1–5). The Southwest and Mountain States have more large beef cow herds than other regions. It is not uncommon for herds in the Mountain States to contain 500 to 1000 cows. These ranches usually include some deeded land with extensive acreages of low-carrying-capacity, government-controlled land, which is leased.

Smaller herds are more prevalent in the North Central and Southeast regions, where the climate is more humid and the land highly productive. In these regions, cropping opportunities are greater, and the cow-calf enterprise may be a secondary source of income. Calving usually occurs in the winter or early spring, with calves being sold as calves, stockers, or feeders to Texas, Arizona, California, and Corn Belt feeders. Pastures are stocked intensively as a result of the more highly productive and improved pasture species, such as coastal Bermuda grass and fescue.

In the Northwest region, beef cows are maintained in small herds on hilly land that is unsuitable for crop production. The majority of the land is privately owned. Cows graze on permanent pasture, aftermath, and other crop residues.

Purebred Enterprise

Purebred cattlemen are the producers of germ plasm (bulls), and their cattle are recorded in a breed association registry. The purebred sire is registered or eligible for registry in the herd book of the breed. He

TABLE 1-5. Structure of U.S. Beef Cattle Raising, 1974

Region	Farms and Ranches		Cows and Heifers That Have Calved		Distribution of Farms and Ranches With Beef Cattle by Cow Herd Size					
	Total	Proportion (%)	Total	Proportion (%)	1–19	20–49	50–99	100–199	200 or more	Total
Southeast	294,444	30.5	9,268	23.1	54.7	30.5	9.8	3.4	1.6	100.0
Southwest	173,104	17.9	9,199	21.0	40.3	33.7	15.0	6.7	4.3	100.0
West	59,704	6.2	3,903	9.7	50.6	21.6	12.2	8.3	7.3	100.0
Great Plains	107,829	11.2	8,395	20.9	23.8	30.3	23.5	14.2	8.2	100.0
North Central	329,435	34.2	9,325	23.3	51.4	33.3	11.5	3.1	0.7	100.0
All regions	964,516	100.0	40,090	47.3	31.4	31.4	13.0	5.4	2.9	100.0

From Boykin, C.C., et al. 1980. Structural characteristics of beef cattle raising in the United States. USDA, Washington, DC. AER No. 450.

usually represents a useful and improved type and is descended from a long line of ancestry of the same type. Type is an ideal or standard of perfection combining all the characters that contribute to the animal's value and efficiency for a specific purpose.

Costs associated with producing purebred animals are higher than those for producing commercial cattle. Calves are not normally sold at weaning, but are fed over the winter to a minimum of one year of age to collect valuable performance information. Therefore, feed and all other costs associated with retaining an animal for 6 months to a year longer than most commercial cattle escalates the cost of production. Promotion and advertising are integral parts of the purebred business and are expensive. The cost of fitting and showing beef cattle has increased at a tremendous rate over the past ten years. Many breeders today are showing their animals at only a few selected shows, rather than a large number, in an attempt to reduce costs.

A registered breeder must be able to do the following to be successful in such a competitive business:

1. To have an excellent sense of the seed stock needs of commercial cattlemen, and to produce cattle that will have a defined purpose.
2. To develop a planned mating system.
3. To think in terms of long-range goals and have the adequate financial resources that such goals necessitate.
4. To merchandise the product efficiently.
5. To stay current on research that might affect the long-range goals.

Stocker Production

A stocker is a weaned young animal grazed or maintained on forage with a limited amount of grain or concentrate feeds. Stocker production is a program of growth, not fattening, which allows for an increase in the skeleton and frame of cattle. Traditionally, stockers were steers and heifers intended for slaughter. In recent years, however, these animals have usually been placed in feedlots for additional weight gain.

Stocker cattle may be grown out in the area in which they were bred, or may be shipped soon after weaning either to grazing areas not fully stocked with cows and young calves or to grain-growing areas, where they are usually grazed on small grain pastures and corn or sorghum stubble, then fattened in feedlots.

In the North Central region, many cattle feeders buy their cattle in the fall and maintain them on stalk fields or aftermath pasture through the winter before putting them in the feedlot.

Stocker operations are most prevalent in the Southwest, Great Plains, and North Central regions of the United States (Boykin et al., 1980). In the Southwest, almost one third of the stocker operators accounted for over half the region's stocker-feeder

cattle, with annual sales of 500 head or more. According to a 1976 survey, the Southwest and Great Plains regions had over 75% of all stocker feeder operations in the United States.

Feedlot Program

This program consists of feeding thrifty calves and well-grown-out older cattle (yearlings) a ration that is moderately high to very high in energy until they are sufficient to yield a carcass of choice quality grade.

Cattle feeding in this country spans over 100 years, but the industry did not reach its current size until recently. Expansion occurred gradually until the late 1950s, then increased rapidly through the 1960s and early 1970s because of abundant and low-

TABLE 1–6. Average Annual Marketings of Fed Cattle from All Feedlots and Farmer Feedlots in Selected States 1964–1967 and 1977–1980

State	Average Annual Marketings (1,000 head) of All Fed Cattle			Average Annual Marketings (1,000 head) of Farmer-Fed Cattle		
	1964-67	1977-80	Change (%)	1964-67	1977-80	Change (%)
Plains States (total)	7,345	15,472	111	3,127	2,530	−19
Texas	1,283	4,437	246	131	83	−37
Kansas[a]	1,093	3,247	197	541	398	−26
Oklahoma	338	721	113	96	25	−74
Colorado	1,170	2,226	90	303	136	−55
New Mexico	198	326	65	12	—[b]	−100
Nebraska[a]	2,680	3,939	47	1,545	1,506	−3
South Dakota	583	576	−1	499	382	−23
Southwest (total)	2,765	2,035	−26	50	8	−84
Arizona	612	625	2	12	1	−92
California	2,153	1,410	−35	38	7	−82
Corn Belt-Lake States (total)	7,602	5,975	−21	7,092	4,714	−34
Michigan	224	246	10	216	173	−20
Minnesota[a]	757	743	−2	700	670	−4
Wisconsin	191	180	−6	183	152	−17
Iowa[a]	3,532	2,921	−17	3,353	2,092	−38
Indiana	466	365	−22	421	320	−24
Ohio	450	332	−26	426	284	−33
Illinois[a]	1,323	930	−30	1,210	806	−33
Missouri	659	258	−61	583	217	−63
Top 23 States	18,942	24,820	31	10,823	7,467	−31

[a] States included wholly or partly in the 1981 survey.
[b] None recorded.
From VanArsdall, R.N., and K.E. Nelson. 1983. Characteristics of farmer cattle feeding. USDA, Washington, DC. AER No. 503.

cost grains, many advances in technology, and strong consumer demand. In the early 1960s, the 23 leading cattle-feeding states were marketing 15 to 16 million head of fed cattle annually, which amounted to more than half of all cattle slaughtered commercially (VanArsdall and Nelson, 1983). Production of fed cattle continued to increase to a peak of 27 million head in 1972, declined, then peaked at about the same level in 1978. After 1972, however, annual production of fed cattle generally stayed in the 20 to 25 million range as a result of several factors, including higher grain prices, which reduced returns from cattle feeding.

Some cattle are fed throughout the country, but the business is now concentrated in an area stretching from the western part of the Corn Belt and eastern Northern Plains through the high plains of Texas and adjacent area.

Fed-beef production continues to shift toward larger, fewer feedlots. Feedlots vary in capacity from 100 to more than 100,000 head. Only 1 to 2% of the feedlots from the 23 major producing states have capacities of 1000 head or more (Gee et al., 1979). Large commercial feedlots predominate marketings in the western States. Production of large quantities of grain and forage resulting from irrigation development was an important factor in commercial feedlot development.

Farmer feeders (having a capacity for less than 1000 head of cattle) accounted for most of fed cattle marketed as recently as the mid-1960s. In 1974, over 219,000 farmer feeders sold 61% of all cattle in the top 23 states; 1564 commercial feedlots turned out the remainder (VanArsdall and Nelson, 1983) (Table 1–6). Total production of fed cattle expanded, but both the number and share of production from farmer feedlots declined during subsequent years. By 1980, the number of farmer feedlots had dropped by nearly half, and their share of total production had fallen below 28%. Commercial feedlots numbered only 2144, but annual sales averaged over 7800 per

feedlot, accounting for nearly three fourths of all fed cattle.

Horizontal integration is the ownership of multiple feedlots. Vertical integration is the ownership of more than one of the functions involved in moving feeder cattle to the consumer as fed beef. Larger feedlots usually involve both horizontal and vertical integration. Functions most frequently associated with fed-beef production are feed processing and slaughtering; only a few feedlots actually sell retail beef.

Custom cattle feeding may be defined as the feeding of cattle owned by individuals other than the feedlot owners. This is usually done to reduce the risks associated with cattle feeding and to minimize the capital requirements of the business. Custom feeding also allows cow-calf and stocker operators the opportunity to retain ownership of their cattle up to the time of slaughter.

REFERENCES

Boykin, C.C., H.C. Gilliam, and R.E. Gustafson. 1980. Structural characteristics of beef cattle raising in the United States. USDA, Washington, DC. AER No. 450.
Byerly, T.C. 1975. Feed use in beef production. A review. J. Anim. Sci. 41:921.
Gee, C.K., R.N. VanArsdall, and R.E. Gustafson. 1979. U.S. fed-beef production costs, 1976-77, and industry structure. USDA, Washington, DC. AER No. 424.
Koch, R.M., and J.W. Algeo. 1983. The beef cattle industry: Changes and challenges. J. Anim. Sci. 57:28 (Supp. 2).
NCA. 1982a. Myths and facts about beef. National Cattlemen's Association. Beef Business Bulletin. Englewood, CO. December 3.
NCA. 1982b. The future for beef. A report by the Special Advisory Committee. National Cattlemen's Association. Beef Business Bulletin. Englewood, CO. March 5.
Ritchie, H.D. 1981. Future and direction of the beef cattle industry in the United States: Cow-calf production. Michigan State University, East Lansing, MI. Beef Cattle Report, AC-LC-8204.
USDA. 1983. Livestock and Meat Statistics. United States Department of Agriculture. Washington, DC.
VanArsdall, R.N., and K.E. Nelson. 1983. Characteristics of farmer cattle feeding. USDA, Washington, DC. AER No. 503.

Selected Readings

Briggs, H.M., and D.M. Briggs. 1980. Modern Breeds of Livestock. 4th Ed. MacMillan, New York.

Cunha, T.J. 1980. The beef cow—forever. Beef Digest. November, p. 18.

Gee, C.K., R.N. VanArsdall, and R.E. Gustafson. 1979. U.S. fed-beef production costs, 1976-77, and industry structure. USDA, Washington, DC. AER No. 424.

Heady, H.F., et al. 1974. Livestock grazing on federal lands in 11 western states. Report of a task force of the Council for Agricultural Science and Technology. J. Range Mgt. 27:174.

Lasley, J.F. 1981. Beef Cattle Production. Prentice-Hall, Englewood Cliffs, NJ.

NCA. 1979. Beef cattle research needs and priorities. National Cattlemen's Association. Beef Business Bulletin. Englewood, CO. November 2.

Osgood, E. 1929. The Day of the Cattlemen. University of Minnesota Press, Minneapolis.

Preston, R.L. 1981. Beef in the 21st century. Beef Digest. January, p. 5.

Rouse, J.E. 1973. World Cattle III: Cattle of North America. University of Oklahoma Press, Norman, OK.

Simpson, J.R., and D.E. Farris. 1982. The World's Beef Business. Iowa State University Press, Ames, IA.

Smith, R.T. 1957. A History of British Livestock Husbandry to 1700. Routledge and Kegan Paul, London, England.

Smith, R.T. 1959. A History of British Livestock Husbandry, 1700–1900. Routledge and Kegan Paul, London, England.

Vaughan, H.W. 1941. Breeds of Livestock in America. College Book Company, Columbus, OH.

Wilson, J. 1909. The Evolution of British Cattle. Vinton and Co., London, England.

Zenner, F.E. 1963. A History of Domesticated Animals. Harper and Row, New York.

QUESTIONS FOR STUDY AND DISCUSSION

1. On what factors will future growth of the beef cattle industry depend?
2. What year was per capita consumption of retail beef at its highest level?
3. What percentage of a beef cow's diet comes from forage?
4. Does beef provide more cholesterol than other animal protein sources? Explain.
5. Name five edible and inedible by-products that are produced from beef.
6. When evaluating the efficiency of production of beef cattle, what are the biggest advantage and disadvantage in comparison with poultry and swine?
7. How much feed grain and oilseed meal protein is required to produce one pound of animal protein in the form of beef? How does this compare with broilers, turkeys, and swine?
8. What effects will the following have on the efficiency of beef production?
   a. Electrical stimulation
   b. Restructured beef products
   c. Feeding bulls instead of steers
   d. Electronic marketing of cattle
   e. Cooperative marketing
9. Today's cattle originated from either or both of two species. Name them and describe the difference(s) between them.
10. When were cattle first domesticated?
11. Name the only member of the *Bos* family that is native to North America.
12. When and by whom were the first cattle brought to the North American continent?
13. How did the Longhorn become established in the western and southwestern United States?
14. When were the first purebred cattle brought to the United States?
15. What was the most important breed of cattle in the United States through the first half of the nineteenth century?
16. What was responsible for the development of cattle trail drives in the Southwest after the Civil War?
17. What lesson(s) did ranchers learn from the severe winters of the late nineteenth century?
18. Name the year in which each of the following events occurred.
   a. Outbreak of cattle tick fever
   b. Importation of the first
      1. Angus
      2. Hereford
      3. Shorthorn
      4. Brahman
      5. Charolais
   c. Formation of the Society of Animal Production
   d. Commercial availability of artificial insemination
19. Define or describe the following beef production systems.
   a. Commercial cow-calf
   b. Purebred
   c. Stocker
   d. Feedlot
20. What are the two most important factors that determine whether a producer will sell his calves at weaning or grow them out over the winter?
21. What qualities must a purebred breeder have to be successful? Why?
22. Define vertical and horizontal integration.

Chapter Two

Breeds of Beef Cattle

A breed of livestock may be defined as a group of animals within a species that share a common ancestry and that as a result of breeding and selection possess common inherited characteristics that distinguish them from another group. Each breed association has developed a structure of animal types that supplies a broad base and a wide variety of genetic material (germ plasm) for use by cattle breeders.

Although the success of a beef cattle herd depends more on the breeding and selection principles followed by the cattle manager than on the breed selected, selection of a breed not adapted to a specific environment can cause economic disaster. Never before in the history of the beef cattle industry has there been such a wide variety of breeds from which to choose. Before selecting a breed or combination of breeds, the cattle producer should consider the following key points:

1. All breeds have strong and weak points.
2. No one breed is adapted to all situations.
3. Hereditary variation exists in all breeds.

When selecting a breed of cattle, the livestock producer should use the following guidelines to aid in making an objective choice:

1. Survey the area to see what breed or breeds are best adapted to local conditions.
2. Study the market demand for the offspring.
3. Compare advantages of a breed or breed cross already produced in your area with those of breeds that have apparent usefulness but that are not being raised extensively in the same area.

Once the breed has been selected, breeding practices that lead to genetic improvement of the herd should be followed.

BREED CHARACTERISTICS*

This section describes those breeds of cattle that have made or may make a significant contribution to the beef cattle industry in the United States. It is not meant to be a complete listing of all breeds of beef cattle.

British Breeds

ANGUS

The Angus breed originated in Scotland from black, polled cattle native to Aberdeen and Angus shires more than 200 years ago. The first herd book was established in 1862. George Grant of Victoria, Kansas, a retired silk merchant, imported the first four Angus bulls into the United States in 1883 and crossed them with native Texas Longhorn cattle. Since then, the breed has become one of the most popular in the United States.

The breed is naturally polled with black skin and hair (Fig. 2–1). When Angus cattle are crossbred with horned breeds of European origin, the crossbred calves are nearly

---

* USDA Bulletin No. 2228 (Putnam and Warwick, 1975) was used extensively in the preparation of this section.

FIG. 2-1. The purebred Angus cow-calf pair. (Courtesy of the American Angus Association.)

always polled. Calves produced from Angus crosses with Charolais (cream colored or creamy white) are usually dark or smoky white. In crosses with red-bodied breeds, the offspring are black but may have white markings characteristic of the other breed. In crosses with white Shorthorns, so-called blue-gray cattle are often produced, having a mixture of white and black hair. Angus cattle transmit maternal characteristics to their offspring, which include fertility, efficiency under minimal management conditions, calving ease, mothering ability, longevity, and early age of puberty. Angus cattle are also known for their ability to transmit marbling to their offspring.

RED ANGUS

Red pigment is inherited as a simple recessive trait in Angus cattle. Red cattle have occurred in the breed since its earliest development. The Red Angus Association of America was founded in 1954. Except for their color, these cattle are similar to the black Angus breeds. Red color absorbs less of the sun's heat, which may be advantageous in hotter climates.

DEVON

The Devon breed originated in Devonshire's grass-covered hills in southwestern England. Colonists brought Devon cattle to America as early as 1623; however, the first record of registered Devon cattle being imported to the United States was in 1817, when cattle were imported by Robert Patterson of Baltimore, Maryland as a gift from the Earl of Leicester of Holkham, England.

Devon cattle are a dark red color (often called "ruby red") with yellow skin and white horns darkening to black tips. The Devon Cattle Association recommends that the breed can be used in crossbreeding programs for its maternal characteristics.

GALLOWAY

The Galloway breed was developed in southwestern Scotland, where the climate is moist and chilly. The breed may have had a common origin with Aberdeen Angus. The two breeds have much in common, but the Galloway can be distinguished from the Angus by its coat of black, long, soft wavy hair and thick mossy undercoat.

The breed was first imported into the United States in 1870. In the early 1880s, 3 bulls and 16 females were imported into the United States from Scotland. The American Galloway Breeders Association was founded in 1902. Today, the breed is most commonly found in Montana, Wyoming, and the Dakotas. The breed has excellent crossing characteristics for adaptability to high altitude and to cold, windy climates. The Galloway has a mild disposition and reaches puberty early.

HEREFORD

Hereford cattle are characterized by horns and by a white face, crest, dewlap, underline, and switch; their legs are white below the hocks and knees, and their bodies are red (Fig. 2–2). They originated in the County of Hereford in England, an area of plains and fertile valleys. In 1817, Henry Clay, the statesman from Kentucky imported the first Hereford cattle. A second importation of this breed in 1840, however, provided for the establishment of Herefords in this country. In the late 1870s, large numbers of cattle were imported, and the breed became popular.

Herefords have superior foraging ability, vigor, and hardiness. Under rigorous

FIG. 2-2. The Hereford (horned). (Courtesy of the American Hereford Association.)

conditions, they tend to produce more calves than do many other breeds. These characteristics, along with their docile nature, account for the breed's popularity in the western United States. The American Hereford Cattle Breeders Association was organized in 1881, and in 1934, the official name was changed to American Hereford Association.

POLLED HEREFORD

The genealogy of the present-day polled Hereford cattle can be traced to the original parent horned Hereford breed imported from England. The first serious breeding program designed to produce polled Herefords was begun in 1898. Warren Gammon, a young Iowa Hereford breeder, conceived the idea after seeing some polled cattle on display at the Trans-Mississippi World's Fair in Omaha, Nebraska in 1898 (Fig. 2–3). In 1901, Gammon sent 2500 inquiries to members of the American Hereford Association. From the replies he received, Gammon located and bought four bulls and ten females. The polled trait resulted from a dominant mutation. Mr. Gammon and others organized the American Polled Hereford Cattle Club in 1900. In 1911, the club combined with the National Polled Hereford Association to form the American Polled Hereford Breeders Association.

SCOTCH HIGHLAND

The Scotch Highland breed of cattle was developed in the Hebrides Islands near the west coast of Scotland. Scotch Highland cattle are hardy and exceptionally good foragers. They have long, coarse outer hair with a soft thick undercoat (Fig. 2–4). Their coat makes them adaptable to extreme cold winds and heavy rainfall. Their color pattern may be black, brindle, red and light red, dun yellow, or silver. They have wide spreading horns, with the bulls' horns curving forward and the cows' horns curving upward.

FIG. 2–3. The Polled Hereford. (Courtesy of the American Polled Hereford Association.)

FIG. 2-4. The Scotch Highland. (Courtesy of the American Scotch Highland Breeders Association.)

SHORTHORN

The Shorthorn breed originated in the late 1700s in northeastern England, principally in the valley of the Tees River. The Coates Herd Book, established in 1822 to record pedigrees of Shorthorn cattle, was the first cattle herd book. Shorthorns were originally bred for the dual purpose of milk and meat. The name "Durham" was often applied, but it is no longer used.

The importation of Shorthorns into the United States between 1820 and 1850 established the breed on a permanent basis in this country. Importations of this breed date back as far as 1783. Amos Cruickshank and other Scottish breeders in the middle 1800s selected intensely within the existing breed for early maturity and for increased compactness and thickness. This eventually led to the separation of the breed into beef and milking types.

Scottish Shorthorns did not attain popularity in the United States until the last two decades of the nineteenth century. Beef-type Shorthorns in the United States today can be traced back almost entirely to Scottish cattle in all lines of their pedigree.

Shorthorn cattle may be red, white, or roan in color. They have short, refined incurving horns. In the late 1800s, polled cattle of predominantly Shorthorn breeding were developed. Polled Shorthorn cattle are registered in the same herd book as horned beef Shorthorns because their ancestors are registered Shorthorns.

SOUTH DEVON

The South Devon breed is reported to have existed in the southwestern part of England for about 400 years. The breed is believed to have originated, in part, from the larger red cattle of Normandy, France, which were imported to England at the time of the Norman Invasions. In England, the breed is classified as dual-purpose (meat and milk) and genetically divergent from the Devon breed.

The first importations of South Devon cattle to the United States occurred in 1936 and 1947. In 1969 and 1970, 215 registered South Devon cattle from England were imported by Beef Hybrids of Stillwater, Minnesota, for crossbreeding with traditional American beef cows.

The South Devon breed is the largest of the British breeds of beef cattle. At maturity, bulls may weigh between 2000 and 2800 lb and cows between 1500 and 1600 lb. They are extremely docile and have a coat of rich medium-red, soft, curly hair with an exceptionally thick hide. They have average-sized horns that curve forward and slightly downward.

RED POLL

Red Poll cattle are light red to very dark red in color with natural white in the tail switch. The breed originated on the eastern coast of England as a result of crossing stocks of the shires of Norfolk and Suffolk. This process began in the early 1800s and was completed by 1846. G.T. Taber of New York imported the first registered Red Polls to the United States in 1873. The Red Poll Club of America was organized in 1883.

The breed is dual-purpose, but also yields carcasses that are lean and low in outer fat covering when fed to slaughter weight. In recent years, Red Poll breeders have been emphasizing beef characteristics more than milk production.

MURRAY-GREY

Murray-Grey cattle originated in Australia from a cross of Angus and Shorthorn. They range in color from silver-gray or gray, to dark gray or dun. Some dun pigment may be found on the underbody, and the breed has a dark muzzle. They are naturally polled. Murray-Grey cattle were first imported into the United States in 1972. The American Murray-Grey Association was formed in August 1970, with offices in Billings, Montana. The Murray-Grey breed is noted for small birth weights, low calf mortality, and excellent carcass quality.

European Breeds

CHAROLAIS

The Charolais breed is one of the oldest of the several breeds of French cattle. The Charolais was developed in the district around Charolles in central France. The King Ranch of Texas is given credit for importing the first Charolais bulls into the United States from Mexico in 1936. The American International Charolais Association was established in 1962.

The coat is white or very light straw in color. Most Charolais cattle are naturally horned, but a growing number of polled animals are also being registered. The horns are white, slender, and tapered. Charolais cattle are large-sized, long-bodied, and heavily muscled. The Charolais breed's outstanding characteristics are its growth rate and carcass cutability.

BLONDE D'AQUITAINE

The Blonde d'Aquitaine breed originated in the Aquitaine Basin of southwestern France. Prior to World War II, the breed was selected for beef and draft; however, after World War II, breeders began emphasizing only beef production traits. These cattle were first imported to the United States in 1971, and the American Blonde d'Aquitaine Association was formed in 1973.

They are solid-colored cattle with a coat of hair that ranges from light to slightly darker shades of yellow-gold. The horns and hooves are light, and mucous membranes are light pink. Blonde d'Acquitaine cattle are similar to Limousin cattle in shape and color; however, they appear to have more pronounced muscling than the Limousin breed.

LIMOUSIN

Limousin cattle range from a golden wheat color in the females to a deep-red gold in the males, darkening somewhat with maturity and age. They are light-horned cattle with no record of polled mutations. Limousin cattle are known for their carcass cutability. Research data collected at the U.S. Meat Animal Research Center (MARC), Clay Center, Nebraska indicate that Limousin-sired calves weigh less at birth and are easier to deliver than calves sired by Charolais and Simmental bulls.

The breed originated in the isolated area of France near Limoges. The breed was introduced into the United States in 1968, and the first 50% Limousin calf was born in the United States in 1969. The North American Limousin Foundation was formed in 1969.

MAINE-ANJOU

The Maine-Anjou breed is the result of continued crossings between English Shorthorns and French Mancelle cows. In France, these cattle are bred for meat and milk production. Maine-Anjou cattle are heavier in weight than any other French breed. They are dark red and white in color, or sometimes roan. They have lightly pigmented skin and medium-sized horns that curve forward. MARC data indicate that the breed excels in growth rate, milk production, and carcass cutability.

SALERS

The Salers breed was developed in the rugged mountains of central France. Although the breed has existed for centuries, the name first appeared in 1840 in Salers, France. The Salers herd book was established in 1905, and in 1925, recording of milk production and growth traits became compulsory. In 1973, Salers semen was imported to North America.

Salers cattle have a solid, deep cherry-red coat that is sometimes spotted with white markings under the belly (Fig. 2–5). Their hair is thick and sometimes curly. Salers have "lyre"-shaped horns, which are light at the base and darker toward the tips. When Salers bulls are bred with Hereford cows, the resulting offspring have red bodies and white faces ("red baldy"). Salers are larger than most English breeds, but slightly smaller than Charolais. The breed is considered to have the easiest calving of European breeds. This is due to a

FIG. 2-5. The Salers. (Courtesy of the American Salers Association.)

small head, slender neck, and long body. Redeeming characteristics of the breed are milk production, calving ease, and carcass quality and cutability.

TARENTAISE

Tarentaise cattle originated in the Savoy Alps in southeastern France near Switzerland and Italy. A Tarentaise bull was first imported to North America in 1972. In 1973, the Canadian and American Tarentaise Associations were founded.

The Tarentaise has a solid light cherry to dark blonde hair coat. Bulls normally darken around the neck and shoulders as they mature. They are adaptable to different climatic conditions. The breed is considered a dairy breed in its native Savoy region. Cows are known for their strong maternal and feminine qualities. They are considered one of the smaller of the European Continental breeds, with mature bulls and cows weighing 1800 and 1100 lb, respectively.

BROWN SWISS

The Brown Swiss breed originated in the Alps of Switzerland. The exact date of the breed's origin is not known. The first Brown Swiss cattle entered America between 1869 and 1870. The breed has been primarily used as a dairy breed; however, in the late 1960s, some American breeders started placing more emphasis on beef performance traits.

Brown Swiss cattle range in color from medium brown to tan. Their dark horns curve upward. Brown Swiss cattle fit into a crossbreeding program by infusing fertility, calving ease, and milking ability.

SIMMENTAL

The Simmental breed originated in the Simmen Valley in western Switzerland. The American Simmental Association was formed in 1968. The first purebred bull entered the United States in 1971, and by 1974, the American Simmental Association had registered nearly 150,000 head.

FIG. 2–6. The Simmental. (Courtesy of the American Simmental Association.)

Simmentals vary from yellowish-brown or straw color to a dark red, combined with white markings (Fig. 2–6). The head, underside of the brisket, and belly are generally white. The legs and tail are usually white and there may be white patches on the body. The hair is soft, and the skin is lightly pigmented. The horns are fine and white in color, and they curve outward from the side toward the front, with the tips turned slightly upward. Simmental cattle are known for their fertility, milk production, growth rate, and carcass cutability.

CHIANINA

The Chianina breed obtained its name and origin from the Chianina Valley in Italy. Italian Chianina cattle are the tallest of all cattle breeds. Mature bulls can stand 72 inches at the withers and weigh up to 4000 lb, while females can weigh up to 2400 lb and stand 60 to 68 inches at the withers. They have porcelain-white hair, a black tongue and palate, a black nose and eye area, and a black switch and anal orifice (Fig. 2–7). Semen of this breed was first imported to the United States in 1971. The Chianina tends to be long-boned and long-muscled, and these characteristics are claimed to offset potential calving difficulties. Use of the Chianina in crossbreeding programs should be as a terminal sire to transmit growth and carcass cutability to the offspring.

GELBVIEH

The Gelbvieh was developed in West Germany. The breed comes from four triple-purpose, yellow breeds—Glan-Donnersbury, Yellow Franconian, Limburg, and Lahn. These four breeds were developed around 1850 and were amalgamated into the Gelbvieh in 1920. Semen was first imported to the United States in 1971, and the first Gelbvieh arrived in North America in 1972.

The Gelbvieh has fine dense hair that is golden red (Fig. 2–8). All mucous membranes and skin are pigmented. The Gelbvieh has a gentle disposition and can tolerate cold weather slightly better than hot weather. Gelbvieh cattle resemble Limousin cattle; however, the Gelbvieh is larger and dual-purpose.

Brahman and Brahman Crosses

AMERICAN BRAHMAN

The Brahman breed was developed in the southern United States in the early 1900s from humped cattle of India (Bos indica), which are often referred to as Zebu. Zebu cattle were first imported in 1849; however, the first significant importation was accomplished by Pierce Ranch, Pierce, Texas and by T.M. O'Connor, Victoria, Texas in 1906. American cattlemen developed the Brahman by combining several Indian breeds and upgrading British

FIG. 2-7. The Chianina. (Courtesy of the American Chianina Association.)

FIG. 2-8.   The Gelbvieh. (Courtesy of the American Gelbvieh Association.)

FIG. 2-9.   The American Brahman. (Courtesy of the American Brahman Breeders Association.)

females. The name "Brahman" was chosen by the American Brahman Breeders Association, which was organized in 1924.

Brahman cattle have a characteristic hump over the shoulders, loose skin (dewlap) under the throat, and large drooping ears (Fig. 2–9). Color may range from light gray or red to almost black with the prevailing color usually light to medium gray. Brahman cattle are resistant to ticks and more tolerant of heat than British or European cattle breeds. Brahman-European crosses have been observed to have distinct production advantages, especially in southern climates. They exhibit a great deal of heterosis or "hybrid vigor," and the offspring often exceed both parental types in growth rate and reproduction.

BRANGUS

Brangus cattle were developed by blending the Brahman and Angus breeds. They possess ⅝ Angus and ⅜ Brahman ancestry. Brangus cattle are black and polled; both are inherited dominant qualities. The recessive red gene is also present in the Brangus breed, which led to the formation of the Red Brangus Association. Brangus is a registered trade name and can only be applied to cattle registered with the International Brangus Association, which was founded in 1949.

BEEFMASTER

Three breeds—the Hereford, Shorthorn, and Brahman—were combined to produce the Beefmaster. Breed development began in 1931 by Tom Lasater on a ranch near Falfurrias, Texas. The foundation herd was moved to Matheson, Colorado in 1949. Approximately 25% Hereford, 25% Shorthorn, and 50% Brahman hereditary material is estimated to have been incorporated into the breed. Disposition, fertility, weight, conformation, hardiness, and milk production were the six criteria stressed in the initial selection program. Any cow that failed to wean a calf or that was classified as a poor milker was culled. Color pattern varies, but reds and duns predominate. Most cattle are horned, but polled animals occur.

BARZONA

Barzona cattle were developed by F.N. Bard for the intermountain desert areas of southwestern and northern Mexico. Development began in Arizona in 1942 with an Africander-Hereford cross. Females resulting from this cross were divided into two herds. Santa Gertrudis bulls were used on one herd, and Angus bulls were used on the other. Progeny were closely culled and crossbacked, with emphasis on fertility, mothering ability, and growth. The Bard herd was closed in 1960.

Barzona cattle are able to perform effectively on a grass-browse combination and were selected for use in the semi-arid regions of the United States and Mexico.

The Barzona breed is nearly solid red, with little white except on the underline and occasionally around the head.

SANTA GERTRUDIS

The Santa Gertrudis was developed by Robert J. Kleberg, Jr., the president of King Ranch in southern Texas, from crosses between beef-type Shorthorn cows and Brahman bulls. Crossbreeding began in 1910, and in 1918, formation of a new breed was initiated. In 1920, a bull calf of ⅜ Brahman and ⅝ Shorthorn ancestry was born and was named Monkey. Monkey became the foundation sire of the Santa Gertrudis breed, and before his death, he had produced over 150 sires. The Santa Gertrudis Breeders International organization was formed in 1951.

The breed is cherry-red in color, and while the majority are horned, polled animals occur and are acceptable (Fig. 2–10). They have loose hides with increased surface area due to neck folds and sheath or navel flap. Mature cows frequently attain a weight of 1600 lb, and mature bulls a weight of 2000 lb.

FIG. 2–10. The Santa Gertrudis. (Courtesy of Santa Gertrudis Breeders International.)

Other Breeds

LONGHORN

Longhorn cattle are descendents of those brought to Mexico by Spanish explorers as early as 1521. These cattle became the foundation stock of the Texas Longhorn that was prominent throughout the Southwest from the late 1500s to the middle 1800s. British breeds of cattle were imported in the 1800s to upgrade the Longhorn to a more beefy type of animal. By the 1920s, the Longhorn breed was on the verge of extinction; however, in 1927, the Federal Government commissioned Will Barnes to round up some of the few remaining Longhorns. These animals were shipped to a wildlife refuge in Oklahoma, and the breed was saved. The Texas Longhorn Breeders Association was formed in 1964.

Longhorn cattle can be black, fawn, brindle, yellow, or red in color and are easily recognized by their huge horns.

Longhorns can endure hunger, thirst, cold, and heat better than most breeds. Recently, interest has revived in using Longhorns in breeding programs because of their small birth weights.

HAYS CONVERTER

The Hays Converter is a three-breed cross developed by Senator Harry Hays of Calgary, Alberta, Canada. Five generations of closed herd development were used to develop the breed from Hereford, Holstein, and Brown Swiss stock. The selection program emphasized a large, strong frame, well-attached udders, sound feet and legs, growth potential, and ease of calving. Hays Converter cattle have white faces. Their bodies may be either black or red; most are black.

BEEFALO

The Beefalo breed is a bison hybrid that was developed by D.C. Basolo, of Tracy,

California. For registry in the Bison Hybrid Herd Book, an animal must have a ⅛ to ⅜ bison bloodline. According to the Bison Hybrid International Association, the breed's attributes are winter hardiness, ability to perform well on low-quality roughage, and small birth weights.

BREED SELECTION

The tasks of characterizing and selecting the correct breeds are difficult. A vast number of cattle breeds and feed resources are available for beef production. In the United States, stocking rates range from one cow per acre to one cow per 300 or 400 acres, because of differences in climate, land, and feed resources (Cundiff, 1981). Traits that are important to efficient beef production differ among the individual breeds. These breed differences can be utilized by crossbreeding, which results in heterosis or "hybrid vigor." It is important to match the crossbreeding system and the characteristics of breeds used with the feed and other available resources on farms or ranches.

An extensive comparative evaluation of breeds of cattle has been conducted at the Meat Animal Research Center (MARC) at Clay Center, Nebraska. The project started in 1969 and included three cycles of sire breeds that were bred artificially to Hereford and Angus cows. To permit comparisons between breeds, Hereford-Angus reciprocal crosses were repeated in each cycle as controls. Breeds evaluated were Hereford, Angus, Jersey, Limousin, South

TABLE 2-1. Breed Crosses Grouped in Biological Types on Basis of Four Major Criteria and Breed Group Means for 200-Day Weaning Weights of F_1 Calves

| Breed Group | Biologic Type Criteria[a] | | | | No. of Calves | 200-Day Weight (lb) | 200-Day Weight Ratio[b] |
	Growth Rate and Mature Size	Lean to Fat Ratio	Age at Puberty	Milk Production			
Jersey-X (J)	X	X	X	XXXX	302	406	94
Hereford-Angus-X (HA)	XX	XX	XXX	XX	962	430	100
Red Poll-X (R)	XX	XX	XX	XXX	214	426	99
South Devon-X (SD)	XXX	XXX	XX	XXX	232	430	100
Tarentaise-X (T)	XXX	XXX	XX	XXX	202	443	103
Pinzgauer-X (P)	XXX	XXX	XX	XXX	376	439	102
Sahiwal-X (Sa)	XX	XXX	XXXXX	XXX	325	432	100
Brahman-X (Br)	XXXX	XXX	XXXXX	XXX	349	456	106
Brown Swiss-X (BS)	XXXX	XXXX	XX	XXXX	263	452	105
Gelbvieh-X (G)	XXXX	XXXX	XX	XXXX	213	461	107
Simmental-X (Si)	XXXXX	XXXX	XXX	XXXX	399	452	105
Maine-Anjou-X (MA)	XXXXX	XXXX	XXX	XXX	222	454	106
Limousin-X (L)	XXX	XXXXX	XXXX	X	371	437	102
Charolais-X (C)	XXXXX	XXXXX	XXXX	X	382	459	107
Chianina-X (Ci)	XXXXX	XXXXX	XXXX	X	238	456	106

[a] Increasing numbers of X's indicate relative differences between breeds.
[b] Ratio relative to Hereford-Angus crosses.
From Cundiff, L.V. 1981. Evaluation of maternal breeds. Proc., The Range Beef Cow Symposium VII. Rapid City, SD. December 7–9.

TABLE 2-2. Breed Group Means for Reproduction and Maternal Performance of F_1 Cows—Cycles I, II and III

Breed Group	No. of Births	Calving Difficulty[a] (%)	Calf Crop		Birth Weight (lb)	Milk Prod.[b] (lb)	200-Day Weight			
			Born (%)	Weaned (%)			Per Calf Weaned (lb)	Ratio[c] (%)	Per Cow Exposed (lb)	Ratio (%)
Cycle I (2- through 8-year-olds)										
Hereford-Angus-X (HA)	738	10	93	85	86	6.6	472	100	401	100
Jersey-X (J)	628	4	92	85	79	9.7	490	104	415	104
Limousin-X (L)	851	9	91	83	88	6.0	481	102	400	100
South Devon-X (SD)	603	12	90	86	91	7.0	489	104	420	105
Simmental-X (Si)	872	14	91	84	91	8.8	518	110	436	109
Charolais-X (C)	693	12	90	81	93	6.0	500	106	408	102
Cycle II (2- through 7-year-olds)										
Hereford-Angus-X (HA)	438	16	91	84	88	6.2	481	100	404	100
Red Poll-X (R)	461	17	90	79	91	7.6	508	106	401	99
Brown Swiss-X (BS)	681	11	92	85	93	8.4	540	112	459	114
Gelbvieh-X (G)	429	14	95	87	92	8.4	539	112	469	116
Maine-Anjou-X (MA)	468	14	94	86	98	6.5	528	110	454	112
Chianina-X (Ci)	475	11	93	86	97	6.2	529	110	455	113
Cycle III (2- through 6-year-olds)										
Hereford-Angus-X (HA)	422	16	89	82	84	5.4	465	100	381	100
Tarentaise-X (T)	306	12	89	82	85	7.2	514	111	421	111
Pinzgauer-X (P)	436	16	91	84	89	7.3	499	107	419	110
Sahiwal-X (Sa)	350	3	94	87	74	7.8	495	106	431	113
Brahman-X (Br)	430	3	93	85	81	8.4	533	115	453	119

[a] Includes calves requiring calf puller or C-section.
[b] Average of three 12-hour milk production measures on a sample of 18 cows per breed group at 3 and 4 years of age in Cycle I, on two sets of 18 cows per breed group at 3 years of age in Cycle III.
[c] Ratio relative to Hereford-Angus crosses.
From Cundiff, L.V. 1981. Evaluation of maternal breeds. Proc., The Range Beef Cow Symposium VII. Rapid City, SD. December 7–9.

Devon, Simmental, and Charolais sires (Cycle I, 1970, 1971, and 1972); Hereford, Angus, Red Poll, Brown Swiss, Gelbvieh, Maine-Anjou and Chianina sires (Cycle II, 1973 and 1974); and Hereford, Angus, Tarentaise, Pinzgauer, Sahiwal and Brahman sires (Cycle III, 1975 and 1976).

Six biologic types were defined using growth rate and mature size, lean to fat ratio, age at puberty, and milk production as a basis for classification (Table 2–1).

The females produced in the program were retained to evaluate reproduction and maternal performance traits. Results on the production of F_1 cows are summarized in Table 2–2. Valid comparisons can be made between breed groups used in the same cycle, which were managed as contemporaries through 8 years of Cycle I, 7 years of Cycle II, and 5 years of Cycle III.

The information presented in Tables 2–1 and 2–2 can be used to determine which breed(s) to use in a crossbreeding program. It must be remembered that these data reflect breed averages, and that within any breed, there are individuals that are higher or lower than the average.

One of the most important conclusions from the MARC data was that the amount and quality of feed resources for the cow herd and for the growing-finishing progeny for slaughter are of critical importance in selecting breeds for a crossbreeding system. Table 2–3 was compiled by the researchers at MARC and matches available feed resources with breed differences.

Table 2–3 classifies feed resources separately for the cow herd and for the growing-finishing progeny for slaughter. In classifying feed resources for the cow herd, "low" represents the energy available from unsupplemented desert ranges or unsupplemented ranges in temperate climates during winter months; "moderate" represents the energy level available from native range pastures supplemented with moderate levels of hay during the winter months; and "high" represents an energy level available from good quality-improved or native range pastures during the growing season in temperate climates, or energy levels that are sustained by liberal feeding of supplemental energy during the winter months.

TABLE 2-3. Synchronizing Genetic Resources With Feed Resources[a]

Feed Resources[b]		General-Purpose Breeds		Maternal Breeds		Terminal-Sire Breeds
Cow Herd	Finishing Progeny	Growth Rate, Mature Size, and Leanness	Milk Prod.	Growth Rate, Mature Size, and Leanness	Milk Prod.	Growth Rate, Mature Size, and Leanness
Low	Low	XX	XXX	X	XXX	XXX
Low	Mod.	XX	XX	XX	XX	XXXX
Low	High	XX	X	XX	XX	XXXX
Mod.	Low	XX	XXXX	XX	XXXX	XXX
Mod.	Mod.	XXX	XXX	XX	XXXX	XXXX
Mod.	High	XXXX	XX	XXX	XXX	XXXX
High	Low	XXXX	XXXXX	XXX	XXXXX	XXXX
High	Mod.	XXXX	XXXX	XXXX	XXXX	XXXXX
High	High	XXXXX	XXXX	XXXX	XXXX	XXXXXX

[a] Increasing X's reflect higher performance levels for growth rate, mature size and leanness or milk production.

[b] Low, moderate, and high levels of energy for the cow herd or for growing and finishing progeny for slaughter.

From Cundiff, L.V. 1981. Evaluation of maternal breeds. Proc., The Range Beef Cow Symposium VII, Rapid City, SD. December 7–9.

In classifying feed resources for the growing-finishing progeny for slaughter, "low" represents the energy available from a quality-improved or native pasture; "moderate" represents the energy available from a good-quality forage diet supplemented with grain or finished on a high-grain diet for a relatively short period (60 to 90 days) prior to slaughter; and "high" represents the high-concentrate diets typically used in feedlots.

Table 2–3 is organized separately for general-purpose, maternal, and terminal-sire breeds to provide alternate breeding strategies. General-purpose breeds are appropriate for straightbreeding or rotational crossbreeding systems (Cundiff, 1981). Maternal and terminal-sire breeds are appropriate for static terminal-sire or rotational terminal-sire crossbreeding programs.

In general, as quantity and quality of feed resources improve, higher levels of size and milk production can be supported in the cow herd. Optimum levels of growth and leanness are higher when feed resources for growing-finishing progeny are of higher quality. The potential for increasing economic efficiency is great when cow productivity is matched with feed resources and other production conditions through systematic crossbreeding (Gosey, 1984).

CATTLE RECORD ASSOCIATIONS

Beef Cattle

American Angus Association
3201 Frederick Boulevard
St. Joseph, MO 64501

Red Angus Association of America
P.O. Box 776
Denton, TX 76201

Barzona Breed Association of America
P.O. Box 1421
Carefree, AZ 56331

Beefmaster Breeders Universal
Suite 350, GPM South Tower
800 North West Loop 410
San Antonio, TX 78216

Foundation Beefmaster Association
200 Livestock Exchange Building
Denver, CO 80216

National Beefmaster Association
817 Sinclair
Ft. Worth, TX 76102

American Blonde d'Aquitaine Association
Rt. 2, Box 21A
Porum, OK 74455

American Brahman Breeders Association
1313 La Concha Lane
Houston, TX 77054

International Brangus Breeders
  Association
9500 Tioga Drive
San Antonio, TX 78230

American Red Brangus Association
404 Colorado
Austin, TX 78701

American International Charolais
  Association
P.O. Box 20247
Kansas City, MO 64195

American Chianina Association
P.O. Box 159
Blue Springs, MO 64015

Devon Cattle Association
P.O. Box 628
Uvalde, TX 78801

American Galloway Breeders Association
302 Livestock Exchange Building
Denver, CO 80216

American Gelbvieh Association
313 Livestock Exchange
Denver, CO 80216

Canadian Hays Converter Association
6707 Elbow Dr. SW, Suite 509
Calgary, Alberta
Canada T2V 0E5

American Hereford Association
715 Hereford Drive
Kansas City, MO 64105

American Polled Hereford Association
4700 East 63rd Street
Kansas City, MO 64130

North American Limousin Breeders
Foundation
309 Livestock Exchange Building
Denver, CO 80216

Texas Longhorn Breeders Association
3701 Airport Freeway
Ft. Worth, TX 76111

American Maine-Anjou Association
564 Livestock Exchange Building
Kansas City, MO 64102

American Murray-Grey Association
1222 N. 27th Street
Billings, MT 59107

American Pinzgauer Association
P.O. Box 1003
Norman, OK 73070

American Salers Association
Suite 101
Livestock Exchange Building
Denver, CO 80216

Santa Gertrudis Breeders International
P.O. Box 1257
Kingsville, TX 78363

American Scotch Highland Breeders
Association
P.O. Box 81
Remer, MN 56672

American Simmental Association
1 Simmental Way
Bozeman, MT 59715-9990

American Shorthorn Association
8288 Hascall Street
Omaha, NE 68124

North American South Devon Association
P.O. Box 68
Lynnville, IA 50153

American Tarentaise Association
123 Airport Road
Ames, IA 50010

Dual-Purpose and Dairy Cattle

The Brown Swiss Cattle Breeders
Association
800 Pleasant Street
Beloit, WI 53511

American Red Poll Association
Box 35519
Louisville, KY 40232

American Milking Shorthorn Society
1722-JJ S. Glenstone
Springfield, MO 65804

REFERENCES

Cundiff, L.V. 1981. Evaluation of maternal breeds. Proc., The Range Beef Cow Symposium VII. Rapid City, SD. December 7–9.
Gosey, J.A. 1984. Fitting cow size and efficiency to feed supply. *In* Beef Cattle Science Handbook. No. 20. Edited by F.H. Baker, and M.E. Miller. Westview Press, Boulder, CO.
Putnam, P.A., and E.J. Warwick. 1975. Beef cattle breeds. USDA, Washington, DC. Agriculture Information Bulletin No. 2228.

Selected Readings

Briggs, H.M., and D.M. Briggs. 1980. Modern breeds of livestock. MacMillan, New York.
Cundiff, L.V., R.M. Koch, K.E. Gregory, and C.M. Smith. 1981. Characterization of biological types of cattle—Cycle II. IV. Postweaning growth and feed efficiency of steers. J. Anim. Sci. *53*:332.
Gregory, K.E., D.B. Laster, L.V. Cundiff, et al. 1979. Characterization of biological types of cattle—Cycle III. II. Growth rate and puberty in females. J. Anim. Sci. *49*:461.
Stephen, L., and W. Sullivan, Jr. 1976. Cattle Breeds Index. Research Communications, Hays, KS.

QUESTIONS FOR STUDY AND DISCUSSION

1. Define *breed.*
2. What criteria should be considered when selecting a breed of beef cattle to use in a breeding program?
3. What are dual-purpose cattle? Name two dual-purpose breeds.
4. Where did Aberdeen Angus originate?
5. What was the first breed of beef cattle imported to the United States?
6. Why are Herefords so popular in Western range country?

7. How do Angus cattle differ from Galloway cattle?
8. Who founded the Polled Hereford breed?
9. What are the Charolais breed's outstanding characteristics?
10. Describe the following European breeds:
    a. Limousin
    b. Salers
    c. Tarentaise
    d. Simmental
    e. Chianina
11. Why was the Brahman breed developed in the United States?
12. Name four breeds of cattle developed in the United States, and discuss their origin.
13. Based on the breeding studies at MARC, identify two breeds of cattle that could be used to improve the following traits within a beef herd.
    a. Growth rate
    b. Milk production
    c. Carcass cutability
    d. Calving ease
14. What breed or breed combination would you select in the following situations?

Feed Resources

Finishing Progeny	Cow Herd
Low	Low
High	High
High	High

Chapter Three

Selection Principles and Breeding Systems

A breeding program is designed to bring about genetic improvement in a herd. To have an effective breeding program, one (1) must have a clearly defined goal, (2) must be able to measure the traits associated with such a goal, and (3) must utilize these measurements in selecting animals.

The overall goal of any breeding program should be improvement in one or more of the following *economically* important traits:

1. Reproductive performance or fertility
2. Calving difficulty and birth weight
3. Nursing or mothering ability
4. Growth rate
5. Efficiency of gain
6. Longevity
7. Carcass merit

Development of an effective breeding program requires an understanding of the basic principles of beef cattle genetics. This chapter outlines some of the basic principles of beef cattle genetics to improve beef cattle and describes breeding systems for use by purebred and commercial cattle producers.

SELECTION PRINCIPLES

Animal differences result from genetic (hereditary) differences transmitted by

USDA Bulletin No. 286 (Cundiff and Gregory, 1977) was used extensively in the preparation of this chapter.

parents and environmental influences. Cattle receive 50% of their inheritance from their sire and 50% from their dam. Cattle have 30 pairs of chromosomes, which are the carriers of genetic material found in the nucleus of the cell. Chromosomes carry inheritance units called *genes*. Genes occur in pairs in most body cells; however, the ovaries and testes produce reproductive cells (*gametes*), which contain only one member of each chromosome pair. One gene of each pair comes from the sire and the other from the dam. A sire or dam transmits one gene from each pair, but not both, to each of the offspring. Pairing of chromosomes at the time of fertilization restores the full complement of chromosomes, and this restoration keeps the chromosome number constant from generation to generation.

The member of a pair of chromosomes that a particular reproductive cell receives is determined by random process. Some reproductive cells contain more genes that are desirable for economically important traits than others. Chance segregation in the production of reproduction cells (gametes) and recombination upon fertilization explain why the same parents produce genetically different offspring.

Beef cattle breeders select animals based on their phenotype. *Phenotype* refers to differences among individuals in a population that can be measured by the senses (e.g., rate of gain, carcass characteristics, and coat color) (Lasley, 1981). Many observed phenotypic differences may be due to a single gene pair with one member of the pair being dominant. The *dominant* gene has the ability to mask the effect of the other member of the pair. The masked gene is referred to as *recessive*. For example, the gene for the polled characteristic (P) masks the gene for horns (p) when both are present. Thus, the polled characteristic is dominant; the horned characteristic is recessive.

Inheritance of most of the economically important traits—for example, carcass characteristics, growth rate, and feed efficiency—is affected by many genes. Environmental effects are usually larger when many genes affect a trait.

Gene Frequency

The objective of a beef cattle selection program is to increase the number or frequency of desirable genes affecting the economically important traits. This is accomplished by culling animals that are below the herd average in genetic merit and retaining those that are above average. *Gene frequency* may be defined as the percentage of available locations that a particular gene occupies in a herd or population.

Genetic Variation

The genetic makeup (*genotype*) of an animal can be divided into two portions: (1) the additive genetic value (breeding value) and (2) the non-additive portion of the genotype.

The breeding value is that portion of the total genotype that is transmitted from parent to offspring and is determined by the additive effects of the genes affecting a particular trait.

Additive gene effects are comparable to adding block upon block in the construction of a building (gene effect upon gene effect). The sum of additive gene effects totaled over all pairs of genes affecting a

particular trait determine an individual *breeding value* for that trait. The results of selection is to increase the frequency of desirable genes that produce additive effects.

The non-additive portion of the genotype is determined by the way in which genes are combined in an individual. When specific pairs of genes produce a response, the result is referred to as *heterosis*, or "hybrid vigor." For traits in which most of the genetic variation is non-additive, the breeding program must be designed to produce specific crosses that produce favorable gene combinations.

Factors Affecting Rate of Genetic Improvement

HERITABILITY

The proportion of total variation (genetic and environmental) caused by additive gene effects is called *heritability*. Stated more simply, heritability is the proportion of differences (measured or observed) between animals that is transmitted to the offspring. The higher the heritability of a trait, the greater the rate of genetic improvement is, or the more effective the selection will be for the trait. For traits of equal economic value, those with high heritability should receive more emphasis in selection than those with low heritability. To ensure that the differences observed were of genetic origin, all animals considered in the selection must have been subjected to the same environmental effects. Average heritability estimates for some of the economically important traits of beef cattle are presented in Table 3–1. A heritability estimate of approximately 40% means that 40% of the variation in a group of individuals is due to heredity and that the remaining 60% is probably due to the environment.

SELECTION DIFFERENTIAL

Selection differential may be defined as the difference between selected individuals and the average of all animals from

TABLE 3–1. Heritability Estimates of Some Economically Important Traits in Beef Cattle

Trait	Heritability (%)
Fertility	0–10
Birth weight	35–40
Weaning weight	25–30
Cow maternal ability	40–45
Post-weaning daily gains	40–45
Post-weaning daily gains (pasture)	30–35
Yearling weight	50–55
Carcass characteristics	
Carcass grade	40–45
Ribeye area	60–65
Marbling score	50–60
Fat thickness	40–45
Percentage retail product	25–30

which they were selected. Selection differential within a beef cattle herd is determined by:

1. The proportion of progeny needed for replacements
2. The number of traits considered in selection
3. The animal differences within a herd

Every effort should be made to obtain the greatest selection differential possible for the economically important traits in a selection program.

GENETIC ASSOCIATION AMONG TRAITS

A genetic correlation among traits results from genes favorable for the expression of one trait tending to be either favorable or unfavorable for the expression of another trait (Cundiff and Gregory, 1977). Genetic correlations may be positive or negative. An example of favorable association is between growth rate and feed efficiency between 7 and 18 months of age. An unfavorable or negative association would be that between outside fat thickness and marbling score.

GENERATION INTERVAL

Generation interval is the average age of the parents when the replacement offspring are born; in most herds, the average is between 4 and 6 years. The speed at

which you could make improvement in a herd depends not only upon the heritability estimate and the selection differential but also upon the generation interval. Keeping the generation interval short and increasing the selection differential are the factors that boost annual improvement. The formula for calculating the annual improvement for any heritable trait is as follows:

Annual Progress
$$= \frac{\text{Heritability} \times \text{Selection Differential}}{\text{Generation Interval}}$$

Because one needs to keep a high percentage of heifers, and just one bull can be saved for each 25 or more heifers, it is possible to have a much higher selection differential on the male side than on the female side. For this reason, most of the genetic improvement needs to come from the sire. For example, assume that the average yearling weight of a herd was 600 lb, and that the breeder selected replacement heifers that weighed 650 lb, giving a selection differential of 50 lb, and a bull with an adjusted yearling weight of 800 lb, giving a selection differential of 200 lb. What would he expect the adjusted yearling weights of the offspring to be?

Selected heifer's
differential 650 − 600 = 50
× heritability estimate = 40%

Expected increase from females	=	20 lb
Contribute ½ to offspring		10 lb
Selected bull differential 800 − 600	=	200
× heritability estimate	=	40%
Expected increase from bull		80 lb
Contribute ½ to offspring		40 lb
Heifer contribution	=	10 more lb
Bull contribution	=	40 more lb
Total		50 lb
+ old herd average		600 lb
Expected yearling weights of progeny		650 lb

Inherited Abnormalities and Lethal Traits in Cattle*

Whenever an abnormal calf is born, the breeder wonders if the condition is due to disease, faulty nutrition, or genetics. Often, the cause is difficult to determine since there are many similar conditions with diverse causes. To be certain that any abnormality is genetic usually requires many test matings and a thorough evaluation of the animal's pedigree, if available. An accurate diagnosis cannot be made from only one calf.

Most genetic abnormalities are recessive and produce their effect in the homozygous genotype. This means that for the calf to show the trait, both the bull and the cow must be carriers of the undesirable gene.

The occurrence of most abnormalities is rare—less than one affected calf per 10,000. When any lethal condition appears in a herd, however, it is likely to show up more than once. This means that the breeder has several carrier cows as well as a carrier bull.

---

* Much of the information in this section was adapted with permission from Christian, R.E. 1980. Inherited abnormalities and lethals in cattle. *In* Cow-Calf Management Guide. Cattleman's Library. Cooperative Extension Service, University of Idaho, Moscow, ID. CL-605.

The best way to eliminate the problem is to buy a new bull that is not related to the one that sired the affected calves. Although the breeder, in this case, cannot be sure that the condition is gone, the chances are quite small that the new bull is also a carrier. Several generations of clean bulls are needed to eliminate the carrier cows from the herd.

The following is a systematic approach for elimination of a genetic abnormality in a purebred herd:

1. Cull all sires that have produced defective offspring.
2. Replace the culled herd sires with animals whose pedigrees are clean for the abnormality in question.
3. Identify all females that have produced defective offspring. They may be culled or retained for progeny testing of future herd sires to determine whether they are heterozygous for the gene responsible for the defect.
4. Cull close relatives of affected animals, including normal offspring of sires and dams that have produced defective individuals.
5. If affected individuals are viable and fertile, retain them for testing prospective breeding animals from among their progeny.
6. Test the progeny of prospective herd sires before using them extensively in the herd.

The following is a partial list of the known genetic abnormalities and lethal traits in beef cattle:

1. Amputation—Short head, bulging eyes, short lower jaw, lower part of legs missing; lethal recessive.
2. Bull dog—Extremely short head, abnormal legs, small overall size, usually born premature; lethal recessive.
3. Defective skin—Defective formation of skin below hocks and around nose and mouth; lethal recessive.
4. Double muscling—Extremely well-developed muscles (particularly in round and rump), reduced fertility in affected cows, gestation length increased about 10 days, difficult calving; incomplete recessive.
5. Dwarfism—Short head, mature-looking body at birth, constant bloating, usually fatal condition; recessive.
6. Flexed pasterns—Front toes turned under, inability to walk; recessive.

7. Hairlessness—calves born without hair, usually fatal; recessive.
8. Mule foot—Only one toe on each foot; recessive.
9. No anus—No exterior opening of anus; lethal recessive.
10. Prolonged gestation—Pregnancy beyond normal term (up to 400 days), calves usually stillborn; recessive.
11. Screw tail—Lower part of tail is kinked; recessive.
12. Cancer eye—Not simply inherited; caused by many pairs of genes. Susceptible cattle are more likely to get cancer eye if exposed to irritating environment. It is a disease of mature cattle. Heritability of cancer eye susceptibility is about 30%.
13. Water on the brain—Accumulation of fluid in the cranial cavity, bulging forehead, usually fatal shortly after birth; recessive.

BREEDING SYSTEMS

The five fundamental types of breeding or mating systems are (Cundiff and Gregory, 1977):

1. Random mating
2. Inbreeding
3. Outbreeding
4. Assortative mating
5. Disassortative mating

A definition of each system follows:

Random mating is mating of individuals without regard to similarity of pedigree or of performance (phenotype).

Inbreeding is mating of individuals that are more closely related than the average of the breed or population. Linebreeding is a form of inbreeding.

Outbreeding is mating of individuals that are less closely related than the average of a breed or population. The term "outcrossing" is also used to mean outbreeding when matings are made within a breed.

Phenotypic assortative mating is the mating of individuals that are more alike in performance traits than the average of the herd or group.

Phenotypic disassortative mating is the mating of individuals that are less alike in performance traits than the average of the herd or group.

Outcrossing

Outcrossing is the mating of unrelated animals within the same breed. Unrelated is interpreted to mean no common ancestors in the first four to six generations of their pedigrees. It is the most common breeding system used by breeders of purebred cattle. Outcrossing selected individuals increases the number of pairs of heterozygous genes in the offspring. Increasing heterozygosity tends to cover up detrimental recessive genes. Also with increased heterozygosity, "hybrid vigor" is expressed when the average of the offspring exceeds the average of the parents.

Inbreeding

Inbreeding is the mating of individuals that are more closely related to each other than the average of the breed. When mates are related through common ancestors, the resulting offspring are inbred.

Inbreeding increases the proportion of gene pairs that are homozygous and decreases the proportion that are heterozygous. Inbreeding of any offspring is one-half the relationship of the parents if the parents are not inbred (Table 3–2). An offspring from a sire-daughter mating would be 25% inbred. This individual would be expected to have 25% fewer heterozygous gene pairs than an animal that is not inbred. Usually, there is no danger of excessive inbreeding in a closed herd if four or more sires are used. If a breeder closes his herd to outside breeding sources, the rate of increase in inbreeding per generation is calculated as follows:

$$\text{Rate of inbreeding per generation} = \frac{1}{8m} + \frac{1}{8f}$$

where: m = number of sires used in each generation

f = total number of females in the herd in each generation

Inbreeding has been shown to depress growth rate, reduce reproductive efficiency, increase calf mortality, and increase

TABLE 3–2. Genetic Relationships and Inbreeding Values

Mating	Relationship of Mate (%)	Inbreeding of Offspring (%)
Parent-offspring	50	25
Full sibling	50	25
Grandparent-grandchild	25	12.5
Half-sibling	25	12.5
First cousin	12.5	6.2
Double first cousin	25	12.5
Uncle-niece	25	12.5

From Brinks, J.S. 1974, 1976. What you can expect from linebreeding. *In* Linebreeding, Simmental Shield, Lindsborg, KS.

the probability that an inherited abnormality may express itself phenotypically.

Inbreeding should be used only as an aid in producing seed stock that with predictable results can serve as parents for outbred or crossbred commercial animals.

Linebreeding

Linebreeding keeps the relationship to a superior animal prominent in the descendants' pedigrees. The primary purpose of linebreeding is to retain within a herd a large proportion of the genes of a designated sire. Linebreeding is also used when there is a high likelihood of reducing the merit of the herd when outside sires are introduced. Only breeders with superior herds should consider linebreeding. The herds should be large enough so that the rate of inbreeding is sufficiently slow to provide opportunity for selection before genetic variation is reduced.

Producers of purebred cattle should consider the following factors when contemplating the implementation of linebreeding within their herd:

1. Linebreeding for its own sake is of questionable value.
2. The breeder should be sure of the merit of his herd and of the direction in which he is headed before instituting linebreeding.
3. The power of linebreeding is usually much less than that of selection.
4. Linebreeding can be used to maintain the genetic contribution from a given animal at a relatively high level instead of allowing it to be cut in half with each generation, as occurs otherwise.
5. Linebreeding builds up homozygosity (sameness) and prepotency within a herd.
6. Market demands can change and leave you dangling with your linebreeding program should you start wrong.
7. Linebreeding can cause an accumulation of undesirable genes as well as desirable ones.

Crossbreeding

Crossbreeding is the mating of animals from different established breeds and is similar in principle to outcrossing. Results from crossbreeding are usually of larger magnitude than outcrossing, since on the average, animals from different breeds are more genetically dissimilar than animals from different families within the same breed.

Crossbreeding increases herd productivity by combining the desirable characteristics of two or more complementary breeds, and by heterosis (hybrid vigor) exhibited by the crossbred calf and cow.

Heterosis is the increase in performance for certain traits attained by crossbred individuals over and above the average performance of their straightbred parents. It is calculated by the following formula:

% Heterosis

$$= \frac{\text{Crossbred Avg} - \text{Straightbred Avg}}{\text{Straightbred Avg}} \times 100$$

TABLE 3–3. Heritability, Heterosis, and Improvement of Various Traits

Type of Trait	Average Heritability	Expected Heterosis	Best Method of Improvement
Reproduction	Low	High	Crossbreeding
Growth	Moderate	Moderate	Crossbreeding and selection
Carcass	High	Low	Selection

For example, if the average weaning weight of the straightbred calves of breed A is 455 lb, and for breed B calves is 445 lb, then the average of the straightbred calves is 450 lb. If the average weaning weight of the crossbred calves is 470 lb, then the percentage heterosis is estimated as:

$$\frac{(470 - 450)}{450} \times 100 = 4.4\%$$

Heterosis is largest for traits of low heritability, such as fertility and livability (Table 3–3). In contrast, traits that are highly heritable exhibit little if any hybrid vigor. Studies at the U.S. Meat Animal Research Center (MARC) have shown that a properly executed crossbreeding program can increase reproductive efficiency of cows and the survival of calves, boost growth rates, and improve milk production significantly.

Crossbreeding is not a substitute for good management, nor is it a cure-all for unproductive cattle. To reap maximum benefits, a crossbreeding system requires better management. Cattle producers should consider the following items before implementing a crossbreeding system:

1. A definite goal and a long-term plan.
2. A planned marketing program.
3. A good management system.
4. Use of superior, performance-tested bulls.
5. Maximum use of the crossbred cow.
6. Accurate records of identification.
7. A herd improvement program that eliminates open cows and replaces low producers with heifers based on their performance records.
8. Artificial insemination (AI) studs and purebred seed stock herds that have bulls with superior genetics.
9. Combination of breeds to fit the management system and market.

CROSSBREEDING SYSTEMS

Two-Breed Crisscross

This system uses only two breeds, rotating purebred bulls from each breed (Fig. 3–1). The present herd of straightbred cows is used as the base. They are bred to bulls of a different breed chosen to complement the first. Then the F_1 heifers are kept for replacements and are bred to a bull of the same breed as the original cow herd. The foundation straightbred cows are gradually replaced with the crossbred heifers. Studies at MARC have shown that utilization of a two-breed system produces a 15% gain in pounds of calf weaned per cow exposed.

During the early generations, the crossbred cows and their calves have varying proportions of the two breeds, but after three generations, it stabilizes at two thirds of one breed and one third of the other breed. From then on, if breed A bulls are used in breeding herd 1, all cows in the herd are 1/3A:2/3B, and the resulting calves are 2/3A:1/3B.

Conversely in herd 2, breed B bulls are mated to cows that are 2/3A:1/3B, and the calves produced are 1/3A:2/3B. The best heifer calves produced in herd 1 can be selected for replacements in herd 2, and vice versa.

Advantages of a two-breed crisscross are that (1) it is simple and that (2) replacement heifers are produced in the herd.

Disadvantages of a two-breed crisscross are that (1) some heterosis is sacrificed in

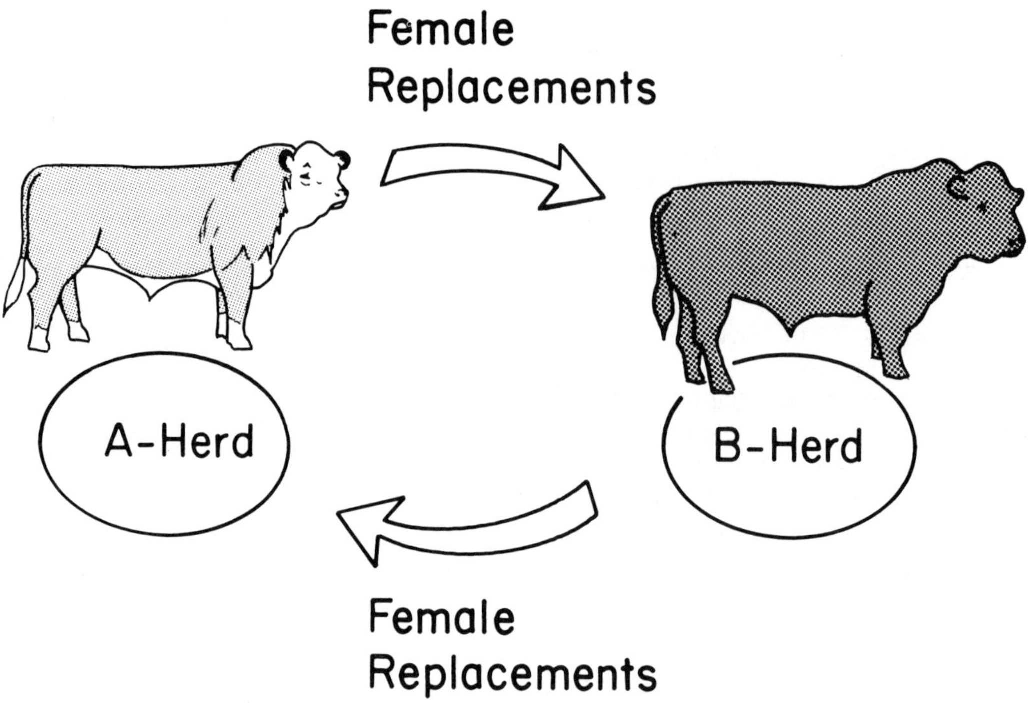

Female
Replacements

FIG. 3–1. Example of a two-breed crisscross. (Drawn by Kathy Dawes Graphics, Moscow, ID.)

the crossbred females and calves since they are not F_1's and that (2) two breeding pastures are required in herds of 50 or more cows. Use of AI would make this system more practical for small herds.

Three-Breed Rotation

In any breeding herd, the crossbred cows are those that are least related to the breed of sire. As with two-breed rotation, heifer replacements are selected from one breeding herd for use in a second breeding herd in which there is a sire breed that has contributed the least to the breed composition of the heifers (Fig. 3–2). For example, herd 1 consisting of breed A bulls mated to cows that are .14A:.29B:.57C produce calves that are .57A:.14B:.29C. Heifers from this group serve as replacements for herd 2 that has breed B bulls. A three-breed cross can add up to 19% more pounds per calf.

Advantages of a three-breed rotation are that (1) it achieves near maximum utiliza-

tion of heterosis and that (2) replacement heifers are produced in the herd.

Disadvantages of a three-breed rotation include the following: (1) Finding three breeds that complement each other is sometimes difficult. (2) The system is practical only in herds of 75 or more cows. (3) Three breeding pastures are required unless AI is used. (4) The system entails the additional cost of maintaining three different breeds of bulls.

Three-Breed Terminal-Cross

This plan involves mating a purebred sire of a third breed to specialized F_1 cows to produce a three-breed cross market animal (Fig. 3–3). The producer buys F_1 females that ideally develop into moderate-sized, highly fertile, good-milking cows. The sire breed should be selected to add growth and carcass merit and to "stamp" a uniform color on the three-breed cross calves that go to market. This

Female Replacements

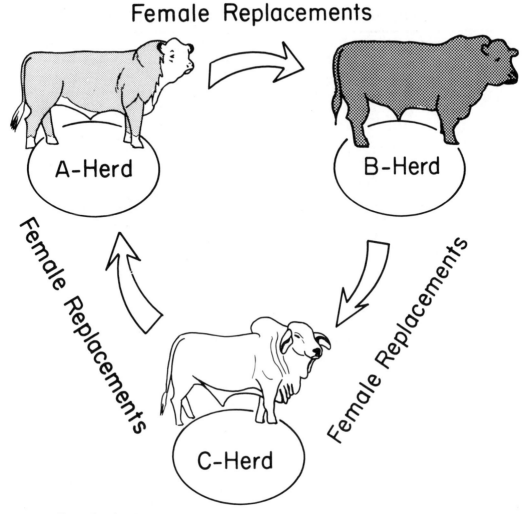

FIG. 3–2. Example of a three-breed rotation. (Drawn by Kathy Dawes Graphics, Moscow, ID.)

program begins once F_1 females are obtained and the sire breed is chosen, and it continues, with the calves being one-half sire breed and one-quarter each breed present in the F_1 cow.

Advantages of a three-breed terminal cross are that (1) it produces maximum hybrid vigor in the F_1 female, that (2) it offers the opportunity to combine the desirable characteristics of sire and dam breeds to produce a specific product, and that (3) only one terminal sire is needed.

Disadvantages of a three-breed terminal cross include the following: (1) F_1 females

must be purchased unless the operation is large enough to produce them. (2) Specific F_1 females are hard to locate, may be expensive, and could introduce disease into the main herd. (3) If a cowman is producing his own F_1 females, three breeds of sires and at least three breeding pastures are required with natural service. AI makes this program much more practical.

Rotational Terminal-Sire-Cross

Rotational terminal-sire crossbreeding involves rotational mating of maternal breeds in a portion of the herd to provide

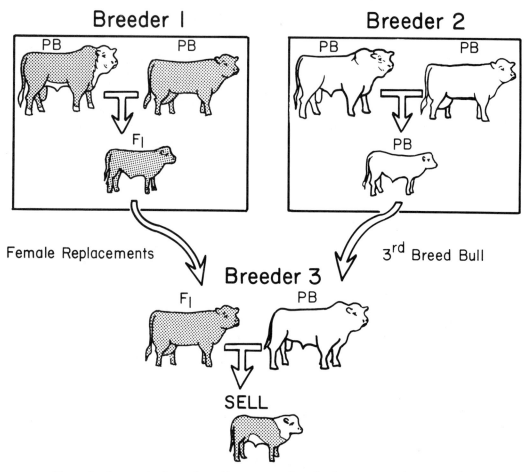

FIG. 3–3. Example of a terminal crossbreeding system. Breeder 3 represents the terminal cross while breeders 1 and 2 are producing the crossbred females and purebred bulls required to achieve a terminal cross. (Drawn by Kathy Dawes Graphics, Moscow, ID.)

female replacements for the entire herd. A portion of the replacements are mated to a terminal-sire breed as outlined in Figure 3–4. In most herds, 45% of the cows must be in the rotational portion of the program to meet requirements for heifer replacements. The other cows in the herd (50 to 55%) would be mated to a terminal-sire breed selected to excel in growth rate and carcass characteristics.

Breeds used in the rotational portion should be comparable in size and milk production to minimize calving difficulty and to stabilize nutrition and management requirements in the cow herd. The youngest cows in the herd would be in the rotational crossing portion. This insures that cows are not mated to a terminal-sire breed of large size until they are 4 or more years of age.

Advantages of a rotational terminal-sire cross are that (1) it offers the opportunity of combining desirable characteristics of sire and dam breeds to produce a specific product, that (2) replacement heifers are produced within the system, and that (3) older cows would be bred only to terminal sires, which would minimize dystocia.

Disadvantages of a rotational terminal-sire cross include the following: (1) Nearly

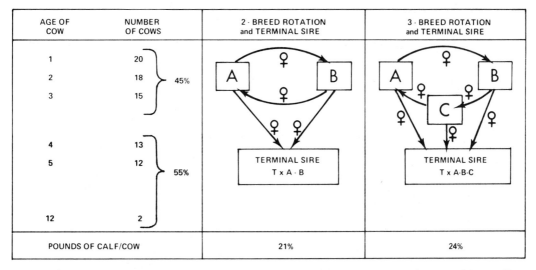

AGE OF COW	NUMBER OF COWS		2 - BREED ROTATION and TERMINAL SIRE	3 - BREED ROTATION and TERMINAL SIRE
1	20			
2	18	45%		
3	15			
4	13			
5	12	55%		
12	2			
POUNDS OF CALF/COW			21%	24%

FIG. 3–4. Rotational terminal-sire crossbreeding system. One hundred cows calving and 20 yearling heifers are assumed. (Adapted from Cundiff, L.V., and K.E. Gregory. 1977. Beef cattle breeding. USDA, Washington, DC. Agriculture Information Bulletin No. 286.)

all heifers produced in the rotational portion of the program are required as replacements. (2) Cows must be identified by breed of sire and birth year. (3) This system requires either three or four breeding pastures unless bulls of the maternal sire breed (A, B, and C) are replaced by bulls of the next breed in the sequence after only two years' use.

Simplified Crossbreeding Systems

Gosey and Linton (1980) have described several modified crossbreeding systems that use the basic principles, while avoiding some of the complexities, of traditional crossbreeding systems.

One of the fastest crossbreeding methods is to purchase two- or three-breed crossbred females and breed them to a different terminal sire breed. Only one breed of bull is needed (the terminal sire), and only one breeding pasture is required. All progeny are marketed, and no identification by sire breed and year of birth is required. The biggest management problem is acquiring high-quality crossbred females each year.

A second method is the equal use of two or more breeds of bulls in the same breeding pasture every year once the crossbred herd is established. A herd of Angus-Hereford cows would be bred to an equal number of Hereford and Angus bulls. This system requires only one breeding pasture, is simple to use, requires no identification, and produces replacement heifers within the herd. Heterosis eventually stabilizes at 50% of maximum for both cows and calves. One disadvantage of this system is a reduction in uniformity of the progeny.

The last system involves the use of only one bull breed for a series of years, then the rotation to a second bull breed. As in the multi-breed bull pasture system, only one breeding pasture is required, and replacement heifers are generated within the system. Straightbred cows of breed A are mated to breed B bulls. When most of the foundation breed A cows have been culled, rotation to breed A bulls is made. The maximum level of heterosis in this system stabilizes at approximately 60% of maximum.

REFERENCES

Brinks, J.S. 1974, 1976. What you can expect from linebreeding. *In* Linebreeding. Simmental Shield, Lindsborg, KS.

Christian, R.E. 1980. Inherited abnormalities and lethals in cattle. *In* Cow-Calf Management Guide. Cattleman's Library. Cooperative Extension Service, University of Idaho, Moscow, ID. CL-605.

Cundiff, L.V., and K.E. Gregory. 1977. Beef cattle breeding. USDA, Washington, DC. Agriculture Information Bulletin No. 286.

Gosey, J., and A. Linton. 1980. Practical breeding programs. *In* Cow-Calf Management Guide. Cattleman's Library. Cooperative Extension Service, University of Idaho, Moscow, ID. CL-1020.

Lasley, J.E. 1981. Beef Cattle Production. Prentice-Hall, Englewood Cliffs, NJ.

Selected Readings

Bogart, R. 1959. Improvement of Livestock. MacMillan, New York.

Carpenter, J.A. 1975. Crossbreeding systems for beef production. *In* Great Plains Beef Cow-Calf Handbook. Cooperative Extension Service, Great Plains States. GPE-8352.

Cundiff, L.V. 1970. Experimental results of crossbreeding cattle for beef production. J. Anim. Sci. *30*:694.

Cundiff, L.V. 1977. Foundations for animal breeding research. J. Anim. Sci. 44:311.

Warwick, E.J., and J.E. Legates. 1979. Breeding and Improvement of Farm Animals. 7th Ed. McGraw-Hill, New York.

QUESTIONS FOR STUDY AND DISCUSSION

1. Why do the same parents produce genetically different offspring?
2. Distinguish between genotype and phenotype of an individual.
3. How many pairs of chromosomes does the normal body cell of beef cattle contain?
4. Describe the difference between a dominant and a recessive gene.
5. Define *gene frequency* and *heritability*.
6. Distinguish between additive and non-additive gene effects.
7. How should one interpret heritability values?
8. What would be the expected weaning weight of the offspring of the following mating?
   Average weaning weight of herd = 400 lb
   Weaning weight of selected females = 450 lb
   Weaning weight of sire used = 550 lb
9. What would be the most practical approach for a commercial cattleman to use to eliminate a genetic defect from his herd? Should a purebred breeder use a more systematic approach?
10. Define the mating systems that have been used to improve beef cattle through breeding.
11. What is the genetic effect of outcrossing? How is it similar to crossbreeding?
12. What is the genetic effect of inbreeding?
13. When should inbreeding be used by a seed stock producer?
14. What is the effect of inbreeding on the economically important trait?
15. When should linebreeding be used?
16. How does crossbreeding improve herd productivity?
17. Define heterosis.
18. Which traits respond more to crossbreeding—those of high heritability or low heritability?
19. Cows of breed A wean 450-lb calves at 205 days of age, and cows of breed B wean calves that weigh 475 lb. If the average weaning weight of the crossbreed calves from the mating breed A and B is 470 lb, what is the percentage of heterosis or hybrid vigor?
20. Outline a two-breed rotation, a three-breed rotation, and a three-breed terminal crossbreeding system. List advantages and disadvantages of each.
21. Would you expect more hybrid vigor from mating Hereford and Charolais, or from mating Hereford and Angus? Why?

Chapter Four

Beef Cattle Improvement Program

The key to genetic improvement through selection is determining which individuals are of superior genetic value. To estimate the genetic potential of an animal, data must be compiled and evaluated. These data include all traits of economic importance as measured by performance records, physical evaluation, and reproduction records. To collect such data, each individual animal must be permanently identified so that the proper information can be matched with it.

IDENTIFICATION METHODS*

Positive animal identification is desirable in all cow herds and is a necessity in purebred herds. Identification makes it possible for the cow-calf producer to keep good records for participation in performance testing programs, for the cattle feeder to keep feedlot-gain and cost-of-gain records, and for any cattleman to verify ownership of stray or stolen cattle.

Requirements of a Good Identification Program

No one identification system is best for all herds. Selection of a system depends on one's objectives. The system selected,

---

* Some of the information in this section was adapted from Nelson, L.A., and W.L. Singleton. 1974. Beef cattle identification methods. Cooperative Extension Service, Purdue University, W. Lafayette, IN. AS-410.

however, should exhibit the following characteristics:

1. It is inexpensive.
2. It is permanent.
3. It is easy to apply.
4. It causes little pain or harm to the animal.
5. It is visible from a short distance.
6. It contains a meaningful numbering system.
7. It uses numbers rather than letters for electronic data processing.
8. It is concise. (It does not use more digits than necessary.)
9. It will not require duplication of numbers within a 10-year period.

Identifying the Cow

All cows in the herd should be permanently identified, and in a form that is easily seen at a distance of several feet. The following sections briefly discuss the various identification systems, their methods of application, and their advantages and disadvantages. Since none of these methods are perfect, use of two systems is recommended to ensure positive and permanent identification. (Manufacturers' instructions on how to use various methods, such as tattoo or eartag, usually accompany the equipment. Additional help can be provided by the local County Extension Office.)

HOT BRAND

This method of identification is permanent and is the best to use if establishment of ownership is the primary purpose. Hot branding burns the hide, and as the burn heals, scar tissue forms, leaving a permanent hair-free outline of the brand (Fig. 4–1).

Disadvantages of hot brands are the initial cost of a set of irons, the labor required for branding, some discomfort to the animal, damage to the hide, possibility of improper application, which can lead to illegible brands, and long winter hair coats that may make hot brands difficult to read. The brand area should be clipped each fall or winter before calving season.

The following are clues to successful hot branding:

1. Use a properly heated iron. The iron is properly heated when it appears silver-gray in the daylight but glows a cherry-red color when held at the bottom of a 5-gallon bucket.
2. Use a properly constructed iron. A face width of approximately ¼ to ½ inch is preferred.
3. Brand only dry cattle. Branding wet cattle results in scalding and excessive scarring.
4. Brand when flies are not a problem. If this is not possible, use an insecticide on the brand.

All ownership hot brands must be registered with a State Brand Board. Brands usually are registered for 5-year periods or for the remaining portion of any 5-year period. A certificate of registration of a brand, as shown in Figure 4–2, is usually issued for each brand registered. All brands must be placed only at such location or locations on the animals as granted by the brand certificate. Wattles, earmarks, dewlaps, or eartags are not registered in most states.

FREEZE BRANDING

This is a relatively new method of branding. It uses extreme cold to selectively destroy the color producing cells (melanocytes) without damaging the hair follicle (Fig. 4–3). The new hair appearing in the brand site should be white.

Freeze branding is relatively painless to the animal, does not leave a thick scar like hot iron branding and the brand is visible year round without clipping. Disadvantages are that in most states, it is not a legal form of ownership; that it is time-consuming (15 to 35 seconds per animal); and that

FIG. 4–1. Hot branding cattle.

since the basis of the brand is the formation of white hair, the method is not easy to use on white-haired animals.

The following procedure should be followed when applying a freeze brand:

1. Prepare the refrigerant in a chilling box. If using dry ice/alcohol, finely crush a quantity of dry ice, and add enough 95% alcohol to cover the branding irons.
2. If using liquid nitrogen, chill the branding irons in the storage tank, or pour the refrigerant into a chilling box.
3. Chill the branding iron in the refrigerant until the vigorous boiling stops and only a slow steady stream of bubbles is seen.
4. Select a firm, fleshy area for the freezing site (an area that does not interfere with state branding laws).
5. Clip the animal closely. Flood the clipped area with 95% alcohol, and scrub with a stainless steel brush.

6. Remove the branding iron from the refrigerant, and shake it vigorously. Apply the iron to the branding site with firm pressure for the proper time as indicated in Table 4–1.

EARTAG

This method utilizes either plastic or metal tags. The flexible, one-piece plastic tags are recommended because they are easy to install, stay pliable in cold weather, and are seldom lost. Metal tags tear out of the ear too easily, are difficult to read from any distance, and can lead to irritation and inflammation if they are placed too close to the head. Plastic tags with a permanently bent bar portion should be put in the center of the ear between cartilage ribs (Fig. 4–4).

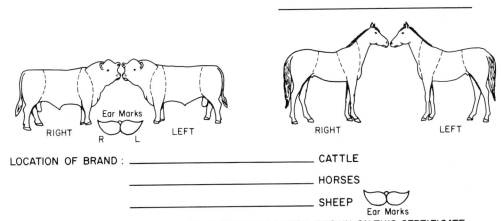

STATE BRAND INSPECTOR   Registration No._____

Date Recorded _____

CERTIFICATE OF BRAND REGISTRATION

The provisions of the Idaho State Brand Board having been complied with, the following brand has been recorded in the State of Idaho for:

NAME AND DESCRIPTION OF BRAND:

RIGHT   Ear Marks   LEFT   RIGHT   LEFT
               R   L

LOCATION OF BRAND : _____ CATTLE

_____ HORSES

_____ SHEEP

Ear Marks

THIS BRAND MUST BE USED ONLY ON THE LOCATION SHOWN ON THIS CERTIFICATE

THIS BRAND EXPIRES JULY 1, 19___

BRANDS MAY NOT BE TRANSFERRED WITHOUT APPROVAL OF THE STATE BRAND INSPECTOR

FIG. 4–2. Certificate of brand registration. (Adapted from Rules and Regulations of Idaho State Brand Board. Boise, ID. April 1, 1977.)

FIG. 4-3. Example of a freeze brand.

Major advantages of the flexible plastic eartag are its low cost, its ease of application, and its legibility when seen at a distance. Also, the cattleman can use several colors of blank tags and different colors of paint to create his own numbers and color-code system. Different colors of tags can be used:

1. In different years.
2. To identify calves from different sires.
3. To differentiate between purebred and grade calves.
4. To designate calves of various parentages in a grading-up program.

TABLE 4-1. Seconds Required for Freeze Branding of Beef Cattle When Dry Ice/Alcohol or Liquid Nitrogen Is Used as a Refrigerant.[a]

Age	Dry Ice/ Alcohol	Liquid Nitrogen
Birth through 1 month	15	10
2–3 months	20	12
4–8 months	25	15
9–18 months	30	18
Over 18 months	35	25

[a] White-haired animals should be branded for 10 seconds longer than shown in table.

Codes for different years, sires, and bloodlines can be placed above the animal's number on the front of the eartag or on the other side.

TATTOO

This method, which involves tattooing a number in the ear with indelible ink, is permanent and is required for registered cattle. The most serious objection is that the animal must be caught to permit a visual inspection of the tattoo in the ear.

It is best to tattoo both ears in the non pigmented, hairless area above the upper ear rib (Fig. 4–4). Tattooing of both ears is especially important when the number involves four or five digits; chances are that the initial or terminal digits will be placed in the hairy area of the ear, and both tattoos may be necessary to read a complete number. Green ink is good on dark pigmented skin. Ear wax must be removed from the tattoo area with alcohol to achieve a legible tattoo.

Black ink in paste form usually works better than liquid ink applied with a roll-on applicator. Regardless of the ink type used, the best tattoos result from thoroughly rubbing the ink into the needle punctures with the thumb and forefinger. New needles can be dulled slightly with a whetstone to produce a better tattoo.

If heifers are vaccinated for brucellosis, the veterinarian's tattoo can be placed in the lower part of the ear below the bottom ear rib, where it will not interfere with the animal's regular tattoo number or eartag.

NECK CHAIN

This method, which consists of a neck chain with a plastic or brass tag attached bearing the animal's number, is used in some herds. Neck chains are easy to use and provide an easily read identification. They are somewhat expensive, however, and many have to be replaced. They must be adjusted as size and condition of the animals change. A neck chain that is too loose may slip over the animal's head and be lost, whereas one that is too tight could choke the animal or result in death by hanging (Fig. 4–5).

EARMARKS

Earmarks supplement brands as an indication of ownership when brands are not easily visible. Earmarks may be recorded with brands. It is one of the best methods for use on range when cattle from several herds are mixed together (grazing in common).

WATTLES

To make a wattle, cut both layers of skin from the connective tissue (fell) that covers the muscle. The wattle should be from ½ to 1½ inches wide and from 3 to 6 inches long. Cut the upper part of the skin entirely free from the animal, leaving the lower end intact, as shown in Figure 4–6. Common places for wattling are the nose, jaw, chin, throat, neck, shoulder, and thigh. Wattles are usually referred to by the part of the anatomy from which they are cut (e.g., nose wattle, jaw wattle, etc.). An exception is the wattle cut on the throat, which is referred to as a bell wattle. Wattles are used

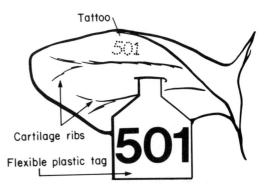

FIG. 4–4. Front view of right ear showing recommended location of tattoo and flexible plastic tag. (Adapted from Nelson, L.A., and W.L. Singleton. 1974. Beef cattle identification methods. Cooperative Extension Service, Purdue University, W. Lafayette, IN. AS-410.)

as a form of identification in grazing associations where cattle from different herds are grazed in common.

In summary, a hot or freeze brand, or a combination of tattoo and a flexible plastic eartag, are the recommended methods of identifying cows. A cow should have only one identification number from birth to the time at which she leaves the herd. To reduce errors in recordkeeping, the eartag

FIG. 4–5. Neck chain holding metal tag identification. (Adapted from Albaugh, R. 1975. How to identify livestock. Cooperative Extension Service, University of California, Leaflet 2297.)

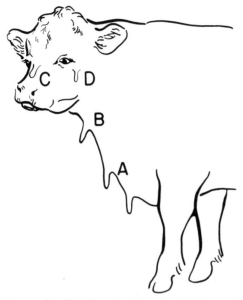

FIG. 4–6. A—Double dewlap. B—Bell wattle. C—Nose wattle. D—Jaw wattle. (Adapted from Albaugh, R. 1975. How to identify livestock. Cooperative Extension Service, University of California. Leaflet 2297.)

number, tattoo number, and herd number of a replacement heifer should be identical.

Identifying the Calf

Performance-testing programs necessitate that calves be individually identified within two or three days after birth. The calf's number can then be recorded in the calving book along with date of birth, birth weight, sex, dam's number, etc. Eartags and tattoos are the two methods best suited for calf identification.

EARTAG

A newborn calf can be easily identified with either a metal or plastic eartag. A flexible plastic tag is recommended until weaning or yearling age. Metal eartags tear out more easily and lead to more infections, especially during fly season. In commercial herds, the eartag may be the only identity because most of the calves are sold after weaning.

TATTOO

In registered herds, calves must be tattooed in the ear to establish the permanent identity used on breed association documents. The same number as is on the eartag should be tattooed in each ear. Then if the eartag is lost, the calf can be identified by the tattoo. The tattoo can include a sire code, but the identification number should be kept to a maximum of four or five digits. Otherwise, the first or last digit will be in the hairy area of the ear and will be difficult to read.

Numbering System

Once a cattleman has determined which method of identification is best suited for his operation, he must develop a meaningful numbering system. The ideal identification system reaches a satisfactory balance between the need for simplicity and the need for providing adequate information.

A numbering system in which the calves are identified consecutively every year is recommended. One advantage of this system is that consecutively numbered tags can be ordered ahead of time for every calf born in a calendar year. Another advantage is that the calf's number contains no more than four digits until 1000 or more calves are born within a herd in a single year. Usually, the last digit of the year is the first digit of the tattoo or tag, designating year of birth. In herds greater than 100 head, the breeding-age females receive a four-digit number. The first digit represents the year of birth (1982 would be 2, 1980 would be 0). The next three digits are the numbers assigned to the cow. The numbers begin with 001 and continue through 999. An example is shown in the following:

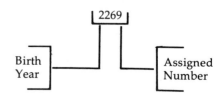

Suckling calves receive a three-digit number that is the same as the mother's numerical number. Each calf should be identified on its day of birth, if possible. When a female calf is selected for replacement, a new number is assigned in the same manner as the existing herd.

For example, a cattleman has a herd of 269 head. Cow 2269 has a calf in 1985. The calf's number will be 269.

2269	269
Cow Number	Calf Number

If the calf is selected for a replacement, the new number will be 5001. The first three digits refer to the order of birth or to the 205-day adjusted weight.

An example follows:

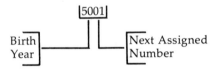

Color coding may also be used in conjunction with a numerical numbering system to provide even more information. For example, color coding may be used to identify different sire or breed progeny groups, sex of the animal, and year of birth, if desired.

Complete and accurate records are the key to good profits in the beef business. Accurate records depend on positive and permanent animal identification. To achieve maximum success in beef cattle identification, the cattleman should:

1. Select a recommended method of identification.
2. Use the proper techniques in identifying the animal.
3. Develop a numbering system that meets his needs, then use it consistently.

RECORDKEEPING

The most important and most difficult part of management is setting goals. Goals should be:

1. Clearly defined.
2. Stated in terms of the results to be accomplished, not methods.

3. Measurable.
4. Challenging but achievable within a realistic time frame.
5. Restricted to desired results.

Records are a tool that can be used to determine progress toward meeting one's goals. They provide the cattleman with a graphic picture of his past management efforts (both good and bad). Without some recordkeeping, it is impossible to handle management decisions effectively. A cattleman cannot improve management until he knows where the opportunities are for improvement.

Records should be kept and used to identify genetically superior animals and cull those that are poor producers. In addition, purebred breeders should remember that record information can be used effectively for promotion and advertising. Figures 4–7 and 4–8 illustrate one type of recordkeeping system that can be used.

BEEF CATTLE IMPROVEMENT PROGRAMS*

Records of performance are useful in that they provide a basis for comparing cattle within a herd as well as within a breed. The first step to identifying high-ranking individuals within a breed is to identify superior individuals within herds. Thus, widespread use of performance testing within herds is the first essential step to beef improvement.

The following are the principal features of keeping good records of performance:

1. All animals are given equal opportunity to perform through uniform feeding and management.
2. Systematic records of economically important traits are maintained for all animals.
3. Records are adjusted for known sources of variation, such as age of dam, age of calf, and sex.

---

* Some of the information in this section was adapted from USDA Extension Service. 1981. Guidelines for Uniform Beef Improvement Program, Beef Improvement Federation Recommendation. Washington, DC. Program Aid 1020.

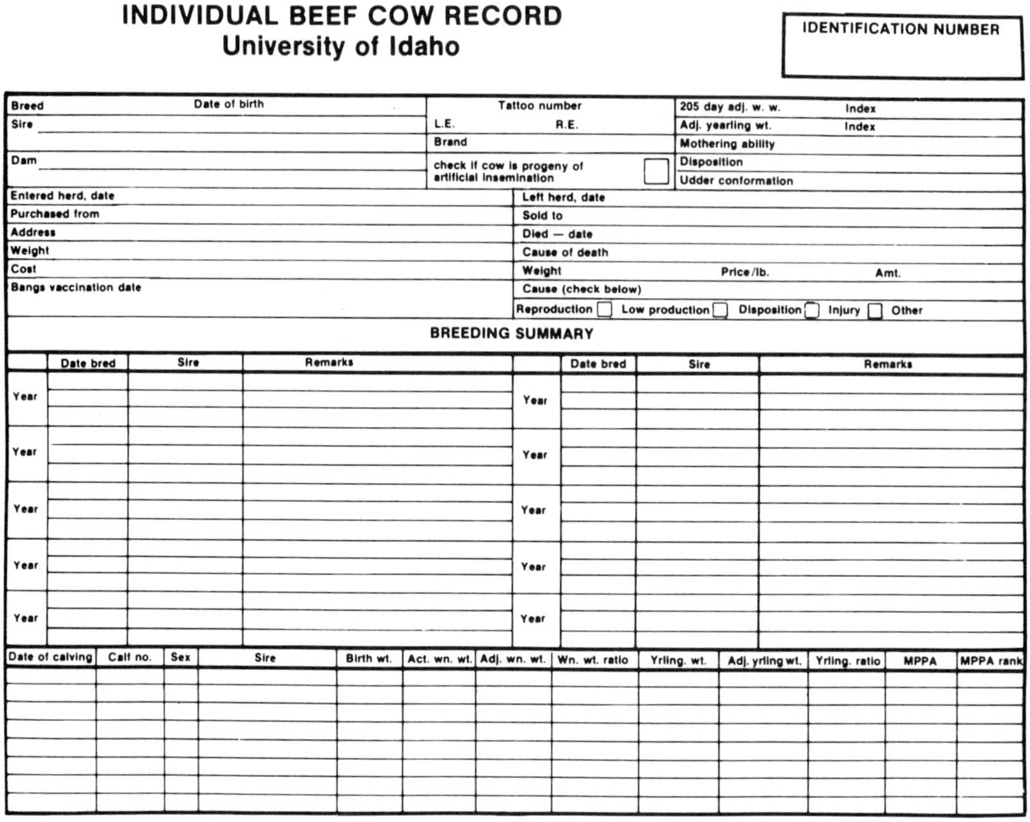

INDIVIDUAL BEEF COW RECORD
University of Idaho

IDENTIFICATION NUMBER

Breed	Date of birth			Tattoo number		205 day adj. w. w.		Index
Sire			L.E.	R.E.		Adj. yearling wt.		Index
			Brand			Mothering ability		
Dam			check if cow is progeny of artificial insemination			Disposition		
						Udder conformation		

Entered herd, date		Left herd, date	
Purchased from		Sold to	
Address		Died — date	
Weight		Cause of death	
Cost		Weight	Price /lb. Amt.
Bangs vaccination date		Cause (check below)	
		Reproduction ☐ Low production ☐ Disposition ☐ Injury ☐ Other	

BREEDING SUMMARY

	Date bred	Sire	Remarks		Date bred	Sire	Remarks
Year				Year			
Year				Year			
Year				Year			
Year				Year			
Year				Year			

Date of calving	Calf no.	Sex	Sire	Birth wt.	Act. wn. wt.	Adj. wn. wt.	Wn. wt. ratio	Yrling. wt.	Adj. yrling wt.	Yrling. ratio	MPPA	MPPA rank

FIG. 4–7. Example of a beef cow record system.

4. Records are used in selecting replacements and in culling poor producers.
5. Nutritional and management programs are comparable to the levels at which the progeny of the herd are expected to perform.

Fertility and the factors that contribute to it have been found to have low heritability. Economically, however, fertility is the most important trait in the beef industry, and in herds where fertility is low, a large selection differential can be achieved. Cattle with poor fertility must be eliminated from the herd for economic reasons. Thus, it is important to maintain complete records on all cows, and fertility records on bulls, in breeding herds. Replacement animals should be selected from parents that have above average fertility.

Production Testing

Production testing is the cow herd phase of a performance testing program in which productivity of each individual cow in the herd is measured in terms of calf weaning weight (Fig. 4–9). A recommended procedure follows:

1. Identify all cows and calves.
2. Record birth weight, sex, and dam of calf.
3. Weigh calves at weaning time (160 to 250 days of age).
4. Make adjustments for age of dam, age of calf, and sex.
5. Record creep-feeding information.
6. Record calf weaning data.
7. Cull cows and replacement heifers on the basis of records.
8. Cull cows and replacement heifers on the basis of physical abnormalities.

CALVING RECORD

Date	Calving score	Cleaned normal	Retained placenta	Other comments	Calving Score
					1 — No assistance
					2 — Hand pull
					3 — Mechanical assistance
					4 — Difficult mechanical assistance
					5 — Cesarean
					6 — Abnormal presentation

Date	Disease or injury	Clinical symptoms	Medications

PREVENTIVE MEDICINE AND VACCINATION RECORD

IBR/PI,	Clostridium	Leptospirosis	Reo Corona	Vitamins A & D	Selenium				

FIG. 4–8. Example of a calving record.

FIG. 4–9. A scale is an important component of a beef cattle performance testing program.

ADJUSTED WEANING WEIGHT

Weaning weights are obtained to evaluate differences in mothering ability and to measure differences in growth potential of calves. For best estimates of genetic worth for weaning weight, it is necessary to adjust the individual calf records to a standard basis. Standardizing weaning weights to 205 days and a mature dam equivalent is recommended. Weight should be recorded within a range of 160 to 250 days. The following formula is used to calculate a 205-day adjusted weaning weight:

205-day wt (lb)

$$= \frac{\text{Actual wt} - \text{birth wt}}{\text{Age in days}} \times 205 + \text{birth wt}$$

To adjust for age of dam, it is recommended that the following adjustment factors be added to the computed 205-day weight for the respective age of dam for each calf:

	Additive Factors (lb)	
Age of Dam	Males	Females
2 years	60	54
3 years	40	36
4 years	20	18
5 to 10 years	0	0
11 years	20	18

There is substantial evidence that these additive adjustment factors for age of dam are not appropriate for all breeds. Therefore, when sufficient evidence is available demonstrating that correction factors other than the ones listed are more appropriate for a given breed, their use is encouraged.

WEANING WEIGHT RATIO

Weaning weight ratios within sex groups are calculated by dividing each calf's 205-day weaning weight that has been adjusted for age of dam by the average of its sex group and expressing it as a percentage of its sex group average. These ratios are useful in ranking individual calves of each sex within a herd for making selections.

205-day adjusted wt ratio

$$= \frac{\substack{\text{205-day adjusted wt} \\ \text{(individual)}}}{\substack{\text{Avg 205-day adjusted wt} \\ \text{herd mates}}} \times 100$$

PRODUCE OF DAM SUMMARY

Keeping a produce of dam summary, a record of lifetime productivity, is recommended. It can provide valuable information for within-herd comparisons. It can be most helpful in identifying both the lowest-producing cows to be culled and the consistently high-producing cows.

A produce of dam summary should include the following information:

Measures relating to reproductive efficiency
1. Age at first calving (days)
2. Current age
3. Number of calves born (lifetime)
4. Number of calves weaned (lifetime)
5. Average age of calves when weaned

Measures relating to productivity
1. Average birth weight
2. Average weaning weight ratio of all calves weaned
3. Average adjusted 365-day weight, weight ratio, and number of contemporaries
4. Most probable producing ability (MPPA)

MOST PROBABLE PRODUCING ABILITY (MPPA)

The MPPA should be included on produce of dam summaries, and ranking of dams should be based on MPPA for the 205-day weaning weight ratio. The MPPA is needed to compare dams that do not have the same number of calf records in their averages. For example, suppose six cows have the following records of production:

Cow	No. Calves	Avg Weaning Wt Ratio	MPPA
A	1	85	94.0
B	2	88	93.2
C	4	90	92.7
D	3	110	106.7
E	4	112	108.8
F	1	115	106.0

In the example, cow A has the lowest lifetime average. This refers to only a single calf, however, for which environmental conditions or the calf's genetic potential for growth might have been below the average of what the cow would normally produce. One or more calves from cows B or C could also have a record of 85 or less. All three cows are probably low producers, but use of MPPA enables more accurate culling and in this example, indicates that cows B and C are slightly lower-producing than cow A.

MPPA for weaning weight ratio is computed by the following formula:

$$MPPA = \overline{H} + \frac{NR}{1 + (N - 1) R} (\overline{C} - \overline{H})$$

where: $\overline{H}$ = 100, the herd average weaning weight ratio

N = the number of calves included in the cow's average

R = 0.4, the repeatability factor from weaning weight ratio

$\overline{C}$ = average of weaning weight ratio for all calves the cow has produced

MPPA of cow D in the previous example is computed as follows:

MPPA cow D

$$= 100 + \frac{3 \times 0.4}{1 + (3 - 1) 0.4} (110 - 100)$$

$$= 106.7$$

Post-Weaning Performance

This phase of the beef cattle improvement program is designed to measure post-weaning performance. Yearling weight (365 days) and long yearling weight (452 and 550 days) are important because of their high heritability and genetic association with growth and efficiency of gain. The following are guidelines to observe when conducting a post-weaning performance test:

1. Identify all animals.
2. Be sure that the adjusted 205-day weaning weights and actual weaning weights are recorded.
3. Take an initial weight before calves are tested. A pretest 21-day period is advised following weaning to compensate for weaning stress.
4. Feed for a minimum of 140 days. Rations should be calculated to provide for an *average* gain of 2.25 lb per day for bulls. Longer test periods are acceptable and advisable for testing on the ranch; however, weights must be recorded at the end of 140 days.
5. Take official on-test and off-test weights on two successive days. An average of those two weights are used to obtain average weight.

The following formula is recommended for calculating 365-day adjusted weights:

Adjusted 365-day wt

$$= \frac{\text{Actual final wt} - \text{Actual weaning wt}}{\text{No. of days between weighing periods}}$$

$$\times 160 + 205 \text{ adjusted wt}$$

Heifers being tested for post-weaning performance are fed a practical ration (high-roughage, low-concentrate) so that they become large enough to begin cycling prior to the breeding season. Some people feel that yearling weights for heifers are not practical and recommend an adjusted 452 or 550-day weight for heifers.

Adjusted 550-day wt

$$= \frac{\text{Actual final wt} - \text{Actual weaning wt}}{\text{No. of days between weighing periods}}$$

$$\times 345 + 205 \text{ adjusted wt}$$

The procedure of using adjusted 365-day weights as a measure of yearling weight applies primarily to herds that develop bulls on a high level of concentrate feeding commencing at weaning time. For herds that develop bulls more slowly, a long yearling weight may be used as an alternative to the adjusted 365-day weight. Yearling weight at 365 days and long yearling weight are particularly important because of their high heritability and their high genetic association with efficiency of gain and pounds of retail, trimmed, boneless beef produced.

WEIGHT RATIOS

Weight ratios for either adjusted 365-day weight (yearlings), adjusted 452-day weight, or adjusted 550-day weight should be computed separately for each sex. The following formula is recommended for computing yearling weight ratio:

Adjusted 365-day wt ratio

$$= \frac{\text{Adjusted 365-day wt (individual)}}{\text{Avg 365-day wt of contemporary group}}$$

$$\times\ 100$$

CENTRAL TESTING STATIONS

Central testing stations are locations where animals are assembled from several herds to evaluate differences in certain performance traits under uniform conditions. Uses of central bull testing stations include:

1. Comparing performance of potential sires to that of similar animals from other herds
2. Comparing sale bulls for commercial producers
3. Progeny-testing sires
4. Educating breeders to familiarize them with performance information
5. Estimating genetic differences between herds or sire progenies

Central bull testing stations are good for promotion and advertising but are too expensive for testing an entire bull-calf crop. Purebred producers, as well as commercial cattlemen, should recognize the limitations of central bull testing stations, that is, that effects of nutritional levels from one stage of development carry over to performance at later stages. Also, cattle have different stages of growth and development; this applies within a breed as well as between breeds. Human children do not all grow at the same rate or at the same age. Some children may have the most rapid growth at 10 years of age, while others grow faster at 12 years of age. This same phenomenon occurs in cattle. Thus, chronologic age may or may not be related to optimum growth rate.

Central testing stations are of greatest educational value when all individuals concerned recognize that only a limited number of traits can be evaluated in them, and that at best, they are merely one phase of a complete performance evaluation program. A primary measure of the effectiveness of central testing stations should be how well they facilitate complete herd testing for the economically important traits. Choices between herds are likely to be made on the basis of diverse information, which may include the results from central testing stations.

The commercial cowman must decide which traits are to receive the most emphasis in his selection program and must select bulls from testing stations accordingly. *Testing does not improve bulls,* it only helps to identify superior ones. Complete herd performance programs in seed stock herds are necessary to achieve satisfactory genetic progress in the beef cattle industry.

The Beef Improvement Federation recommends the following procedures and policies for central testing of bulls:

1. Age of calves at time of delivery to time of testing should be at least 180 days and not more than 305 days.
2. Herds from which bulls are consigned should be on herd testing programs for pre-weaning and post-weaning performance. Calves should have completed the weaning phase of the performance records program, and the following information should be submitted to the testing station:

    Sire, dam, birth date, actual weaning weight and date, adjusted 205-day weight, within herd weaning weight ratio (based on average of all bull calves in same weaning season and management group), and the number of calves making up this average.
3. Cattle should be brought to the testing station for an adjustment or pre-test period of 21 days or more immediately prior to the actual testing.
4. The duration of the feeding test may be influenced by feeding conditions and/or breed goals. It should be 140 days or more.
5. Bulls should be examined by a competent veterinarian for reproductive and structural soundness at the beginning and end of the test.

6. Test rations vary according to locally available feeds and test objectives. Feeding should be allowed on a free-choice basis, with no restrictions on quantity of feed consumed. Rations between 60 and 70% total digestible nutrients (TDN) should be adequate for the expression of genetic differences in growth.
7. Testing of sire groups bulls is more desirable than individual testing since it provides more information to the breeder and to the prospective buyers.

Figure 4–10 illustrates the form recommended for reporting testing station results and gives definitions of these measurements.

PROGENY EVALUATION

Progeny evaluation is the testing of a sire based upon the merits of his offspring. The progeny test, which uses reference sire progeny as the common base of comparison, provides a fair comparison of the breeding value differences among bulls. The basics of a sound progeny test are as follows:

Comparable cows: All bulls to be compared must be mated to a comparable set of cows, to eliminate cow differences between sire progeny averages.

Equal progeny treatment: The resulting progeny from all bulls must be given equal treatment, to reduce environmental differences resulting from the differences between sire progeny averages.

NATIONAL SIRE EVALUATION

Sire selection and evaluation are basic to all beef breeding programs. The performance of the individuals, their ancestors, and collateral relatives, and of their progeny, can all be used to estimate differences in breeding value among sires. Breeding value is twice the difference between the average performance of a large number of progeny by one sire (or dam) and that of the population average when the same sire (or dam) and other sires (or dams) are mated randomly in the population, with all progeny being handled alike. The difference is doubled because parents transmit only one half of their genes (one gene of each pair) to their offspring.

The goal of national sire evaluation is to increase the number of sires whose breeding value differences obtained from all sources can be fairly compared. It is an attempt to provide comparative information on important traits of sires across herds and throughout breeds. Breed associations have developed national sire evaluation programs following the guidelines of the Beef Improvement Federation. The guidelines are based on the use, in different herds, of certain bulls designated as reference sires. When reference sires are used in several herds through artificial insemination, comparisons of sires can be made through the tie provided by reference sires common to each herd. This process is outlined in Figure 4–11.

Data generated are expressed as *expected progeny difference* (EPD) and *possible change* for each trait tested. The EPD is an estimate from existing progeny data of how future progeny are expected to perform relative to the performance of progeny of reference sires. Possible change is an estimate of the accuracy of the EPD value. It is a measure of the range of EPD values that could be expected from use of a specific sire when the EPD values for two sires for the same trait are similar. A bull with an EPD of +10 lb for yearling weight is expected to sire calves that at one year of age will be 20 lb heavier than calves sired by a bull with an EPD of −10 for the same trait. Some breed associations report EPD as a ratio rather than as a positive or negative number. A trait ratio compares the performance of a bull's progeny to that of all other sires in the breed for a specific trait. A ratio of 100 would mean that his progeny are exactly average for the breed. An example of a sire summary from a breed association is shown in Figure 4–12.

Most sire summaries also list estimated breeding values (EBVs) for some traits. EBVs are estimates of an individual animal's true breeding value for a trait based

MEASUREMENTS RECOMMENDED FOR ALL TEST STATIONS

					W.W. Ratio							
	WEANING						GAIN TEST				YEARLING	
Lot No. (1)	Birth Date (2)	Actual Wt (3)	Weaning Date (4)	Adj. 205- Day Wt (5)	W/in Herd and No. (6)	(Date) Initial Test Wt (7)	(Date) Final Test Wt (8)	Age in Days (9)	ADG (10)	Test Gain Ratio (11)	Adj. 365- Day Wt (12)	365- Day Wt Ratio (13)

OPTIONAL MEASUREMENTS FOR ALL TEST STATIONS

YEARLING

Wt Per Day of Age (14)	Conf. Score (15)	Index (16)	Fat Thick. (17)	Est. Yield Grade (18)	Adj. Feed Conv. (19)	Initial Cond. Score (20)

Owner, address, breed, and sire. (Inserted between sire groups, or in a column at the left.)

Each test group (i.e., breed and age group) should be listed together on the report and averaged. (Age range in each group should not exceed 90 days and breed should be averaged separately within age group.)

Sire group averages are shown for 3 or more progeny of same sire.

If sire groups include calves from different age groups, data may be listed together by sires, but with only the average of ratios shown.

(1) Ear tag test number. Tattoo should be recorded elsewhere and may be put on this report if space permits.

(2) Month/day/year of birth. Ex. 2/15/71 for Feb. 15, 1971. If all in the same year, may omit year.

(3) Actual weight used to compute 205-day adj. weight.

(4) Month/day/year when weights were taken to compute 205-day adj. weight.

(5) Weaning weight adjusted to 205 days and for age of dam according to BIF. If creep fed, add C after weight.

(6) Adj. 205-day wt divided by average of all bull calves in same herd in same weaning season group and same management code. Minimum entrance requirement is optional with test management. The number of calves making the average is listed in parentheses. Ex. 105 (17)

(7) &

(8) Average of at least 2 full weights taken on different days. May be more than 1 day apart if desired.

(9) Age at end of test.

(10) Final weight − initial weight + days on test. Minimum length, 140 days, no maximum.

(11) Average Daily Gain + test group average of average daily gain. (Breed within age group average.)

(12) $\dfrac{\text{Final test weight} - \text{Actual weaning weight}}{\text{Days between weights}} \times 160 +$ adj. 205-day wt (adj. for dam's age).

(13) Adj. 365-day wt + test group average of adj. 365-day weights. (Breed within age group average.)

OPTIONAL

(14) Test wt + days of age when weighed.

(15) Any locally adopted scoring system.

(16) Indexes will vary with individual test objectives. They should all be based on ratios to the group average of a trait multiplied by some percentage figure, thus resulting in values ranging below and above a mean of 100.

(17) Fat thickness may be measured by sonoscope and should be expressed per cwt. *and* the absolute value.

(18) Cutability estimates based on sonoscope readings of ribeye area and fat thickness may be classified into the market yield grades of 1, 2, 3, 4, or 5.

(19) Feed conversion of any group fed together in one pen should be expressed as pounds of feed per 100 pounds of gain. The actual amount of feed should be adjusted to a common body weight to eliminate differences in maintenance requirements.

(20) Initial degree of fatness may be visually estimated and scored on a scale of 1 to 5, with 1 being very thin; 5, excessively fat; and 3, average in condition.

FIG. 4–10. Recommended form for reporting test station results. (From USDA Extension Service. 1981. Guidelines for Uniform Beef Improvement Programs. Beef Improvement Federation Recommendation. Washington, DC. Program Aid 1020.)

on the performance of the individual and its close relatives. The simplest EBVs are calculated from an animal's own record by multiplying the heritability estimate for the trait by the animal's deviation from the contemporary group mean. Heritability reveals what portion of the deviation is genetic. An example of how to calculate an EBV for weaning weight is shown in the following:

EBV = Heritability × (Individual's records − Contemporary group mean)

$$= .3 \text{ (Heritability for weaning weight)} \times (550 \text{ lb} - 500 \text{ lb})$$

$$= +15 \text{ lb}$$

In this example, the contemporary group averaged 500 lb for weaning weight, and the individual animal's record was 550 lb for the same trait. The calculated EBV is +15 lb for weaning weight. EBV may be expressed in units of weight or as a ratio. Accuracy figures are attached to EBVs and reveal how closely the EBV predicts the true breeding value.

Breeding values are usually calculated for growth traits (weaning and yearling weight), maternal performance, and occasionally birth weight. Maternal EBV is an estimate of the eventual milking abilities of a bull's daughters.

Commercial Cattle Performance Program

Commercial cattlemen need performance record programs that are affordable

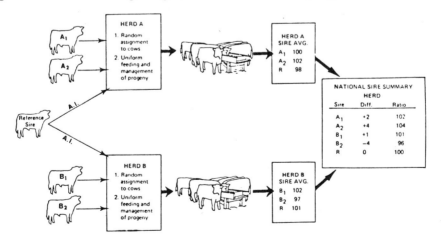

FIG. 4–11. Example of national sire evaluation format. (From Cundiff, L.V., and K.L. Gregory. 1977. Beef cattle breeding. USDA, Washington, DC. Agriculture Information Bulletin No. 286.)

All the Data that prints for the bulls in the Sire Selector is located in this section. The bulls are arranged in alphabetical order by their official name.

* = Reference Sire

Horned Polled Status (3)	NAME OF BULL (1) / BULL'S SIRE / DAM'S SIRE (2)	Country (5) / Currently Registered To City (8) / State	Number (4)	Birthdate (7)		CALVING EASE FIRST CALF	CALVING EAST 2ND CALF & OLDER	BIRTH WEIGHT	CALVING EASE INDEX	WEANING WEIGHT	YEARLING WEIGHT	DAUGHTER'S 1ST CALF CALVING EASE	DAUGHTER'S 1ST CALF BIRTH WEIGHT	DAUGHTER'S 1ST CALF WEAN WEIGHT	USDA QUALITY GRADE	CARCASS - YIELD GRADE	RETAIL CUTS PER DAY AGE
* H	A-A DINDER / HEROD	ENGLAND / FINLEY JR, JAMES H / CLAUDE, TX	101539	04-27-72	RATIO	109.32	100.23	99.52	101.50	103.02	101.53	106.32	100.21	98.43	104.46	108.96	107.34
					ACCURACY	5.32	2.75	.79	(12) 2.15	1.06 ★	3.22	11.90	3.09	3.57	8.40	9.34	4.15 ★
					PROGENY	214	431	679	717	558	76	40	41	39	37	37	37
* H	GALANT / GLADIS	SWITZERLAND / SIMMENTAL BREEDERS CARDSTON 10 / CARDSTON, ALBERTA CANADA		01-25-68	RATIO	93.92	99.39	99.72	98.57	100.67	100.50	98.93	99.87	101.81	101.06	90.63	101.23
					ACCURACY	1.43	.47	.16	.47	.19	.48	1.20	.31	.32 ★	2.68	2.99	1.33
					PROGENY	3468	13619	18230	22590	20380	3974	4506	4623	5668	273	276	276
* H	UMHAU / SIMON	GERMANY / GENETICS DIVISION-CARNATION / HUGHSON, CA	19044	03-10-70	RATIO	89.34	99.64	97.61	96.66	101.75	99.36	105.44	99.59	101.98			
					ACCURACY	10.51	2.88	1.13	3.31	1.32	3.10	8.90	3.22	2.50			
					PROGENY	41	359	331	411	341	86	65	36	69			

1. **Bull's Official Name.** Remember many bulls have have nicknames but will appear only under their official registered name including prefixes if they were used.

   A star (*) by a bull's name also indicates he is a Reference Sire.

2. **Names of his Sire and his Dam's Sire**, along with a Reference Sire designation and their horned polled status.

3. **Bull's Horned-Polled Status.**

4. **Bull's ASA Registration Number**

5. **Bull's Country of Origin.**

6. **These are all 12 Traits,** that are summarized in the Sire Selector. See Appendix A for explanation of terms.

7. **Bull's Date of Birth.**

8. **Bull's Current Owner, City & State.**

9. **Ratio.** A trait ratio compares the performance of a bull's progeny to progeny of all other sires in the breed for this trait. A ratio of 100 would mean his progeny are exactly average for the breed. Ratios over 100 indicate that his progeny are above average. Obviously, half of all Simmental sires will be above 100 and half below 100 in each trait; this does not mean that bulls with ratios of less than 100 are not worth using.

10. **Accuracy.** Random changes (+ or -) in a trait ratio should be expected as additional progeny are reported. The majority of trait ratios should change less than the accuracy value. When a bull has had 300 progeny reported for any trait, his performance in that trait is rather firmly established and it's not likely to change much in the future.

11. **Progeny.** Number of progeny sired by this bull that were compared to a Reference Sire in this trait.

12. **Genetic Trait Leader.** A Star, ★, under a specific trait, indicates that this sire is a 1985 Genetic Trait Leader for that trait.

FIG. 4-12. Example of a sire summary. (Courtesy of the American Simmental Association.)

and workable. A performance breeding program for a commercial producer differs from that of a purebred breeder in that the commercial producer sells pounds whereas the purebred breeder sells breeding value.

A commercial breeder should first document the current level of production and quality in the herd and set goals for 5 to 10 years in the future. These goals should reflect changes in production and quality that individual breeders consider necessary for establishing the most profitable system based on conditions imposed within their own operation. Tables 4–2 and 4–3 can serve as a guide to determining the existing conditions of a herd, the desired improvements in the herd, and the changes that can be made toward the realization of these goals through the use of a performance program.

The types of records kept by commercial producers and the particular traits that are documented vary, depending on the extent to which producers plan to utilize them in the decision-making process.

The ways in which cattle respond to any situation are limited by their genetic capability and by their environment. There are four environmental responses that can be easily measured:

1. Percentage of calf crop
2. Length of calving season
3. Weaning weight
4. Yearling weight

Figure 4–13 indicates the expected relative herd improvement for these four major beef production measures when producers select for weaning weight or yearling weight within different environments.

TABLE 4–2. The Planning Process

The following questions can be used in beginning the planning process for positive genetic change on the farm or ranch.

	Yes	No
1. Are you in business for a profit?		
2. Are beef cows the best use of your resources?		
3. Are you happy with your present breed or cross?		
4. Are your numbers adequate for your land and management?		
5. Are you managing your forages properly?		
6. Is nutrition a problem for:		
Dry cows?		
Cows and calves?		
Replacement heifers?		
First-calf heifers?		
Bulls?		
7. Is your percentage of calf crop too low?		
8. Is your death rate of calves too high:		
At birth?		
From birth to weaning?		
After weaning?		
9. Are too many cows failing to breed?		
10. Are your weaning weights too low?		
11. Is calving difficulty a major problem?		
12. Are your cows permanently identified?		
13. Do your herd bulls have performance records?		
14. Do those records indicate superiority?		
15. Do you crossbreed?		
If yes, is your program well planned?		
16. Do your calves earn high economic return?		

From USDA Extension Service. 1981. Guidelines for Uniform Beef Improvement Programs, Beef Improvement Federation Recommendation. Washington, DC. Program Aid 1020.

TABLE 4-3. Positive Planning

	Calves Weaned (%)	Calving Difficulty	Pregnancies by Exam (%)	Cow Weights or Size	Calf Weaning Weights	Calf Quality or Acceptability	Planned Breeding Program (Straightbred or Crossbred)	Use of Superior Bulls
Where are you?								
Where do you want to be? (goals)								
Major problems to overcome								
Plan to correct problems								
Net return if goal is achieved								

From USDA Extension Service. 1981. Guidelines for Uniform Beef Improvement Programs, Beef Improvement Federation Recommendation. Washington, DC. Program Aid 1020.

TABLE 4-4. Goals and Performance Testing Practices for Beef Producers

Management Changes or Goals[a]	Practice	Description of Practice	Possible Effects
1. No change	Buy bulls.	Bulls performance-tested for desired trait.	Increase in performance of offspring in desired trait; amount depends on heritability of trait and reach.
2. No change	Buy replacement heifers.	Heifers performance-tested for desired trait.	Greater increase in performance compared with herd to which they are added if they are from a higher-performing herd and the top end of the heifer crop.
3. No change	Cull cows.	Culling done on the basis of breeding soundness.	Elimination of cows with poor reproductive performance and poor physical structure; improvement in calf crop and weaning weights; elimination of cows with poor calves.
4. Numbering	Individually identify all animals.	Identification by eartag, tattoo, neck chain, freeze brand, etc.	Collection of individual performance records; individual identification to pinpoint outstanding or poor performance; allowance for more precise management; inventory of cattle.
5. Collection of performance data	Individually weigh animals; keep individual calving records.	Records of actual and adjusted 205-day weights; actual and adjusted yearling weights; birth date of calves; days from last calf; time exposed to bull.	Increased weaning and yearling weights; pinpointing of cows not calving every 365 days; identification of cows that breed late in the season; identification of cows that are poor performers by breeder's chosen production standards.
6. Change in size by skeletal measurements	Record linear measurements.	Measurement of individual animals to obtain hip height.	Selections based on hip height at any age resulting in change of skeletal size on a herd basis.
7. Decrease in number of open cows	Perform pregnancy test.	Elimination of open cows.	Increase in fertility; decrease in winter feed costs per calf raised.
8. Shortening of calving season	Perform pregnancy test; shorten breeding season.	Elimination of late breeders; removal of bulls after a set time.	Increase in average actual weaning weight; increase in reproductive rate.
9. Maintenance of only healthy cows	Physically examine cows.	"Chuting" of cow; physical examination of eyes, mouth, udder, and reproductive tract; pregnancy test.	Maintenance of physically strong cows in good condition, capable of raising large calves; decrease in calving interval; decrease in labor involved with problem cows.
10. Use of only highly fertile bulls	Perform physical examination and semen evaluation of bulls.	"Chuting" of bull; physical examination of bulls for breeding capability; evaluation of semen for quality and quantity.	Settling of more cows within a shorter time; more active bulls; possibly fewer bulls.
11. Advantageous use made of hybrid vigor and genetic variation	Crossbreed cattle.	Mating of cows of one breed to bulls of another breed.	Gains in weaning weight and in fertility of heifers; gains in yearling weight; increase in variation of offspring; limited use of replacement heifers.
12. Increase in efficiency of selection procedure and reduction of cattle handling	Select replacement heifers.	Selection of replacement heifers from the oldest and largest heifers in the contemporary group.	Selection of heifers from cows (1) that are early breeders, (2) that are good milkers, and (3) that transmit good growth impulse; easier maintenance of heifers in winter and ease in developing to breeding weight.
13. Collection of performance data	Individually identify replacement heifers.	Performance records on first and second calf heifers; culling of cows based on basis of first and second records.	Increase in productivity by elimination, based on relationship of first and second calf production to all other calves produced by a cow and culling on combined average of first and second calf production.

[a] Large commercial herds can participate in herd-management changes 1, 3, 7, 8, 9, 10, 11, 12, and 13 without disrupting large-herd management to any major extent. From USDA Extension Service. 1981. Guidelines for Uniform Beef Improvement Programs, Beef Improvement Federation Recommendation. Washington, DC. Program Aid 1020.

Level of environment and management limit the response that can be expected in a herd improvement program in which emphasis is placed on factors affected by environment (Fig. 4–13). If the environmental level is poor, the response in herd improvement programs in those traits that are environmentally influenced will be poor.

Performance records in commercial herds provide producers with information for maximizing growth management of their herds as well as of individuals within herds. These records help identify environmental factors that can be changed so that growth potential can be maximized and profit potential increased.

Table 4–4 shows different levels of participation in performance testing. Progress to be made depends on accuracy of the records, superiority of the animals under consideration, heritability of the traits, number of traits considered, and level of management in the herd.

Major Components of a Performance Testing Program for Commercial Herds

Sire Selection. This is the single most important factor in making genetic progress. In a commercial operation, the breed of sire should be selected based on the crossbreeding program that has been determined for the herd. Traits of importance are birth weight, yearling weight and ratio, post-weaning gain, scrotal circumference, and maternal ability as evaluated through the production and reproduction of the dam.

Individual Identification. Identification of all animals within a herd is a prerequisite for a performance program.

Replacement-Female Selection. At least 40% and not more than 80% of all heifers produced should be retained as replacements at weaning time. Performance information should be combined with visual appraisal in making selections.

Cow-Culling. Ruthless culling of the cow herd is essential to improving herd

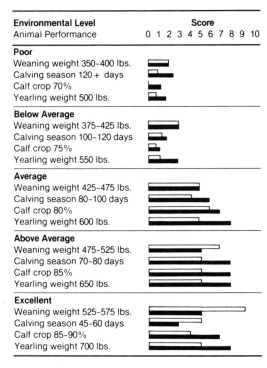

Weaning weight ☐
Yearling weight ■

FIG. 4–13. Estimation of relative expected improvement in four herd production measures using performance records for weaning weight and yearling weight at different environmental levels. (From USDA Extension Service. 1981. Guidelines for Uniform Beef Improvement Programs, Beef Improvement Federation Recommendation. Washington, DC. Program Aid 1020.)

performance. Cows that fail to become pregnant or cows that lose calves should be culled automatically. Cows that produce lightweight calves or poor-quality calves should also be culled.

REFERENCES

Nelson, L.A., and W.L. Singleton. 1974. Beef cattle identification methods. Cooperative Extension Service, Purdue University, W. Lafayette, IN. AS-410.

USDA Extension Service. 1981. Guidelines for Uniform Beef Improvement Programs, Beef Improvement Federation Recommendation. Washington, DC. Program Aid 1020.

Selected Readings

Absher, C., F. Thrigt, and N. Gay. 1976. Beef: Individual identification of cattle. Cooperative Extension Service, University of Kentucky, Lexington, KY. SR-3000.

Albaugh, R. 1975. How to identify livestock. Cooperative Extension Service, University of California. Leaflet 2297.

Cundiff, L.V., and K.L. Gregory. 1977. Beef cattle breeding. USDA, Washington, DC. Agriculture Information Bulletin No. 286.

Farrell, R., R.I. Hostetler, and J.B. Johnson. 1978. Freeze marking farm animals. Cooperative Extension Service, Washington State University, Pullman, WA. PNW Bulletin 173.

Gosey, J.A. 1983. Breeding values and linear measurements. Proc., The Range Beef Cow Symposium VIII, Sterling, CO. December 5–7.

QUESTIONS FOR STUDY AND DISCUSSION

1. List the requirements of a good identification program.
2. List the different methods of identification and discuss their advantages and disadvantages.
3. Outline a meaningful numbering system for a commercial cattle operation.
4. What are the principal features of a good record-of-performance operation?
5. Describe and explain the importance of the following:
   a. Production testing
   b. 205-Day adjusted weight
   c. Calf index
   d. Most probable producing ability
   e. Performance testing
   f. Adjusted 365-day weight
   g. Progeny evaluation
   h. National sire evaluation
   i. Expected progeny difference
   j. Estimated breeding value (EBV)
6. Describe the advantages and disadvantages of a central bull testing station.
7. Diagram a management program for performance testing.
8. Outline a recommended production and performance testing procedure.
9. Calculate the adjusted 205-day weight of a calf that weighed 90 lb at birth and 575 lb at 230 days of age.
10. Why do purebred and commercial cattle performance programs differ?
11. What are the major components of a performance testing program for commercial herds?

Chapter Five

Selection and Management of Bulls

The major genetic decision that a beef breeder makes is the selection of a sire. Sire selection can account for 80 to 90% of the progress made in a breeding program. The breeder wants a sire that does the following:

1. Finds and impregnates cows in heat.
2. Sires calves of high value.
3. Works successfully over numerous breeding seasons.
4. Requires little extra management or care.
5. Has a high salvage value.

This chapter discusses the factors to consider when selecting a bull and the ways to manage the bull to maximize his use within a herd.

SELECTION

Performance

Select a bull of which performance information has been kept, paying close attention to traits of high heritability and economic importance.

BIRTH WEIGHT

The heritability for birth weight is high (48%), and birth weight is positively correlated with future growth rate. Therefore, future growth rate can be increased by selecting cattle for heavier birth weights. Birth weight is highly related to calving difficulty, however, and selection for heavier calves to increase growth rate may increase the incidence of dystocia. Ideally, breeders should select bulls with moderate birth weights and rapid post-natal growth.

WEANING WEIGHT

Approximately 30% of a bull calf's superiority at weaning (205 days of age) is passed on to his progeny. Weaning weight is related more to the milk production of the dam and environmental factors than to the genetic makeup of the calf. Weaning weight should receive minor emphasis in sire selection for growth potential because of its low heritability in comparison to yearling weight.

YEARLING WEIGHT

Yearling weight is the most valuable trait for predicting the genetic growth potential of a herd sire because of its high heritability (40 to 50%). Therefore, it should receive major emphasis in bull selection programs. A bull's performance should be at least equal to the herd average in which he was produced, or to the average of a group of bulls on testing programs. His superiority or inferiority with respect to the average is indicated by a weight ratio. A bull with a weight ratio above 100 is heavier than the group average. Individuals with adjusted weights below average have weight ratios less than 100. The following table illustrates these examples:

Bull	365-Day Weight (lb)	365-Day Weight Ratio	Deviation from the Average
A	1,090	109	9% above
B	1,010	101	1% above
C	900	90	10% below
Average	1,000	100	

Breeding Soundness

Studies in Texas and Colorado, involving over 12,000 bulls of service age, have demonstrated that approximately one in every five beef bulls is a questionable or unsatisfactory breeding bull.

At present, there is no accurate way to predict that a bull will settle 50% or 80% of the cows exposed to him; however, through a breeding soundness examination prior to the breeding season, those bulls of questionable or unsatisfactory breeding potential can be identified. They should be eliminated to reduce both the time lost in the breeding pasture and the resulting number of lighter calves at weaning time.

A complete breeding soundness evaluation consists of the following:

1. Physical examination
2. Scrotal measurement
3. Semen evaluation

In addition to these factors, some assessment should be made, if possible, of the bull's desire and ability to breed a female in heat.

PHYSICAL EXAMINATION

Initially, the physical examination consists of a general evaluation of the bull's health and well-being (Fig. 5–1). Good feet and legs are essential if the bull is expected to walk and mount females in estrus. Most structural faults such as sickle-hocks and post-legs are heritable, lead to lameness of the individual, and should be discriminated against in the selection process (Fig. 5–2). In addition, the bull needs to be able to see, eat, and smell. Any factor that lowers the efficiency of one of these activities (e.g., cancer eye) will have a negative effect on the bull's breeding ability. Overall body condition of the bull should also be noted, as bulls that are thin or, conversely, overconditioned may not achieve their optimum potential. A history of recent illness is also helpful since factors that affect semen quality can occur months before any change is detectable in the semen sample.

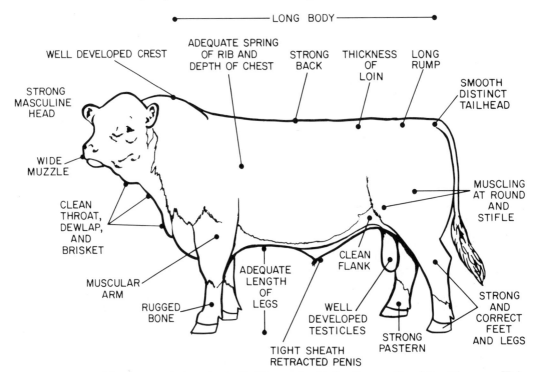

FIG. 5–1. Desirable characteristics of a bull. (Drawn by Kathy Dawes Graphics, Moscow, ID.)

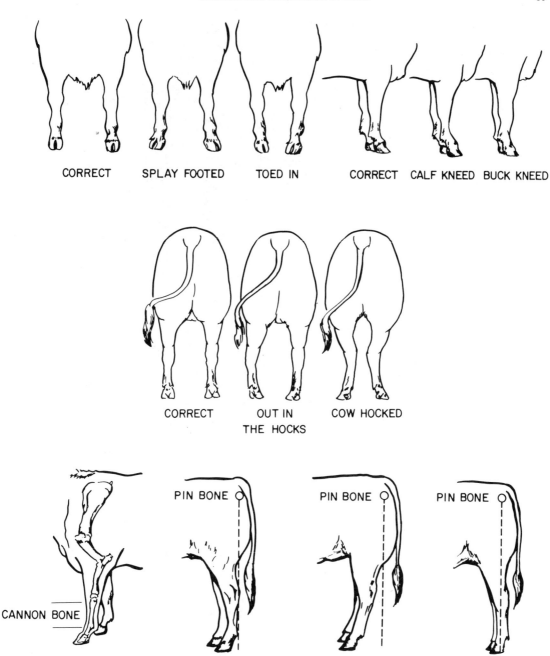

CORRECT   SPLAY FOOTED   TOED IN   CORRECT   CALF KNEED   BUCK KNEED

CORRECT   OUT IN THE HOCKS   COW HOCKED

CANNON BONE   PIN BONE   PIN BONE   PIN BONE

CORRECT   SICKLE HOCKED   POST LEGGED

FIG. 5–2. Structural characteristics of feet and legs. (Adapted from Livestock Production. 1981. A curriculum guide for Idaho vocational agriculture instructors. Idaho State Board for Vocational Education. Moscow, ID.)

A thorough examination of the bull's reproductive system follows the general health examination (Fig. 5–3). The vesicular glands, ampullae, and prostate can be examined by rectal palpation, whereas the spermatic cord, scrotum, testicles, and epididymides can be palpated externally. The penis and prepuce are usually examined

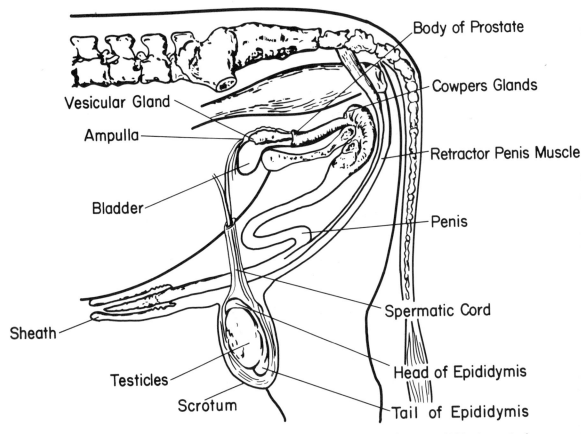

FIG. 5–3. Reproductive tract of a bull. (Adapted from Livestock Production. 1981. A curriculum
guide for Idaho vocational instructors. Idaho State Board for Vocational Education. Moscow, ID.)

during electro-ejaculation for the semen
sample. A tight sheath and strong retractor
muscle of the penis are important when
selecting bulls to breed in low brush
country.

Degenerative change in the testicle is a
frequent cause of infertility in bulls. This
condition is a common occurrence following inflammatory reactions, but many
other factors contribute to this condition.
Regardless of the cause, the effect is reduced fertility. Changes in testicular tone
associated with this degenerative process
can be detected by testicular palpation.

HEIGHT MEASUREMENTS

Some linear measurements (especially
height measurements) have received in-

creased emphasis in beef cattle selection
programs; however, the relationship between linear measurements and net profits
has not been well defined.

Frame scores may be used to monitor
skeletal development in cattle (Massey,
1975) (Table 5–1). Currently, frame scores
describe cattle of the same age according to
wither height. Hip measurements, however, are favored over wither height measurements because the hip matures earlier,
and the hip provides a standard reference
point for measurement (Brown et al., 1956).

Height measurements may also be used
to describe and to market feeder cattle in
commercial herds (see Chapter 12), and to
monitor changes in skeletal size in seed
stock herds. Height measurement should

TABLE 5–1. Frame Size Based on Hip Height Measurements (inches)

Age (mos)	Frame Size							
	1	2	3	4	5	6	7	8
6	35	37	39	41	43	45	47	49
7	36	38	40	42	44	46	48	50
8	37	39	41	43	45	47	49	51
9	38	40	42	44	46	48	50	52
10	39	41	43	45	47	48	51	53
11	40	42	44	46	48	50	52	54
12	41	43	45	47	49	51	53	55
14	42	44	46	48	50	52	54	56
16	43	45	47	49	51	53	55	57
18	44	46	48	50	52	54	56	58

Sex adjustments are estimated to be minus 1 inch for steers and minus 2 inches for heifers. Height at shoulder is approximately two inches less. Adapted from shoulder height measurements at University of Missouri.

be used as a supplement to weight performance data to help describe compositional maturity and optimum slaughter weight of feedlot cattle.

The ranking of cows on height and weight/height ratio is not indicative of cow productivity (Gosey, 1983). Height measurements early in life are associated with mature skeletal size.

Cattle producers should not select directly for height. If they do, they run the risk of encountering unfavorable correlated responses in such traits as delayed reproductive maturity, delayed compositional maturity, and greater mature size. Height should be used only as a supplement to other performance information in a comprehensive performance testing program.

SCROTAL SIZE

Measurement of scrotal circumference with a scrotal tape gives a relatively accurate estimate of the semen-producing ability of a young bull (Fig. 5–4). Scrotal circumference is highly correlated with weight of testes and sperm output in growing bulls (Coulter, 1982, and Lunstra et al., 1978).

Scrotal circumference has been reported to be a highly heritable trait, with most estimates around .60 (Brinks et al., 1978, and Coulter, 1982). There appear to be breed differences, however, as well as considerable variability among bulls within breeds. The relatively high heritability coupled with large within-breed variation indicates that selection would be effective in increasing scrotal circumference as well as in changing traits genetically as they correlate with scrotal circumference. Brinks and associates (1978) reported a cor-

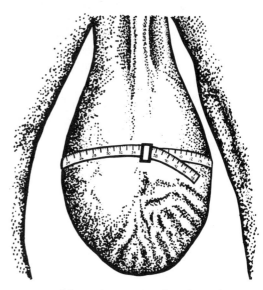

FIG. 5–4. Measuring scrotal circumference. (Drawn by Montana State University Graphics Department, Bozeman, MT.)

relation of .58 between scrotal circumference and percentage of normal sperm.

Scrotal circumference has been shown to be a more accurate indicator of when a bull reached puberty than either age or weight, regardless of breed or breed cross (Lunstra, 1982). Scrotal circumference also influences age at puberty in females (Lunstra, 1982, and Brinks et al., 1978). In general, as scrotal circumference increases, age at puberty in heifers decreases.

SEMEN EVALUATION

A semen sample may be collected from bulls by a variety of methods, but electro-ejaculation is preferred under normal field conditions. This procedure is harmless to the bull, and semen collected with the electro-ejaculator has not been shown to differ in fertility from semen collected by other methods.

The two semen characteristics shown to be the most reliable in evaluating fertility under field conditions are *initial sperm motility* and *sperm morphology* (form and structure of individual sperm cells). Unfortunately, a great deal of variation is possible in estimating semen motility, resulting from differences between technicians and within the work of a single technician. Environmental factors can also have a severe effect on this estimate; thus, the environment of the semen sample needs to be closely controlled.

There is a definite relationship between sperm morphology and infertility in the bull (Beverly, 1979). A system has been developed whereby abnormal spermatozoa are categorized as either "primary" or "secondary" abnormalities; this has proved useful in evaluating bulls. Primary abnormalities (abnormal head and midpiece shapes, abaxial attachment of midpieces, tightly curled tails) are thought to indicate defects in sperm production. Secondary abnormalities (separated head, distal droplets, bent tails) are thought to be defects of the sperm storage system. Although these designations may not be

clear-cut, they are fairly easy to visualize and are indicative of the bull's breeding potential.

BULL CLASSIFICATION

The system adopted in 1976 by the Society for Theriogenology classifies bulls as one of the following:

1. Satisfactory potential breeders
2. Questionable potential breeders
3. Unsatisfactory potential breeders

Flexibility is needed, however, in applying these assessments to all bulls. For example, some allowances are needed for bulls that are very young or very old or that are recovering from illness. In general, a bull should not be culled on the basis of one breeding soundness evaluation. If poor semen is the cause of a less than satisfactory rating, another sample should be collected. Usually, if a bull does not measure up, he should be retested in 30 to 60 days. The sperm cells produced by the bull take approximately 2 months to move through the reproductive system. If an injury to his testicles occurred, it would take approximately 2 months for him to have a normal ejaculate. Often, the physical exam indicates the causes of a poor sample of semen and provides a guideline as to when to reevaluate the bull.

Obviously, bulls not in good overall health or those with physical or reproductive system abnormalities resembling those previously discussed cannot have a satisfactory rating following examination. If the bull has the required physical capabilities, he is then scored for each of the criteria listed in Table 5–2.

Scrotal size constitutes 40% of the total score, spermatozoal motility 20% of the total score, and spermatozoal morphology 40% of the total score. In general, a bull with a total score of 60 or higher is classified as satisfactory, 30 to 59 as questionable, and less than 30 as unsatisfactory. A bull scoring 85, however, does not necessarily have better fertility than a bull scor-

TABLE 5-2. Breeding Soundness Evaluation (scrotal size and spermatozoal scoring system)

Criteria	Classifications			
	Very Good	Good	Fair	Poor
Spermatozoal morphology				
Primary abnormalities	<10	10–19	20–29	>29
Total abnormalities	<25	26–40	41–59	59
Score	40	24	10	3
Scrotal circumference (cm)[a]				
12–14 months of age	>34	30–34	<30	<30
15–20 months of age	>36	31–36	<31	<31
21–30 months of age	>38	32–38	<32	<32
30+ months of age	>39	34–39	<34	<34
Score	40	24	10	10
	Rapid Linear	Moderate Linear	Slow Linear to Erratic	Very Slow and Erratic
Spermatozoal motility Score	20	12	10	3

[a] These values do not apply to Brahman bulls.

ing 65. The score simply indicates that both are satisfactory potential breeders.

Pedigree

Record of ancestry is important to a breeder of registered cattle. Although it is less important to a commercial producer, pedigree should not be ignored. Performance information from close relatives, especially carcass data from steer progeny and mothering ability of daughters, is extremely useful in determining breeding value of a sire. Pedigrees also help in selecting against such genetic abnormalities as dwarfism and "double muscling." When using pedigree information, however, only the closest relatives (progeny, half-siblings, sire, and dam) should receive much consideration.

Recently, more emphasis has been placed on performance pedigrees. A performance pedigree includes performance information of the individual and at least its sire or dam. Estimated breeding values (EBVs) and their accuracy are usually reported. EBV is a mathematical calculation that expresses the value of an animal as a parent for certain traits. To calculate EBV, the performance records of the individual, all of its naturally raised sibs and half-sibs, its sire and dam, the naturally raised progeny of sire and dams, and the naturally raised progeny of parental (sire of sire) and maternal (sire of dam) grandsires are utilized. Accuracy (ACC) of EBVs is a correlation between the EBV and the animal's true breeding value. The ACC value depends on the heritability of the trait and on the amount and sources of information included in its calculation. Accuracy values may range from 0 to 100%; a higher ACC value indicates a more accurate EBV. An example of a performance pedigree is shown in Figure 5-5.

Selection goals listed in Table 5-3 represent a plan for "specification buying" of bulls by commercial cattlemen (Gosey, 1983). Breeding values form an important component of the specifications. Purebred breeders should provide maternal, weaning, and yearling EBVs and their corresponding accuracies on all bulls provided

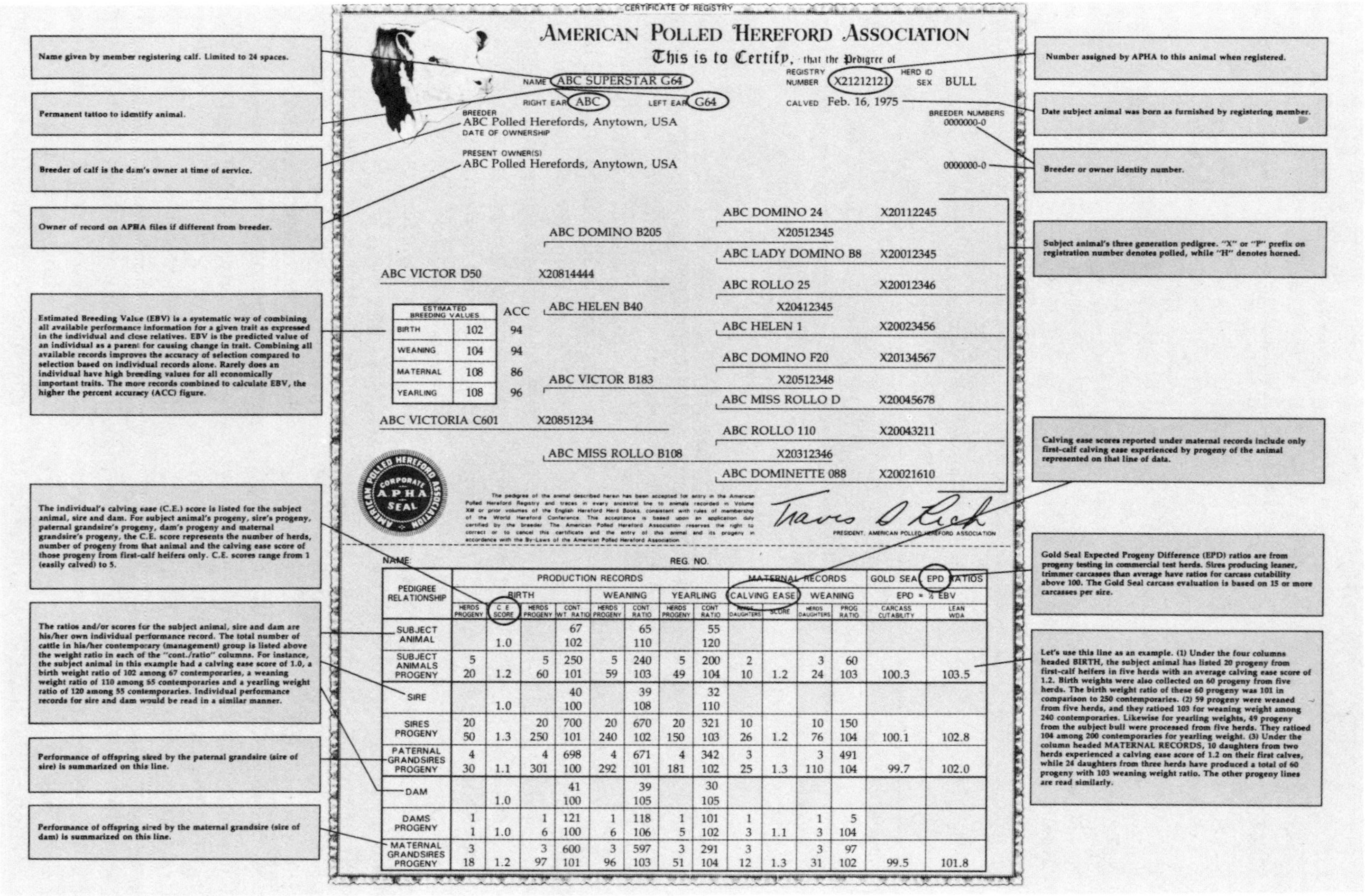

CERTIFICATE OF REGISTRY

AMERICAN POLLED HEREFORD ASSOCIATION

This is to Certify, that the Pedigree of

NAME ABC SUPERSTAR G64 — REGISTRY NUMBER X21212121 — HERD ID — SEX BULL

RIGHT EAR ABC — LEFT EAR G64 — CALVED Feb. 16, 1975

BREEDER ABC Polled Herefords, Anytown, USA — BREEDER NUMBERS 0000000-0

DATE OF OWNERSHIP

PRESENT OWNER(S) ABC Polled Herefords, Anytown, USA — 0000000-0

ABC VICTOR D50 X20814444	ABC DOMINO B205 X20512345	ABC DOMINO 24	X20112245
		ABC LADY DOMINO B8	X20012345
	ABC HELEN B40 X20412345	ABC ROLLO 25	X20012346
		ABC HELEN 1	X20023456
ABC VICTORIA C601 X20851234	ABC VICTOR B183 X20512348	ABC DOMINO F20	X20134567
		ABC MISS ROLLO D	X20045678
	ABC MISS ROLLO B108 X20312346	ABC ROLLO 110	X20043211
		ABC DOMINETTE 088	X20021610

ESTIMATED BREEDING VALUES		ACC
BIRTH	102	94
WEANING	104	94
MATERNAL	108	86
YEARLING	108	96

CORPORATE A.P.H.A. SEAL — AMERICAN POLLED HEREFORD ASSOCIATION

The pedigree of the animal described herein has been accepted for entry in the American Polled Hereford Registry and traces in every ancestral line to animals recorded in Volume XIII or prior volumes of the English Hereford Herd Books, consistent with rules of membership of the World Hereford Conference. This acceptance is based upon an application duly certified by the breeder. The American Polled Hereford Association reserves the right to correct or to cancel this certificate and the entry of this animal and its progeny in accordance with the By-Laws of the American Polled Hereford Association.

PRESIDENT, AMERICAN POLLED HEREFORD ASSOCIATION

NAME — REG. NO.

PEDIGREE RELATIONSHIP	PRODUCTION RECORDS: BIRTH				WEANING		YEARLING		MATERNAL RECORDS: CALVING EASE		WEANING		GOLD SEAL EPD RATIOS (EPD = ½ EBV)	
	HERDS PROGENY	C.E. SCORE	HERDS PROGENY	CONT WT. RATIO	HERDS PROGENY	CONT RATIO	HERDS PROGENY	CONT RATIO	HERDS DAUGHTERS	SCORE	HERDS DAUGHTERS	PROG RATIO	CARCASS CUTABILITY	LEAN WDA
SUBJECT ANIMAL		1.0		67 102		65 110		55 120						
SUBJECT ANIMALS PROGENY	5 20	1.2	5 60	250 101	5 59	240 103	5 49	200 104	2 10	1.2	3 24	60 103	100.3	103.5
SIRE		1.0		40 100		39 108		32 110						
SIRES PROGENY	20 50	1.3	20 250	700 101	20 240	670 102	20 150	321 103	10 26	1.2	10 76	150 104	100.1	102.8
PATERNAL GRANDSIRES PROGENY	4 30	1.1	4 301	698 100	4 292	671 101	4 181	342 102	3 25	1.3	3 110	491 104	99.7	102.0
DAM		1.0		41 100		39 105		30 105						
DAMS PROGENY	1 1	1.0	1 6	121 100	1 6	118 106	1 5	101 102	1 3	1.1	1 3	5 104		
MATERNAL GRANDSIRES PROGENY	3 18	1.2	3 97	600 101	3 96	597 103	3 51	291 104	3 12	1.3	3 31	97 102	99.5	101.8

Name given by member registering calf. Limited to 24 spaces.

Permanent tattoo to identify animal.

Breeder of calf is the dam's owner at time of service.

Owner of record on APHA files if different from breeder.

Estimated Breeding Value (EBV) is a systematic way of combining all available performance information for a given trait as expressed in the individual and close relatives. EBV is the predicted value of an individual as a parent for causing change in trait. Combining all available records improves the accuracy of selection compared to selection based on individual records alone. Rarely does an individual have high breeding values for all economically important traits. The more records combined to calculate EBV, the higher the percent accuracy (ACC) figure.

The individual's calving ease (C.E.) score is listed for the subject animal, sire and dam. For subject animal's progeny, sire's progeny, paternal grandsire's progeny, dam's progeny and maternal grandsire's progeny, the C.E. score represents the number of herds, number of progeny from that animal and the calving ease score of those progeny from first-calf heifers only. C.E. scores range from 1 (easily calved) to 5.

The ratios and/or scores for the subject animal, sire and dam are his/her own individual performance record. The total number of cattle in his/her contemporary (management) group is listed above the weight ratio in each of the "cont./ratio" columns. For instance, the subject animal in this example had a calving ease score of 1.0, a birth weight ratio of 102 among 67 contemporaries, a weaning weight ratio of 110 among 65 contemporaries and a yearling weight ratio of 120 among 55 contemporaries. Individual performance records for sire and dam would be read in a similar manner.

Performance of offspring sired by the paternal grandsire (sire of sire) is summarized on this line.

Performance of offspring sired by the maternal grandsire (sire of dam) is summarized on this line.

Number assigned by APHA to this animal when registered.

Date subject animal was born as furnished by registering member.

Breeder or owner identity number.

Subject animal's three generation pedigree. "X" or "P" prefix on registration number denotes polled, while "H" denotes horned.

Calving ease scores reported under maternal records include only first-calf calving ease experienced by progeny of the animal represented on that line of data.

Gold Seal Expected Progeny Difference (EPD) ratios are from progeny testing in commercial test herds. Sires producing leaner, trimmer carcasses than average have ratios for carcass cutability above 100. The Gold Seal carcass evaluation is based on 15 or more carcasses per sire.

Let's use this line as an example. (1) Under the four columns headed BIRTH, the subject animal has listed 20 progeny from first-calf heifers in five herds with an average calving ease score of 1.2. Birth weights were also collected on 60 progeny from five herds. The birth weight ratio of these 60 progeny was 101 in comparison to 250 contemporaries. (2) 59 progeny were weaned from five herds, and they ratioed 103 for weaning weight among 240 contemporaries. Likewise for yearling weights, 49 progeny from the subject bull were processed from five herds. They ratioed 104 among 200 contemporaries for yearling weight. (3) Under the column headed MATERNAL RECORDS, 10 daughters from two herds experienced a calving ease score of 1.2 on their first calves, while 24 daughters from three herds have produced a total of 60 progeny with 103 weaning weight ratio. The other progeny lines are read similarly.

FIG. 5–5. Example of a performance pedigree for beef cattle. (Courtesy of the American Polled Hereford Association.)

TABLE 5–3. Bull Selection Goals Listed by Breed Type

Trait	Breed Type	
	Maternal	Terminal
Function		
Fertility		
Semen examination	Live sperm	Live sperm
Reproductive tract	No defects	No defects
Scrotal circumference	Minimum: 34 cm at 1 yr	Minimum: 32 cm at 1 yr
Calving ease score	Unassisted delivery	Minor assistance accepted
Birth weight	65 to 85 lb	70 to 100 lb
Structural soundness	Excellent	Adequate
Milk Production		
Maternal EBV	102 to 110	Not important
Growth		
Weaning EBV	98 to 102	104+
Yearling EBV	98 to 102	104+
Market Acceptance		
Frame score (1 yr)	4 to 5.5	5 to 6
Fat thickness (1100 lb)	.2 to .4	<.2

From Gosey, J.A. 1983. Breeding Values and Linear Measurements. Proc., The Range Beef Cow Symposium VIII, Sterling, CO. December 5–7.

for sale. All abbreviations and traits should be defined in a prominent place in the sale catalog. The same information should appear in the same format for all cattle in the catalog. Accuracy values should be provided with all breeding values. A page should be devoted to the breeder's selection, culling philosophy, and goals.

Price

"How much can I afford to pay for a superior, performance-tested bull?" In general, one can expect to pay at least two to three times the value of a market steer. Thus, if a finished steer is worth $700 at 15 months of age, a producer should expect to pay at least $1400 to $2100 for a performance-tested bull of the same age that will noticeably improve his herd.

"How much more valuable is a superior, performance-tested bull over a mediocre herd bull in terms of economic value of the calves produced?" The following is an actual example of prices paid for two bulls of the same breed, one having above average

growth potential and the other having below average potential:

Bull	365-Day Values		
	Weight (lb)	Ratio	Price
A	1,143	107	$1,400
B	923	98	$1,000
Difference	220		$ 400

The price differential of only $400 is by no means an accurate measure of Bull A's true economic superiority to Bull B, as can be seen from the following. (For purposes of illustration, assume that bulls A and B were raised under comparable environments and management levels.)

1. Bull A will transmit about half his superiority over Bull B to his progeny (about 88 lb) because differences in yearling weights are about 40 to 50% heritable.

220 lb difference $\times$ .4 heritability = 88 lb

2. Because a sire contributes only half of the genotype (genetic makeup) of his progeny, we must again halve the heritable difference of 88 lb.

88 lb × .5 sire's contribution = 44 lb

3. Assuming that each additional pound of yearling weight is worth $0.60, each yearling-age progeny of Bull A should be worth $26.40 more than progeny of Bull B, not including possible improvement in grade as well.

44 lb difference × $0.60 per lb
= $26.40 difference per calf

4. If Bull A sires 25 progeny per year, his value over Bull B is equal to $660.00 per year.

$26.40 per calf × 25 calves/year = $660.00

5. Assuming that Bull A is used for 4 years, his actual value over Bull B would be equivalent to 4 times $660.00, or $2640.00. However, if a breeder pays $2640.00 more for Bull A than for Bull B, he will not gain financially, but will merely be trading dollars. Alternatively, he may pay up to $2640 more and make money. Also, Bull A's daughters that he keeps in the herd as replacements should be superior to those of Bull B.

MANAGEMENT

The bull's year can be divided into three seasons:

1. Pre-breeding or conditioning period (2 months)
2. Breeding season (2 to 3 months)
3. Post-breeding season or rest and recuperation period (7 to 8 months)

Actual length of each segment varies from one ranch to another, but the basic management during each period remains about the same.

Pre-Breeding Period

At the start of the conditioning period, the bull battery should be well established. The breeding soundness of all bulls should be evaluated. Unsatisfactory breeders should be removed from the battery and replaced with satisfactory breeders. New bulls should be acquired at least 60 days prior to the start of the breeding season. This provides time for new bulls to adjust to the environment, become familiar with other bulls, and develop a social structure. All bulls should be included in a complete health program. Hoof trimming should be done at the start of the conditioning period to allow time for adequate regrowth prior to turnout.

EXERCISE

Bulls must exercise daily to get into proper breeding shape. Supplemental feeding or salting and watering areas should be located as far apart as possible in the bull pasture to encourage exercise. Bulls that have been conditioned properly are less prone to fighting and breeding injuries than unconditioned bulls.

NUTRITION

Nutrient requirements of bulls as adapted from the National Research Council (NRC, 1984) are shown in Table 5–4. These requirements should be used as guidelines in developing a sound nutritional program.

Bulls should be in moderate condition when they are put into the breeding pasture. Yearling bulls should weigh approximately 1000 lb prior to the breeding season. An 800-lb bull requires a daily minimum of 18.5 lb of dry matter (DM) containing 65.5% total digestible nutrients (TDN). A ration of 21 lb of a good quality hay plus 5 lb of grain provides adequate energy. The following procedure can be used to determine the amount of feed required to satisfy a bull's nutrient requirements:

1. Determine the nutrient requirements from Table 5–4.

Minimum dry matter intake = 18.5 lb
TDN requirement   = 12.1 lb
Crude protein (CP)   = 1.76 lb

TABLE 5-4. Nutrient Requirements of Medium-Framed Bulls Gaining 2.0 lb Daily

Weight (lb)	Dry Matter Intake (lb)	Protein Intake (lb)	Protein (%)	TDN (%)	Calcium (%)	Phosphorus (%)
600	14.9	1.61	10.8	65.5	.43	.24
700	16.7	1.69	10.1	65.5	.38	.22
800	18.5	1.76	9.5	65.5	.33	.21
900	20.2	1.83	9.1	65.5	.31	.21
1,000	21.8	1.90	8.7	65.5	.28	.19
1,100	23.4	1.97	8.4	65.5	.26	.19

Adapted from NRC. 1984. Nutrient requirements of domestic animals. Vol. 6. Nutrient Requirements of Beef Cattle. National Academy of Sciences, Washington, DC.

2. Determine amount of grain fed and nutrient contribution of the grain.

Feed 4 pounds barley daily (as-fed basis).

Nutrient content barley:
    % DM   = 89.0
    % TDN   = 82.0 (DM basis)
    % Crude protein = 10.9 (DM basis)

Four pounds of barley (as-fed) provides:
    4 lb × .89 (% DM)   = 3.6 lb DM
    3.6 lb DM × 0.82 (% DM)   = 2.9 lb TDN
    3.6 lb DM × 0.109 (% CP)   = 0.4 lb CP

3. Determine nutrients that must be provided by roughage portion of ration:

	DM (lb)	TDN (lb)	CP (lb)
Animal requirements	18.5	12.1	1.76
Amount from barley	3.6	2.9	0.4
Difference	14.9	9.2	1.36

4. How much roughage (alfalfa hay) must be fed to provide 9.2 lb TDN?

Nutrient composition of alfalfa hay:
    % DM = 88.0
    % TDN = 55.0 (DM basis)
    % CP = 18.0 (DM basis)

$$\frac{9.2 \text{ lb TDN required}}{55 \text{ (\% TDN alfalfa hay)}} \times 100$$

    = 16.7 lb alfalfa hay DM

Therefore, 16.7 lb of alfalfa hay DM provides 9.2 lb of TDN. 16.73 lb of alfalfa hay

DM is equal to 19.0 lb of alfalfa hay *as fed* [16.78 ÷ 88 (% DM) × 100].

5. The adequacy of the ration can be determined as follows:

	DM (lb)	TDN (lb)	CP (lb)
Animal requirements	18.5	12.1	1.76
Amount from barley	3.6	2.9	0.4
Amount from hay	16.7	9.2	3.01
Difference	+1.8	+0.0	+1.65

6. The hay-barley ration provides adequate DM, TDN, and protein. DM intake is not critical provided the TDN and CP requirements are met. DM intake becomes important if a poor-quality roughage (e.g., wheat straw, meadow hay, corn stalks) that limits dry matter intake is fed. Protein supplementation is not required in the previous example; however, if corn silage or a poor-quality roughage that contains a low protein content is fed, then protein supplementation may be required. Calcium (Ca) and phosphorus (P) intake should be calculated in a similar manner, and an appropriate mineral supplementation program should be designed.

The nutrient requirements of 2-year-old bulls are not as critical as those of yearlings because they have attained more of their mature size. A 1400- or 1500 lb 2-year-old would need to gain about 1 lb per day during the conditioning period.

Overconditioning of newly acquired bulls (2-year-olds or bulls fitted for show and sale) can be a problem if not handled properly. Starting them on a ration that is similar to the one they were on previously is a good practice. Intake should be restricted to 60 to 70% of their previous intake. Grain intake should be reduced at about 10% per week until the desired level is achieved.

Yearling bulls that have been on a performance test should be fed to continue growth. It is good management to remove bulls from the performance test 30 to 45 days prior to the sale and reduce their concentrate consumption to approximately 5 lb by sale day.

The amount of grain fed to older bulls during the conditioning period depends on their physical condition. If bulls have been wintered in good condition, grain feeding is probably not required. If bulls are very thin at the start of the conditioning period, a good quality hay should be fed *ad libitum* for 60 to 80 days. The biggest mistake most cattlemen make in handling mature bulls is keeping them on a low plane of nutrition until 30 days prior to the start of the breeding season. This is a costly and inefficient way to manage bulls. If grain is fed, the following precautions should be taken:

1. Provide at least 2 feet of feeder space per bull.
2. Make sure that all bulls have gathered at the feeding area prior to feeding.
3. If more than 10 lb of grain is fed daily, it should be divided into two equal feedings.

4. When 10 lb or less of concentrate is needed daily, corn or barley may be fed as the energy supplement.
5. If more than 10 lb of concentrate is fed daily, the feed should be given bulk by adding one part of oats, beet pulp, or wheat bran to two parts of corn or barley.
6. Wheat is not a good feed for bulls and should be avoided.

Breeding Season

BULL TO FEMALE RATIO

Many factors influence the number of cows one bull can impregnate during a breeding season. Factors related to the environment such as terrain, water availability, carrying capacity of the land, and pasture size play a major role. Factors related to the bull include age, condition, mating ability, libido, fertility, sperm reserve, social behavior, and injury. Management factors such as length of the breeding season, reproductive diseases, breeding intensity, and amount of observation must also be taken into consideration.

Obviously, it is difficult to standardize a bull to female ratio for all management situations. An estimate of the number of cows that the average bull can serve during a 60- to 90-day breeding season is presented in Table 5–5.

SOCIAL BEHAVIOR AND DOMINANCE

Social ranking of bulls can influence sexual activity when they are mated in groups. A South African study, represented by Table 5–6, showed that the oldest or second oldest bull in the group sired 60% or more of the calves per year (Chenoweth, 1978). Social behavior of bulls appears to be controlled by age and seniority within groups. Social behavior can detrimentally affect genetic progress by allowing genetically inferior bulls to breed too many cows in a breeding season. To avoid dominance problems, bulls of the same age and size should be used together if possible. Yearling bulls should not be expected to compete with older bulls in the same breeding pasture. Older bulls do not "teach" younger bulls how to breed cows,

TABLE 5–5. Estimated Average Mating Capacity of Bulls Within a 60- to 90-Day Breeding Season

	Number of Cows	
Age of Bull	Pasture Breeding	Range Breeding
15–18 months	20–25	15–20
2 years	25–30	20–25
3–5 years	25–35	25–30
Over 6 years	25–35	25–30
Over 8 years	20–25	15–20

and in fact may prevent them from breeding cows and cripple them in the process.

BREEDING MANAGEMENT TIPS

1. In general, do not expose a yearling bull to more than 20 to 25 cows during a breeding season of 45 to 60 days.
2. Two-year-old bulls are capable of breeding 25 to 30 cows. Some Kansas data indicate that having more than 25 cows per bull is likely to result in a calving period that extends over 4 to 6 months.
3. Provide a satisfactory breeding area. Good footing is a must. Clear pens and pastures of boards, wire, and other debris that could cause injury.
4. Observe the cow herd closely, and keep accurate records to assure that the bull finds cows in estrus and services them and that a large percentage of cows conceive at the first service.

Post-Breeding Season

The objectives during this time period are to keep feed costs at a practical minimum, to keep bulls in modest condition, to minimize the chance of injury, and to allow for adequate growth in young bulls.

Bulls should be sorted into three groups after the breeding season is completed: (1) mature bulls in good condition, (2) young bulls that are still growing, and those that are extremely thin or need special care, and (3) old or crippled bulls that will be marketed.

NUTRITION

Mature bulls in good condition require only a maintenance ration. An all-roughage (hay) diet fed daily at about 2% of their body weight in dry feed is satisfactory. Corn silage fed to wintering bulls must be limited to prevent them from becoming overconditioned. Approximately 63 lb of corn silage (35% DM) on an as-fed basis provides adequate energy for a mature 2200-lb bull during the winter. Protein supplementation is required with corn silage to provide adequate protein intake (1 to 1.5 lb of a 30 to 40% crude protein in supplement).

All bulls require adequate phosphorus and vitamin A for successful reproduction. If bulls are being wintered on a poor-quality

TABLE 5-6. Reproductive Performance of Three or Four Bulls Exposed to a Group of Cows Over a 5-Year Period

Bulls Used	Percentage of Calves Sired by Each Bull				
	1964	1965	1966	1967	1968
A	70.4 (10)[a]	76.0 (11)	12.2 (12)	0[b]	0[b]
B	16.7 (4)	18.0 (5)	63.4 (6)	72.5 (7)	25.1 (8)
C	7.4 (3)	6.0 (4)	12.2 (5)	12.5 (6)	62.5 (7)
D	5.5 (2)	0[c] (3)	12.2 (4)	15.0 (5)	12.4 (6)

[a] Age of bull in years.
[b] Bulls absent from the herd.
From unpublished data by Osterhoff as adapted by Chenoweth, P.J. 1978. Bull behavior, selection and management. Charolais Bull-O-Gram. April/May.

roughage or corn silage, vitamin A may need to be added to the mineral mixture or fed with a supplement. Alfalfa hay of good quality usually has adequate vitamin A for wintering bulls. A common mineral mixture fed to bulls is 2 parts of a trace mineralized salt to one part of mineral. The type of mineral supplement fed with the salt depends on the type of roughage fed, specifically on its calcium and phosphorus content. If alfalfa hay is being fed, adequate calcium is probably already provided by the hay; therefore, in that situation, a phosphorus supplement such as monoammonium or monosodium phosphate (22% phosphorus) would be satisfactory. If calcium and phosphorus are required, steamed (feed grade) bone meal (38% Ca; 12% P) or dicalcium phosphate (22.5% Ca; 18.5% P) may be used.

Yearling bulls should be used for only 60 days or less during the breeding season. Using them for a longer period of time may have detrimental effects on future growth. Yearlings should be kept separate from older bulls at least through their second winter and fed the best roughage available and some grain. During their second winter, yearling bulls still grow rapidly and replace all their physical conditioning lost during breeding.

REFERENCES

Beverly, J. 1979. Fertility in bulls. Proc., Western Beef Symposium. Boise, ID. October 29–30.

Brinks, J.S., M.J. McInerney, P.J. Chenoweth, et al. 1978. Relationship of age at puberty in heifers to reproductive traits in young bulls. 29th Ann. Beef Cattle Improvement Report. Colorado State University Experiment Station, Fort Collins, CO. General Series 973:12.

Brown, C.J., M.L. Ray, W. Gifford, and R.S. Honea. 1956. Growth and development of Aberdeen-Angus cattle. Arkansas Agricultural Experiment Station, Fayetteville, AR. Bulletin 571.

Chenoweth, P.J. 1978. Bull behavior, selection and management. Charolais Bull-O-Gram. April/May.

Coulter, G.H. 1982. This business of testicle size. Proc., Ann. Conf. on Artificial Insemination and Embryo Transfer in Beef Cattle, Denver, CO. National Assoc. Animal Breeders, Columbia, MO.

Gosey, J.A. 1983. Breeding values and linear measurements. Proc., The Range Beef Cow Symposium VIII. Sterling, CO. December 5–7.

Lunstra, D.D., J.J. Ford, and S.E. Echternkamp. 1978. Puberty in beef bulls: Hormone concentrations, growth, testicular development, sperm production and sexual aggressiveness in bulls of different breeds. J. Anim. Sci. 46:1054.

Lunstra, D.D. 1982. Testicular development and onset of puberty in beef bulls. In Beef Research Programs Rep. No. 1. U.S. Mgmt. Animal Res. Center, Clay Center, NE. ARM-NC-21. p. 26.

Massey, J.W. 1975. On-the-farm performance testing—Missouri beef cattle improvement programs. Missouri Cooperative Extension Service, Columbia, MO. MP-474.

NRC. 1984. Nutrient requirements of domestic animals. Vol. 6. Nutrient Requirements of Beef Cattle. National Research Council, National Academy of Sciences, Washington, DC.

Selected Readings

Linton, A. 1980. Bull management before the breeding season. In Cattleman's Library. Cooperative Extension Service, University of Idaho, Moscow, ID. CL-435.

Linton, A. 1980. Tips on bull buying. In Cattleman's Library. Cooperative Extension Service, University of Idaho, Moscow, ID. CL-430.

Linton, A. 1980. Bull management after the breeding season. In Cattleman's Library. Cooperative Extension Service, University of Idaho, Moscow, ID. CL-430.

Mankin, J.D., and L. Ickes. 1980. Breeding soundness of bulls. In Cattleman's Library. Cooperative Extension Service, University of Idaho, Moscow, ID. CL-425.

Ney, J.J., V.M. Thomas, and G.W. Gibson. 1979. A feed-planning guide for wintering beef cattle. Cooperative Extension Service, University of Idaho, Moscow, ID. CIS-498.

Singleton, W.L., and L.A. Nelson. 1971. Selection and management of herd bulls. Cooperative Extension Service, Purdue University, W. Lafayette, IN. AS-395.

QUESTIONS FOR STUDY AND DISCUSSION

1. Discuss the performance traits that should be taken into consideration when selecting bulls.

2. Why is yearling weight emphasized more than weaning weight in bull selection programs?

3. List the components of a breeding soundness evaluation and discuss their importance.

4. Distinguish between primary and secondary abnormalities of spermatozoa.

5. What is a performance pedigree?

6. What records are used when calculating an estimated breeding value (EBV) for a bull?

7. Why is expected progeny difference equal to one half of the breeding value for a given trait?

8. How much could you afford to pay for a bull if 1000-lb market steers are selling for $650, or $0.65 per pound?

9. Given the following information, calculate the actual economic value of Bull A over Bull B.

	365-Day Adjusted Weight
a. Bull A	1075 lb
Bull B	1025 lb

   b. Each additional pound of yearling weight is worth $0.60 per pound.

   c. Bull A sires 20 calves per year and is used for 4 years.

10. How many cows should a yearling bull be exposed to during a breeding season of 2 to 3 months?

11. Outline an annual bull management program for a newly purchased yearling bull.

12. a. Would 25 lb (as-fed) of the following alfalfa-grass hay satisfy the TDN requirement of a 1000-lb yearling bull gaining 2.0 lb per day?

Alfalfa hay

% DM	% Crude Protein (DM Basis)	% TDN (DM Basis)
86	15	54

   b. How many pounds of barley in addition to the hay already fed would be required on an as-fed basis to satisfy the bull's TDN requirements?

Barley

% DM	% Crude Protein (DM Basis)	% TDN (DM Basis)
87	11	80

Chapter Six

Female Reproduction

The most important factor affecting profit in a cow-calf enterprise is reproduction. Reproductive efficiency dramatically influences the pounds of calf weaned per cow exposed to the bull. An understanding of the anatomy and physiology of reproduction is a necessary prerequisite to developing a sound reproductive management program.

ANATOMY

The primary products of the cow's reproductive tract are hormones and the ovum (egg). A well-coordinated passageway is necessary for successful completion of the reproductive process. A normal egg must be released from the ovary, picked up by the infundibulum, and passed to the oviduct, where fertilization occurs. The fertilized egg is passed to the uterus, where growth and development take place. Finally, delivery of the calf through the cervix, vagina, and vulva occurs. The reproductive tract then returns to a condition that permits another normal pregnancy. A diagram of the organs of reproduction for the female is shown in Figure 6–1. The major functions of the reproductive organs are given in Table 6–1.

The *ovary* is the primary reproductive organ of the female and has two important functions:

1. Production of the female reproductive cell, the ovum.

2. Production of two major hormones, estrogen and progesterone.

The paired ovaries of the cow are oval or bean-shaped, are 1 to 1½ inches long, and are located in the abdominal cavity.

The ovary contains several thousand tiny structures called primordial follicles. These can be quiescent for several years, but with each reproductive cycle, a few develop into primary follicles, each of which consists of a germ cell surrounded by a layer of nurse cells (Fig. 6–2). This germ cell has the potential to mature into an egg if the follicle continues its development to a tertiary stage. Most follicles never complete their development, however. They die and are absorbed by the ovary and replaced by newly formed primary follicles. Follicles also produce the hormone estrogen.

Primary follicles that complete their development do so through a series of developmental phases. Nurse cells start to grow and multiply around the follicle, and a central cavity forms (antrum). As the follicle and the cavity grow larger, the ovum is attached by a stalk of cells opposite the site of ovulation. The side opposite the ovum bulges with rapid follicular growth from the surface of the ovary, and the wall becomes thin. This mature follicle is called a Graafian follicle. The mature follicle ruptures, and the follicular fluids flow out carrying the ovum with it.

Following ovulation, the cells lining the follicle become luteinized and grow into a

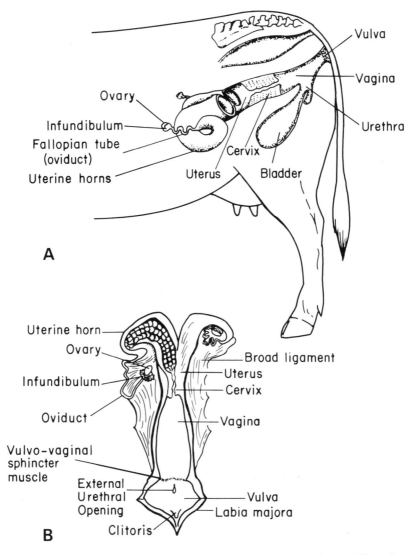

FIG. 6–1. Reproductive system of the cow as it appears in the natural state (*A*) and excised (*B*). (*A*, Adapted from material provided by the National Association of Artificial Breeders; *B*, Adapted from Bearden, J.H., and J. Fuquay. 1980. Applied Animal Reproduction. 2nd Ed. Reston Publishing Co., Reston, VA.)

corpus luteum. The corpus luteum secretes progesterone, a hormone important to maintaining pregnancy.

If the newly ovulated egg is not fertilized and pregnancy does not occur, the corpus luteum degenerates, and a new 21-day estrous cycle begins. If pregnancy occurs, however, the corpus luteum remains functional until delivery of the fetus.

At ovulation, the egg enters the *infundibulum* and passes into the 8- to 10-inch-long *oviduct*. Fertilization occurs in the upper portion of the oviduct. The egg remains capable of fertilization for only a few hours; thus, it is important that fertile sperm be present near the time of ovulation. The egg moves through the oviduct into the uterus within the next 3 to 4 days.

TABLE 6–1. Major Functions of the Female Reproductive Organs

Organ	Function(s)
Ovary	Production of oocytes Production of estrogens (Graafian follicle) Production of progestins (corpus luteum)
Oviduct	Gamete transport (spermatozoa and oocytes) Site of fertilization
Uterus	Location for retaining and nourishing the embryo and fetus
Cervix	Prevention of microbial contamination of uterus Reservoir for semen and transport of spermatozoa Site of semen deposit during natural mating in sows and mares
Vagina	Organ of copulation Site of semen deposit during natural mating in cows and ewes Birth canal
Vulva	External opening to reproductive tract

From Bearden J.H., and J. Fuquay. 1980. Applied Animal Reproduction. Reprinted by permission of Reston Publishing Company, a Prentice-Hall Company, 11480 Sunset Hills Road, Reston, VA 22090.

If fertilized, the egg begins embryologic development in the uterus. If it is not fertilized, it degenerates and disappears.

The horns and body of the uterus are usually about 14 to 18 inches long. On the inner surface of the uterus are 70 to 120 button-like projections called *caruncles.* During pregnancy in beef cattle, fetal *cotyledons* fuse with caruncles to form *placentomes.* The caruncle-cotyledon attachments (placentomes) bring fetal blood vessels into proximity with maternal blood vessels to provide a site for exchange of nutrients and waste for the fetus.

The *cervix* is about 4 inches long and 2 inches in diameter. The walls of the cervix are thick and form a small-diameter lumen, which connects the uterus and the *vagina.* In natural service, semen is deposited in the forward part of the vagina. In artificial insemination, however, semen is deposited in or through the cervix.

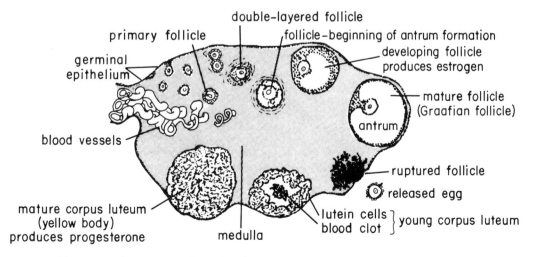

FIG. 6–2. Changes in the ovary. (Adapted with permission from Patten, B.M. 1964. Foundations of Embryology. 2nd Ed. McGraw-Hill, New York.)

The vagina is about 10 inches long and has relatively thin and elastic walls. It extends from the cervix to the *urethral opening*. The reproductive and urinary systems share a common passageway from this point to the exterior. This area is called the *vestibule*. The *vulva* is the external genitalia and forms the opening of the reproductive and urinary tracts.

HORMONAL REGULATION OF FEMALE REPRODUCTION

The step-by-step occurrence of the events in the estrous cycle are determined chiefly by hormones. Hormones are specific chemical substances produced by specialized glands, called endocrine glands. Hormones are secreted into the bloodstream and transported to various parts of the body, where they exert several special effects.

The female hormone, estrogen, is produced by the Graafian follicle. A second hormone of the ovary is progesterone, which is produced by the corpus luteum. *Estrogen* has the following actions:

1. Development and maintenance of the secondary sex organs.
2. Stimulation and maintenance of estrus, the period of sexual receptivity.
3. Alteration of rate and type of growth, especially muscle growth, via growth hormone.

Progesterone is the hormone of pregnancy. It suppresses secretion of pituitary hormone and thus development of follicles and subsequent secretion of estrogen. As a result, the female does not experience estrus. Progesterone is necessary for preparing the uterus to receive the fertilized egg, and it produces the proper uterine environment for the maintenance of pregnancy.

Uterine development is initiated by estrogen and completed by progesterone. The fertilized egg does not implant and survive in the uterus unless the tissue has been properly prepared by estrogen and progesterone. Estrogen causes rhythmic contractions of uterine muscles and aids in gamete transport near the time of estrus. Progesterone has a quieting effect on the uterus so that contractions do not disturb pregnancy.

The *anterior pituitary gland,* located at the base of the brain, produces gonadotrophic hormones, which control the production of the hormones of the ovary. Follicle stimulating hormones (FSH) stimulate development and growth of the follicle, and luteinizing hormone (LH) causes the rupture of the follicle and development and maintenance of the corpus luteum.

The *posterior pituitary gland* releases a hormone called *oxytocin,* which affects contractions of the uterus. Oxytocin is released in response to the mating stimulus and thus aids in gamete transport. This hormone also causes milk let-down.

THE ESTROUS CYCLE

The estrous cycle is defined as the time from one period of estrus to another (Bearden and Fuquay, 1980). During the cycle, ovulation occurs, and the reproductive tract is prepared for conception. The estrous cycle of the cow averages 21 days in length (ranges from 17 to 24 days). The sequence of events and hormones involved for a typical 21-day cycle in which pregnancy does not occur is outlined in Figure 6–3.

Stages of the Estrous Cycle

The four stages of the estrous cycle are estrus, metestrus, diestrus, and proestrus.

ESTRUS

Estrus is the period of time in which the female is receptive to being bred by the male (standing heat). Estrus lasts for 12 to 18 hours in the cow. A cow in estrus, displaying mating behaviors:

1. Attempts to mount other cows.
2. Solicits mounts from other cows.
3. May smell vulva of other cows.
4. Raises and switches tail.
5. Leaves herd in search of a bull.
6. Has clear mucus streaming from vulva.

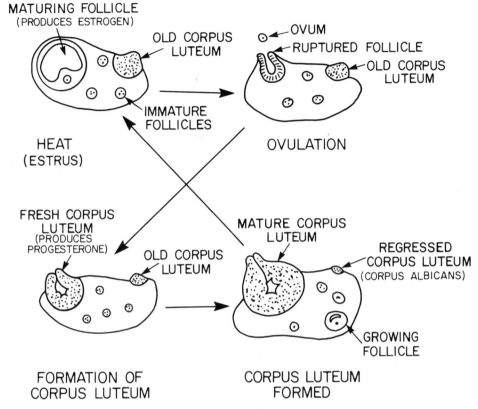

FIG. 6–3. Sequence of events in a typical 21-day estrous cycle in which pregnancy does not occur. (Adapted from National Association of Artificial Breeders.)

7. Stands to be mounted.
8. Shows signs of nervousness.
9. Bawls.
10. Urinates frequently.

METESTRUS

Metestrus begins with the end of estrus and lasts for approximately 3 days (ending on day 2 to 4 of cycle). During this period, the mature Graafian follicle releases the ripened egg, or ovum. Ovulation occurs 4 to 16 hours after the end of estrus. Ovulation is triggered by an LH surge from the anterior pituitary gland. Usually, only one egg is released during ovulation in the cow, but sometimes more than one egg is released, which could produce multiple births.

The cells lining the follicle change and become the lutein cells of the corpus luteum (CL). This change is primarily caused by LH. The CL develops rapidly.

DIESTRUS

The mature CL is functional during *diestrus* and produces the hormone progesterone (day 5 to 17 of cycle). Progesterone prevents the growth of a new follicle, maintains pregnancy in the female, and blocks LH release by the pituitary gland.

PROESTRUS

During *proestrus*, a prostaglandin ($PGF_{2\alpha}$) from the uterus causes a regression of the CL along with a sharp decline in blood levels of progesterone (day 17 to 21 of cycle). The decline in blood progesterone releases the blocking action on follicular growth. This results in the production and release of FSH and LH by the anterior pituitary gland. Follicular growth begins, and the Graafian follicle develops and produces estrogen. Estrogen causes the behavioral and physiologic signs of estrus, triggers LH release, and initiates the development of the uterus to receive the fertilized ovum, if mating occurs.

GESTATION

The gestation period begins at the time of fertilization of the ovum by the sperm in the oviduct. Average length of the gestation period is 281 days, with a range from 270 to 290 days. It ends when the calf is born (parturition).

The fertilized egg descends from the oviduct to the horn of the uterus within a period of about 4 days after breeding. The ovarian hormones, estrogen and progesterone, condition the uterus for receiving the fertilized ovum. The walls of the uterus are thickened, and the blood supply increases. During this time, the fertilized egg begins to divide. On about the fifth day after breeding, the egg has undergone 4 or 5 divisions and consists of 16 to 32 united, similar cells. Attachment to the wall of the uterine horn occurs no later than 30 days after conception.

The *placenta*, which consists of a series of membranes, develops around the embryo and serves to supply it with nutrients, oxygen, and carbon dioxide from the blood of the mother. It is filled with fluid, which protects the embryo. The membranes of the fetus are connected to the mother's uterus through placentomes. Retained placentas after parturition are due to the failure of the maternal and fetal membranes to separate at the placentomes.

The period from the 46th day of pregnancy until calving is called the fetal period. Increased size and weight of the fetus is the most important characteristic of this period. Figure 6–4 shows the relationship between the days of pregnancy and the weight of the reproductive tract and fetus.

As shown in Figure 6–4, the entire tract and fetus weigh only about 45 lb at 150 days of gestation. Therefore, the general requirements of energy, protein, and minerals to nourish the calf adequately during the first 5-month period of pregnancy are not much greater than normal maintenance requirements. Since approximately 110 lb of weight is added to the tract and fetus from the 5th month through termination of pregnancy, however, the energy, protein, and mineral requirements are greater during the last 4 to 5 months of gestation.

During pregnancy, the uterus greatly enlarges to accommodate the fetus. The uterus pushes into the abdominal cavity, and the vagina may stretch to twice its normal length. During delivery, the elastic

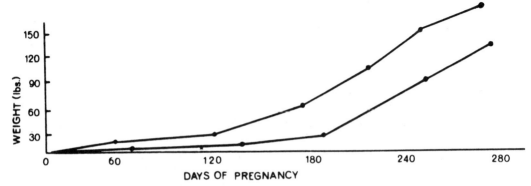

FIG. 6–4. Weight change in fetus and reproductive tract during pregnancy. (Adapted from Singleton, W.L., and L.A. Nelson. 1975. Assisting the beef cow at calving time. Cooperative Extension Service, Purdue University, W. Lafayette, IN. AS-405.)

walls of the vagina expand to the limits of the pelvic cavity in response to the uterine contractions, which are brought about by specific hormones.

PARTURITION AND LACTATION

During the latter stages of pregnancy, preparations are made for the next two steps in reproduction, *parturition* (calving) and *lactation* (milk production).

Parturition occurs about 281 days after conception, but varies with breed, sex of calf, and several other factors. After the signs of labor have appeared, the fetus is usually expelled within 4 to 5 hours. If birth has not occurred within 12 hours after labor has started, help by an experienced person should be given to the cow. The placental membranes should be expelled within 16 hours after delivery. Cows should be observed at least three times a day near time of parturition.

Near calving time, another hormone, *prolactin*, is secreted from the anterior pituitary gland and promotes initiation of lactation. Once initiated, lactation is maintained by several hormones along with the suckling stimulus.

Milk production increases rapidly after parturition, reaching a peak in 2 to 4 weeks. From 4 weeks following parturition to weaning, milk production gradually declines.

After parturition, the cow must produce milk as well as prepare the reproductive tract for another conception. During pregnancy, the uterus increases greatly in size and weight. Before another successful pregnancy can occur, the female reproductive tract must be adequately repaired. The process of uterine repair, *involution*, usually takes less than 50 days. This means that there should be a lapse of at least two months between birth and rebreeding.

PUBERTY

Puberty in females is defined as the time of the first estrus accompanied by ovulation. Puberty in heifers occurs as a result of

TABLE 6–2. Age and Weight at Puberty of Different Breeds and Crosses

Breed or Cross	Age at Puberty (days)	Weight at Puberty (lb)
Hereford	415	603
Angus	366	561
HA and AH	371	585
Jersey crosses	322	482
S. Devon crosses	364	603
Limousin crosses	398	642
Charolais crosses	398	667
Simmental crosses	372	629

From U.S. Meat Animal Research Center. April 1976. Progress Report No. 3. Clay Center, NE. ARS-NC-41.

endocrine activity. Table 6–2 gives the ages and weights at puberty of different breeds and crosses. Most beef heifers should not be bred at puberty because they would have difficulty with parturition. Factors influencing age and weight at puberty are:

1. Genetics—larger breeds of cattle (Charolais, Limousin) are usually older and heavier at puberty than smaller breeds (Angus, Hereford, Shorthorn).
2. Nutrition—inadequate nutrition reduces growth rate, therefore delaying the onset of puberty.
3. Environmental effects—elevated environmental temperatures have been shown to delay puberty in heifers as well as produce poor herd health and sanitation.

CALVING MANAGEMENT

Calving Season

The best season for calving is determined by weather, facilities, method of winter feeding, labor availability, and markets of the individual beef producer.

The most common calving season is the spring. Cows are dry (non-lactating) during most of the winter, and therefore, the total amount of feed required for a spring calving herd is less than for a fall calving herd, where the cows lactate through the winter months. A lactating cow requires approximately 50% more feed than a dry cow. Spring calving cows are bred on pasture while fall calving cows are usually bred in drylot. Traditionally, conception rates of cows bred on pasture are higher than those of cows bred in drylot. Milk production is usually higher with pasture than with dry, winter feed. Consequently, spring calving cows who are just beginning their peak milk production at that time should produce more total milk than fall calving cows who feed on dry forage. Often, fall calving does not fit in well with other farming operations. If beef cattle are a supplemental enterprise, calving should take place at a time when the least conflict occurs with the primary farm enterprise.

At the time of turnout onto pasture in the spring of the year, calves born in the fall are 3 to 5 months older and 100 to 200 lb heavier than calves born in the spring. Peak milk production by fall-calving cows has already occurred; consequently, their calves depend more on forage than on milk to satisfy their nutrient requirements. Fall-born calves are also better equipped than spring-born calves to consume more forage dry matter. Spring-born calves are usually too light and too dependent on milk at turnout onto pasture to maximize the use of available forage when forage nutritional quality is at its peak.

Some beef cattle producers calve in the fall as well as in the spring. Purebred breeders sometimes split calving seasons to ensure that they are able to show or sell calves in different age groups. Some commercial cattle producers select heifer replacements from their fall born calves for the main herd that calves in the spring. These heifers would calve at 2.5 years of age rather than at 2. Herd replacements for the fall herd come from the spring herd.

Time of Calving

The calving season is the most labor-intensive time for the cow-calf producer. It often involves either the hiring of additional labor to handle the night calving chores, or long days and sleepless nights if hiring additional labor is not economically feasible. Recently, several studies have shown that a beef cattle producer may be able to increase the percentage of calves born during the daylight hours by changing the time at which they are fed.

Gus Konefal, a purebred breeder from Manitoba, Canada, was one of the first individuals to investigate the possibility of changing calving time by manipulating feeding time. He established two different feeding programs for his cows and continued those feeding regimens from about 1 month prior to the start of calving until the completion of calving. He recorded the time of day when each calf was born, and the results are shown in Table 6–3. Cows fed later in the day had 80% of their calves born during the daylight hours, compared with 38% of those fed earlier in the day.

Iowa State University conducted a survey of 15 cattle producers that could be clearly classified as feeding either early in the day (before noon) or late in the day (5 to 10 P.M.). Cows fed late had 85% of their calves born during the day while only 15% were born at night (Table 6–4). Only 49.8% of the cows in the morning feed group calved during daylight hours.

In a three-year study conducted at the Livestock and Range Research Station

TABLE 6–3. Influence of Feeding Time on Calving Time

Group	No. Calvings	Calving Time (%) 7 A.M. to 7 P.M.	Calving Time (%) 7 P.M. to 7 A.M.
Fed 11 A.M. to noon and 9 to 10 P.M.	44	80	20
Fed 8 to 9 A.M. and 3 to 4 P.M.	39	38	62

Reported by R.B. Stagmiller and R.A. Bellows (1981), from data by Gus Konefal, Arborg, Manitoba.

(LARRS) at Miles City, Montana, the effect of time of feeding on calving time was recorded (Table 6–5). Approximately 67% of the cows fed early (7 to 9 A.M.), calved from 6 A.M. to 10 P.M., and 33% calved at night (10 P.M. to 6 A.M.). In the cows fed late, 78.1% calved during the day and early evening hours, and 22.8% calved at night. In the Konefal study and Iowa survey, feeding occurred as late as 9 to 10 P.M. whereas cows in the LARRS study were mostly fed at 5 to 6 P.M. in the late feeding group. This 3- to 4-hour difference may account for more cows calving during the daylight hours in the earlier studies.

Feeding cows in the evening increases the number of cows calving during the daylight hours; however, *this has not eliminated nighttime calvings.* Therefore, beef cattle producers still need to observe their cows during the late night and early morning hours.

Reducing Calving Difficulty in Beef Cattle

Calf deaths at or near birth represent a major economic loss for beef cattle producers. As factors contributing to reductions of the net percentage of calf crop, calf losses are second in importance only to cows that cannot become pregnant. Calving data compiled over 15 years at the Livestock and Range Research Station at Miles City, Montana, which involved 13,000 calvings, revealed that of all calves lost, 68% were lost during the first 3 days after birth (Table 6–6). Of calves lost during the first 3 days, 61% were lost as a result of dystocia (calving difficulty). Thus, dystocia represents the single largest cause of calf losses. Seventy-five percent of the calves lost were anatomically normal at autopsy.

In recent years, increased attention has been given to calving difficulty. One reason for this is the mating of breeds containing larger bulls to British breeds of cows (Angus, Hereford, Shorthorn, etc.). Increased calving problems are also being encountered within purebred herds, because

TABLE 6–4. Time of Feeding— Time of Calving

Group	No. Calvings	Calving Time (%)	
		6 A.M. to 6 P.M.	6 P.M. to 6 A.M.
Morning-fed only	695	49.8	50.2
Evening-fed only (5 to 10 P.M.)	1331	85.1	14.9

Reported by R.B. Stagmiller and R.A. Bellows (1981), from data by Iowa State Extension Service, C. Iverson.

genetically large bulls are often mated to cows of only average size.

Beef cattle producers need to understand the numerous factors influencing calving difficulty if they are to take steps to reduce it. The following section discusses the factors influencing calving difficulty, the calving process, preparation for calving, and potential post-delivery problems.

Factors Associated With Calving Difficulty in Beef Cattle

The major factors contributing to calving difficulty, listed in the order of most important to least important are:

1. Birth weight
2. Age of dam

TABLE 6–5. Effect of Feeding Time on Time of Calving at the LARRS, Miles City, MT[a]

Group	No. Calvings	Calving Time (%)	
		6 A.M. to 10 P.M.	10 P.M. to 6 A.M.
Early-fed (7 to 9 A.M.)	334	66.9	33.3
Late-fed (5 to 6 P.M.)	347	78.1	22.8

[a] Summary of 3 years of data.
From Stagmiller, R.B., and R.A. Bellows. 1981. Daytime calving. Proc., Range Beef Cow Symposium VII, Rapid City, SD. December 7–9.

TABLE 6–6. Calf Losses by Days Following Calving

Days after Calving	Percentage of All Losses	Major Category	Percentage of Losses from Major Category
1–3	68	Dystocia	61
4–6	6	Accidents	31
7–9	5	Scours	38
10–42	12	Pneumonia	32
43–weaning	9	Missing	33

From Stagmiller, R.B. 1980. Managing the beef cow herd to minimize calving difficulty. Proc., 2nd Ann. Western Beef Symposium. Boise, ID. October 28–29.

3. Gestation length
4. Sex of calf
5. Precalving nutrition and environment
6. Shape of the calf

Cattlemen usually have little control over items 3, 4, and 6, whereas they can influence the other four factors. A discussion of these factors follows.

BIRTH WEIGHT

Heavy birth weights account for most of the problems related to calving difficulty. Average heritabilities and the genetic relationship of some measurements of growth rate with birth weight are listed in Table 6–7.

A genetic correlation is simply the degree to which traits are transmitted or inherited together. A high positive genetic

TABLE 6–7. Heritabilities of Growth Traits and Their Genetic Correlations With Birth Weight

Trait	Heritability (%)	Genetic Correlation With Birth Weight
Birth weight	44	—
Weaning weight	32	0.58
Yearling weight	58	0.61
Mature weight	84	0.68

From Petty, R.R., and T.C. Cartwright. 1966. A summary of genetic and environmental statistics for growth and conformation traits of young beef cattle. Texas Agriculture Experiment Station Department, College Station, TX. Technical Report No. 5.

correlation (relationship) indicates that two traits increase or decrease together; that is, selection to increase one trait increases the second trait. Birth weight is positively related to weaning, yearling, and mature weights. Therefore, selection for any of these traits also causes some increase in birth weight. A major challenge to the beef industry is to find a way to minimize this correlated response in birth weight.

Stagmiller (1980) reported that bulls with the highest rate-of-gain indexes at LARRS had a wide variation in their own birth weights. Among the top ten gaining bulls of their Line 1 herd for each year during a 9-year period, differences in birth weight ranged from 17 to 35 lb (Table 6–8). The importance of these data is that it may be possible to select for increased growth rate and slow the rate of increase in birth weight.

Dickerson and associates (1974) developed a selection index that places negative emphasis on birth weight but still allows selection for rapid growth rate.

Selection index =
Yearling weight − (3.2 × Birth weight)

This index should reduce the expected increase in birth weight by 55%, reduce the expected increase in mature cow size by 25% and reduce the expected increase in yearling weight by only 10%. The index is currently being used in the Line 1 herd at the Northern Agriculture Research Center in Havre, Montana (Burfening, 1981). Ta-

TABLE 6–8. Range in Birth Weights Among the Ten Bulls With Heaviest Yearling Weights (lb)

Year	Heaviest Wt	Lightest Wt	Range[a]
1968	110	73	28
1969	100	77	23
1970	95	78	17
1971	98	80	18
1972	106	71	35
1973	102	68	34
1974	108	85	23
1975	112	78	34
1976	105	80	25

[a] If birth weight heritability is .48, then approximately one half of this difference is passed on to the progeny.

From Stagmiller, R.B. 1980. Managing the beef cow herd to minimize calving difficulty. Proc., 2nd Ann. Western Beef Symposium. Boise, ID. October 28–29.

TABLE 6–10. Percentage of Calving Difficulty by Age of Dam

Age of Dam (yr)	No. Calvings	Percentage of Calving Difficulties
2	437	29.7
3	475	10.5
4	427	7.2
5–10	1478	2.7
11–13	134	3.7

Based on Brinks, J.S., et al. 1973. Calving difficulty and its association with subsequent productivity in Herefords. J. Anim. Sci. 36:11.

ble 6–9 shows the difference between selecting with Dickerson's index and using yearling weight. The difference was 7.1 lb in birth weight and 17 lb in yearling weight. Therefore, calves selected by Dickerson's index would be 1.6 lb lighter at birth and 4.9 lb lighter as yearlings.

AGE OF DAM

First-calf heifers have the highest percentage of calving difficulty of any age group of dams (Table 6–10). Heifers are small and still growing. They have grown to about 75% of their mature size, but produce calves 90% as large as those of older cows.

Pelvic area (birth canal) increases as the female develops to sexual maturity. Thus, a higher proportion of calving difficulty in 2- or 3-year-old dams is due to smaller pelvic openings. Heifers and mature cows with small pelvic areas are likely to require assistance at calving. Even heifers with large pelvic areas may need help, however, when giving birth to calves with a heavy birth weight.

The calf's birth weight and the cow's pelvic area have a combined effect on dystocia. Two-year-old heifers tend to have either a pelvis that is too small or a calf that is too large to allow them to deliver without assistance. Therefore, calving problems could be reduced by decreasing birth weight through bull selection and increasing pelvic areas by selecting larger, better-growing heifers.

TABLE 6–9. Effect of Selection of Yearling Weight or Index

Selection Criteria	No. Bulls	Birth Wt (lb)	Yearling Wt (lb)
Yearling weight	12	94.9	1007.0
Index[a]	12	87.8	990.0
Difference (Y − I)		7.1	17.0
Predicted difference[b]		1.6	4.9

[a] Index × Yearling wt − 3.2 (Birth wt).
[b] (Predicted difference) × (Heritability of trait) × (0.5).
Data based on Line 1 herd, Northern Agriculture Research Center, Havre, MI. From Burfening, P.J. 1981. Genetic dilemma. Beef Digest. August. p. 35.

OTHER FACTORS

Calves normally gain about one pound per day near calving time. Therefore, a gestation period that is one week longer than average adds about seven pounds to the calf's birth weight. Data from Montana State University indicate a positive genetic correlation of .30 between weaning weight and gestation length (Burfening, 1981). Therefore, selection for heavier weights results in longer gestation lengths and heavier birth weights.

Bull calves weigh approximately 5 lb (or 7%) more than heifer calves at birth. Bellows and co-authors (1971) reported in a study with first calf heifers that about 2 to 2½ times as many bull calves were given assistance at birth as heifer calves. Brinks and associates (1973), in a study of dams from 2 to 14 years of age, found that about 50% more bull calves were given assistance than heifers.

Overfeeding has a greater effect on dystocia than underfeeding (Falk, 1981). Arnett and associates (1971) reported that 7 of 12 obese heifers required assistance at parturition, compared with 1 of 12 of the "normal" group. Similar data have been reported by Bond and Wiltbank (1970), and by Wiltbank and co-authors (1965).

Calf shape is generally believed by cattlemen to influence calving ease. Laster and co-authors (1973), however, reported that calf shape (shoulder width, hip width, chest width, body length, and wither height), independent of birth weight, is minimally related or not related to calving difficulty. Although the literature shows a lack of agreement regarding the relationship between calf shape and dystocia, one should consider the skeletal shape of bulls purchased or produced in attempting to minimize calving difficulty.

Stages of Calving (parturition)

A general understanding of the birth process is needed to save the maximum number of calves. Normal calving can be divided into three stages—preparation, fetal expulsion, and expulsion of the placenta or afterbirth (Table 6–11).

RECOGNIZING NORMAL CALVING

Most animals give birth unassisted if they are given a chance to do so, and if the calf is normally positioned (Fig. 6–5). Recognizing normal calving is as important as recognizing abnormal calving. In this way, one avoids giving unwanted help.

The first sign of calving is development of the cow's udder, which occurs 4 to 6 weeks prior to calving. Within 12 hours of calving, the udder becomes engorged with milk, and the teats become smooth and

TABLE 6-11. Stages of Parturition[a]

Stage	Events	Duration
Preparation	1. Calf rotates to upright position. 2. Uterine contractions begin. 3. Cervix begins to dilate, allowing water sac to protrude.	2 to 6 hr
Fetal expulsion	1. Cow usually lies down, and abdominal contractions begin. 2. Fetus enters birth canal. 3. Front feet and head protrude first. 4. Calf is delivered.	1 hr or less
Placental expulsion (cleaning)	1. Caruncle-cotyledon (button) attachments relax. 2. Uterine contractions expel membranes.	2 to 12 hr

[a] Heifers take approximately 50% more time than cows.

From Singleton, W.L., and L.A. Nelson. 1975. Assisting the beef cow at calving time. Cooperative Extension Service, Purdue University. W. Lafayette, IN. AS-405.

extended. The pelvic ligaments relax 24 hours prior to parturition, and the area between the tailhead and pin bones become loose and sunken.

Stage 1—Preparation (2 to 6 hours)

The cow shows signs of nervousness and slight pain. She kicks at her belly and twitches her tail. Usually, the cow stands away from the rest of the herd. The calf rotates to an upright position with its forelegs and head pointed toward the birth canal.

The cervix dilates, and uterine contractions begin. Uterine contractions are not visible except for the discomfort of the cow. Initially, uterine contractions occur at approximately 15-minute intervals, and as labor progresses, they become more frequent and occur every few minutes. These contractions cause the water bag surrounding the calf to press against and enlarge the cervix. As the contractions become more frequent, occurring about every 1½ to 3 minutes, cervical dilation is almost three-quarters complete, and the water bag starts to protrude through the cervix. The last few contractions during Stage 1 produce large amounts of manure and urine. This ensures that there is as much room as possible for the passage of the calf through the birth canal. Near the end of Stage 1, the pains of uterine contractions force most cows to lie down, although some heifers do not. The water bag may become visible near the end of this stage.

Stage 2—Delivery (1 hour or less)

This stage begins when the fetus enters the birth canal, and usually occurs while the cow is lying down. She appears to be oblivious of her surroundings and concentrates solely on expelling the fetus with abdominal pressing.

Uterine contractions occur about every 2 to 3 minutes and last from ½ to 1½ seconds. These contractions are accompanied by the voluntary contractions of the diaphragm and abdominal muscles. Straining

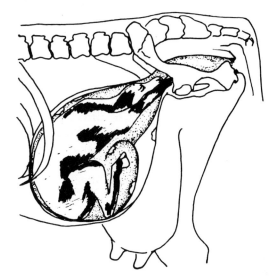

FIG. 6–5. Reproductive tract of a cow near normal calving. (Adapted from Singleton, W.L., and L.A. Nelson. 1975. Assisting the beef cow at calving time. Cooperative Extension Service, Purdue University, W. Lafayette, IN. AS-405.)

or abdominal pressing is stimulated by the feet and head of the calf coming in contact with the birth canal. The pressing of the head and feet against the cervix causes the cervix to relax and open further.

If the calf's head or feet are not in proper position to put pressure on the birth canal, the cow may exhibit weak or nonexistent abdominal pressing and may not attempt to deliver the calf. If a cow is apparently in Stage 1 but does not progress to Stage 2, the position of the calf should be checked.

The water bag moves into the vagina, after which the calf's feet, then its head, begin to pass through the cervix. The water bag may rupture at this stage, but it usually ruptures when the feet reach the vulva, and by that time, the calf's head has entered the front part of the vagina. Once the feet have appeared, the intensity of the contractions increase, occurring every 15 seconds to 1½ minutes. The tongue appears first, followed by the nose and head. After the head has appeared, the dam exerts maximum straining to push the shoulders and the chest through the pelvic girdle. Once the shoulders have passed, the abdominal muscles relax, and the calf's hips and hind legs extend back to permit easier passage of the hip region.

Once the feet and head are in the birth canal, and are subject to direct pressure from the abdominal contractions, delivery should not be delayed. Australian workers have found that normal delivery in heifers should be completed in 2 hours after the membrane sac is first noticed. Examination should be made under either of the following circumstances:

1. The heifer has been in intense labor for 2 hours without making progress toward delivery.
2. The water sac or feet are observed, and delivery is not completed in 1 hour.

Stage 3—Cleaning (2 to 12 hours)

The placenta (fetal membranes) is expelled by continued uterine contractions after the separation of the fetal cotyledons from the maternal caruncles attached to the uterus. The cotyledon, or button attachment, between uterus and placenta relaxes and separates. Retained placentas occur most frequently in cows that have had a

premature delivery, twin calves, milk fever, vitamin A deficiencies, mineral imbalances, and fatigue due to dystocia. A retained placenta is an ideal growth medium for contamination by bacteria. Prompt veterinary assistance must be given to the cow to prevent metritis (infection) and allow for normal rebreeding.

PREPARING FOR CALVING ASSISTANCE

Normal delivery in cattle should be completed within 2 to 3 hours, because the flow of maternal oxygen to the fetus decreases, owing to separation of the maternal caruncles from the fetal cotyledons. Calves can live for up to 8 hours in the uterus of a cow if delivery does not progress beyond the initial phases of Stage 2. Once the feet and head are in the birth canal, however, delivery should not be delayed for a prolonged period of time. If prolonged, the calf may be stillborn or may undergo permanent brain damage because of lack of oxygen. Assisted deliveries should not be attempted without proper preparation of facilities and equipment. A checklist for calving is given in Table 6–12.

TABLE 6–12. Calving Equipment Checklist

_____ Clean, lighted maternity pen (10' × 10')	_____ Ear tags and applicator
	_____ Tattoo equipment
_____ Pulling chains or straps (nylon) and handles	_____ Towel
	_____ Uterine bolus
_____ Disinfectant	_____ Scour vaccine
_____ Lubricant	_____ Vitamins A, D, and E
_____ Syringe and needles	
_____ Antibiotics	
_____ Scour bolus & gun	
_____ Flashlight	
_____ Record book	Optional
_____ Halter and lariat	_____ Dehorning paste
_____ Iodine	_____ Bucket heater
_____ Oxytocin	_____ Respirator
_____ Piece of hose (8" long)	_____ Epinephrine
_____ Paper towels	_____ Scales
_____ Bucket and scrub brush	_____ Pulling jack

Adapted with permission from American Breeders Service. 1983. A.I. Management Manual. W.R. Grace & Co., DeForest, WI.

Steps in Calving Assistance

1. Determine extent of cervical dilation by palpation. Final dilation of the birth canal is completed in stage 2. *Assistance can be given too early.* Ensure dilation is complete before rendering assistance.
2. Determine the position of the calf within the first hour of labor.
3. Examine size of calf in relation to the birth canal. If this examination can be made before the head and front feet are in the pelvis, successful cesarean section is possible. *If the calf is small enough to fit in the birth canal, proceed.*
4. If examination indicates a dry fetus and birth canal, add lubrication such as a commercial obstetric lubricant (K-Y jelly) or soap. Ivory soap and warm water are acceptable.
5. Attach obstetric pulling chains or straps to the front legs of the calf, placing the loop around each leg. Slide chains up the cannon bone 2 to 3 inches above the ankle joints and dewclaws, and make a half hitch below the ankle (Fig. 6–6). Do not attach the chains by making one loop around each front leg, because of the danger of breaking one or both legs. Make sure chains pull from the bottom of the leg.
6. Alternately pull one leg and then the other a few inches at a time (Fig. 6–7). This allows the shoulders to pass through the pelvis opening one at a time. If the shoulders happen to "lock" at the opening, apply traction to the calf's head both to reduce its compaction against the sacrum (or top of the birth canal) and to reduce the dimensions of the shoulder and chest region.
7. Once the shoulders and rib cage are exposed, pull the calf downward at a 45° angle, or nearly parallel with the legs of the cow. This causes the hips of the calf to be elevated in the birth canal and pass through the widest portion of the birth canal.
8. Tearing or lacerations of the uterus and cervix occur most frequently when the calf's head and shoulders come through the birth canal. The rule of thumb is to apply gradual pressure. Slow gradual pressure increases dilation of the cervix and ensures maximum expansion of the pelvic region.
9. "Hip locks" may cause loss of the calf. If the hips lock when a cow is lying down, push the fetus back a short distance, then apply traction to the front legs in the direction of the cow's udder flank. This rotates the calf enough so that one hip bone goes through the pelvic opening ahead of the other.

FIG. 6–6. Proper attachment of pulling chains or straps. (Adapted from Singleton, W.L., and L.A. Nelson. 1975. Assisting the beef cow at calving time. Cooperative Extension Service, Purdue University, W. Lafayette, IN. AS-405.)

10. After calf removal, check the uterus for tears or lacerations that may require suturing. Insert a suitable uterine antibiotic to prevent uterine infection.
11. After all assisted deliveries, inject the cow with oxytocin to aid in cleaning of the reproductive tract and to help prevent prolapse.
12. All breech positions (anterior) should be considered an emergency since the umbilical cord is cramped between the fetus and pelvis early in the delivery (Fig. 6–8). Blood circulation to the fetus is consequently decreased, and the fetus may die or sustain

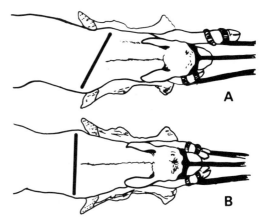

FIG. 6–7. Applying traction at delivery—properly (*A*) and improperly (*B*). (Adapted from Singleton, W.L., and L.A. Nelson. 1975. Assisting the beef cow at calving time. Cooperative Extension Service, Purdue University, W. Lafayette, IN. AS-405.)

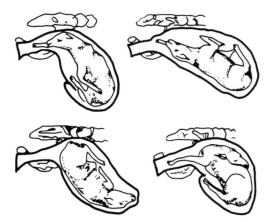

FIG. 6–8. Some routine abnormal positions of the calf for delivery. (Adapted from Singleton, W.L., and L.A. Nelson. 1975. Assisting the beef cow at calving time. Cooperative Extension Service, Purdue University, W. Lafayette, IN. AS-405.)

brain damage unless delivery is rapid. Rupture of the umbilical vein occurs many times in backward presentations, prior to complete expulsion of the calf. This stimulates the calf's respiratory reflex; consequently, mucus enters the lungs, which interferes with the calf's breathing.

13. Mechanical calf pullers can be of great help in an assisted birth; however, their use by inexperienced individuals can do more harm than good.

Starting the Calf

1. Clear mucus from the calf's mouth and throat by hand and by suspending the calf briefly by its rear legs to allow drainage of fluid from the lungs. Stimulate respiration by rubbing it briskly, tickling its nose with straw, or slapping it with the flat of the hand.

2. If artificial respiration is required, use mouth-to-nostril resuscitation. Use a commercial respirator, or place a short section of ¾-inch garden hose into one nostril, hold mouth and nostrils shut so that air enters and leaves only through the hose, then literally blow into the hose to allow expiration of air. Repeat at 5- to 7-second intervals until the calf begins to breathe. If a commercial respirator is used, make sure that breathing-quality air, and not straight oxygen, is used.

3. Disinfect the umbilical cord with a weak tincture of iodine solution to prevent infection.

4. Make sure that the calf consumes "colostrum" milk within 4 hours after birth to ensure that it receives important antibodies to prevent disease.

5. Collect birth-weight data and ease of calving records.

INDUCED CALVING

Induced calving in beef cattle is a management tool that has been used by beef cattle producers to promote parturition in heifers so that a sufficient work force is available during calving. Dexamethasone, flumethasone, and estradiol cypionate are examples of drugs that have been used to induce calving. In general, inducing calving 0 to 15 days early may decrease growth rate of calves and reduce the subsequent pregnancy rate of cows (Moody and Han, 1976). Decreased pregnancy rates are due to a higher percentage of retained placentas in cows induced to calve early. The use of induced calving as a management tool should be carefully considered before its use. It should be limited to such special situations as the treatment of cows determined to calve during the last 10 days of the projected calving season, when observation is not practical. Oxytocin therapy must be an integral part of an induced calving program to aid in "cleaning" of the cow and reduce problems with retained placentas.

MANAGEMENT OF A BEEF COW HERD AI PROGRAM*

Artificial insemination (AI) is the act of depositing semen into the cow's reproductive tract by means other than a bull. By using semen from superior sires, AI can profitably increase the quality and gaining ability of beef cattle.

---

* Adapted from Singleton, W.L., and D.C. Petritz. 1975. Management and economics of a beef cow herd AI program. Cooperative Extension Service, Purdue University. W. Lafayette, IN. ID-100.

After a rather slow beginning, AI has been gaining in popularity among beef producers because of the following:

1. Relaxation of breed association rules
2. Introduction of exotic breeds
3. Greater realization of the importance of using performance tested bulls
4. Increased interest in crossbreeding
5. Availability of practical synchronization programs

The following is a discussion of the advantages and disadvantages of AI, the conditions necessary for its success, and the mechanics of starting and carrying out an AI program.

Management Considerations

The benefits of AI include:

1. *More widespread use of genetically superior sires.* Once superior bulls are identified, their use can be maximized. Through AI, up to 20,000 cows per year can be inseminated with the semen from one bull, regardless of where the bull or cows are located.
2. *Services of proven sires at a lower cost.* Through AI, a producer's cow can be mated to bulls that he could not otherwise afford to own.
3. *Elimination of the cost, care, and danger of keeping bulls in small herds.* In most herds, one or more cleanup bulls are still utilized, however.
4. *Improved reproductive health of the herd.* When properly used, the separate sterile pipette for breeding each cow does not spread disease from cow to cow. AI can be helpful in controlling vibriosis and trichomoniasis.
5. *A tool for crossbreeding.* Several breeds of bulls can be used in the same year without ownership. Cows can be maintained in one pasture and still be selectively mated to bulls of the producer's choice.
6. *Continued use of a valuable sire in the case of his injury or death.* Semen can be collected from injured bulls with an electroejaculator without requiring them to mount a cow.

7. *More accurate records.* Records kept on heat detection and conception rate focus more attention on reproductive efficiency in a herd, thus providing a base for improvement. Breeding dates provide a good estimation of calving time, thereby enabling the producer to watch cows at a time nearer to calving.

The limitations of AI include:

1. *Requirement of good herd management.* A herd with previously low reproductive performance due to disease, inadequate nutrition, extended calving season, or other fertility problems, is not necessarily helped by AI.
2. *Requirement of a well-trained and interested operator.* The person in charge must be convinced that AI will benefit his herd. He must be trained in heat detection, and also in insemination techniques, to administer a within-herd AI program.
3. *Requirement of handling facilities.* Availability of facilities for corralling and restraining the cows is essential for AI. It is also desirable for other herd management practices, such as identification, performance testing, health programs, and pregnancy testing.

CONDITIONS FOR A SUCCESSFUL AI PROGRAM

An artificial insemination program is not feasible for every beef cattle producer. In fact, the cowman should be able to answer "yes" to each of the following questions before he gives it further consideration:

1. Does the producer sincerely believe that the genetic gains made by using proven sires is worth the extra time and effort?
2. Is the producer willing to spend time with his herd, observing the cows in estrus during early morning and late evening hours?
3. Is the producer willing to keep records and analyze them in culling and selection?
4. Does his herd already have a relatively short breeding season (60 to 90 days)?
5. Is the current calf crop weaned in the 80 to 90% range?
6. Is the herd provided with proper nutritional levels before, during, and after calving?

7. Are pastures or feed available where the herd can be reasonably well concentrated for the 30- to 45-day breeding period?
8. Are facilities available to handle the cows with ease and without exciting them?
9. Will net farm income be improved by adopting an AI breeding program?

Economic Considerations

While the cow-calf producer can probably visualize the benefits and limitations of artificial insemination, he must also ask himself, "What would be the change in net returns if I used AI instead of a natural mating system?" Because of the many differences that exist among beef herds, it is impossible to give a firm answer to this question. Instead, each producer should determine the economic feasibility of AI for his own situation. Good management requires careful planning and analysis of a proposed change before it is made.

A producer needs to consider three separate areas of cost-return data when analyzing an AI program:

1. He must estimate the added value of the feeder calves and replacement heifers resulting from the improved breeding program.
2. He must estimate the cost savings from keeping fewer breeding bulls.
3. He must estimate the added costs that an AI program involves, including semen, storage tank rental or purchase, liquid nitrogen, extra labor for detecting cows in heat and for handling cows, expense for a technician or for attending insemination school, breeding supplies, and other special handling facilities.

Starting and Maintaining a Beef AI Program

PLANNING THE PROGRAM

If after carefully analyzing the management and economic considerations, the producer is convinced that artificial insemination will work in his herd, he should plan a sound breeding program for his particular situation. The following might serve as a checklist for formulating such a program.

1. Use quality semen. An important factor contributing to success or failure of an AI program is quality of the semen. Quality semen can be obtained from:
   a. Commercial bull studs
   b. Independent semen distributors
   c. Individual purebred breeders
   d. A producer-owned bull, in either partnership or full ownership.

   In most instances, the bull is housed in a stall, and the semen is collected under quality-controlled conditions. Semen from privately owned bulls is usually collected on the farm and transported to a laboratory for processing and freezing. Many diseases can be transmitted through semen; therefore, it is recommended to use semen from bulls that have been tested and are known to be free from disease.
2. Begin with a well-managed herd that already has a short breeding season and high fertility. Cows should be 50 to 60 days postpartum (after calving) before the breeding season begins.
3. Begin with a healthy herd. Venereal diseases, such as vibriosis, cause reproductive failure even under natural breeding conditions.
4. Provide proper facilities for quiet and easy handling of the cows during insemination.
5. Individually identify the cows with either eartags, neck chains, or brands.
6. Ensure that both competent labor and adequately trained personnel are available when needed. Help is needed to detect estrus and bring cows to the insemination area. Training and expertise are crucial for successful heat (estrus) detection and insemination. Several commercial AI firms conduct excellent insemination schools for beef producers at a reasonable cost.

Mechanics of an AI Program

Once high-quality semen from proven bulls is purchased, and the reproductive status of the cow herd is good, the AI program can proceed. The following is a brief explanation of the mechanics of an AI program.

The most important factor affecting the success of an AI program is the use of cows with normal reproductive cycles so that those in heat can be identified.

HEAT DETECTION

The cow has an 18- to 22-day estrus (heat) cycle, i.e., she sheds an egg (ovulates) once every 18 to 22 days. From 18 to 30 hours before ovulation, she shows signs of estrus and accepts the bull. The herd should be observed for estrus twice daily—in early morning and in late evening.

Heat detection aids may be helpful. One of the most effective has been the chin-ball marker. This device, worn beneath the detector animal's chin, consists of a dye reservoir with a steel ball, which is attached to a strong leather halter. When the detector animal mounts the cow in heat, the device acts as a giant ballpoint pen, leaving a mark on the cow's back and rump. Chin-ball markers are commonly used on bulls that have been surgically altered to prevent contact of the penis with the vulva. Current methods for surgical alteration include complete removal of the penis and relocation of the penis right or left of its normal position ("sidewinder bulls"). These bulls retain their sex drive but cannot have sexual contact.

A newer technique used by some cattlemen is to inject a female or mature steer with a synthetic male hormone, testosterone, to provoke male-like behavior. This method is easy to use. Use of one androgenized cow for every 30 breeding females is recommended.

Kamar heat detector pads are sometimes used instead of chin-ball markers. The plastic device is glued to the tailhead of cows eligible for breeding. Prolonged pressure from the mounting animal's brisket turns the originally white detector red. At least 3 seconds of full weight are necessary to change the detector's color.

INSEMINATION

For maximum conception, only cows that have exhibited standing heat should be inseminated. Cows should be inseminated

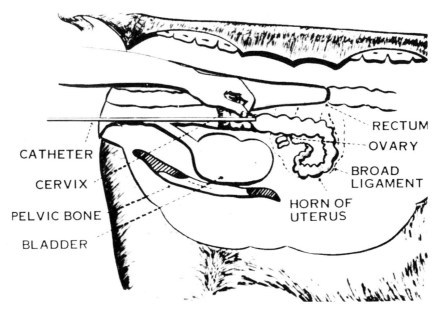

FIG. 6–9. Retrovaginal method of artificial insemination. (Courtesy of the American Breeders Service. 1983. A.I. Management Manual. W.R. Grace & Co., DeForest, WI.)

from the last half of standing heat to a few hours after standing heat for maximum fertility. This schedule is important because it insures that the sperm reaches the egg at the most optimum time for fertilization.

AI semen is stored at $-320°F$ in liquid nitrogen and may be packaged in ampules or in straws. It is thawed and placed into the cow through a 14- to 16-inch plastic pipette by the recto-vaginal method (Fig. 6–9). AI is best accomplished while the cow is standing in a blind chute with a pole behind her.

Heat Synchronization

Artificial insemination has not been widely used with beef cattle, mainly because of the time and labor required to detect cows in estrus accurately. It has long been felt that if estrus could be synchronized, AI would be facilitated, and more general use of genetically superior sires would be possible. Heat synchronization also allows heifers to breed and calve as a group, which helps to concentrate labor during the calving season.

Estrus and ovulation can be controlled by several hormonal products. Two products contain prostaglandins (Lutalyse and Estrumate), and one is an estrogen-progesterone product (Syncro-Mate-B) that is administered to cycling (reproductively active) females.

To understand how estrus is synchronized by prostaglandins or estrogen-progesterone, one must first understand the estrous cycle. After a cow demonstrates estrus and ovulates, a corpus luteum (CL, or "yellow body") is formed on the ovary. The newly formed CL produces progesterone, which prepares the uterus for pregnancy. If the cow does not conceive, the CL starts to regress 16 to 18 days after estrus, and a new egg-producing follicle begins to grow and mature.

Prostaglandins are drugs prescribed only by veterinarians. They are administered by intramuscular injection in the rump. Dosages for Lutalyse and Estrumate differ.

Prostaglandins cause premature regression of the CL (from days 6 to 17 in their estrous cycle). The cow comes into heat and ovulates within 2 to 5 days. Early in the cycle (days 1 to 5), as well as late in the cycle (days 18 to 21), prostaglandins have no effect.

Syncro-Mate-B is a non-prescription drug. It is made up of an ear implant containing a synthetic progestin called norgestomet, or an intramuscular injection containing estradiol valerate and norgestomet.

All cycling animals can be treated with Syncro-Mate-B regardless of where they are in the estrous cycle. Heifers are implanted and injected with the hormone solution on the same day. Nine days later, the ear implant must be removed.

Treatment early in the estrous cycle with Syncro-Mate-B (days 1 to 5) resets the ovary and causes a rapid decline in progesterone production. The progestin in the implant prevents the female from coming into heat until the implant is removed. If administered mid-cycle (days 6 to 17), Syncro-Mate-B causes CL regression and shortens progesterone production by a few days. This prevents the heifer from coming into heat until the desired time. Treatment during the later stages of the cycle (days 18 to 21) prevents a normal heat period from occurring for approximately two days, thereby extending the cycle. Synchronization of heat occurs 24 to 48 hours after implant removal.

SYSTEMS FOR USING PROSTAGLANDINS

Prostaglandins synchronize estrus by causing premature regression of the mature corpus luteum; therefore, they are only effective with cycling heifers and cows. It is important to recognize cycling from non-cycling females. Ovarian palpation is one method that can be used to determine if the female has a functional

TABLE 6–13. Outline of One-Injection System Using Prostaglandin

Days 1 to 5	Heat detection and artificial insemination of animals exhibiting estrus are conducted.
Day 6	If approximately 25% of cows exhibit heat, all other females are injected with prostaglandin.
Days 7 to 11	Heat detection and artificial breeding continues.
Days 27 to 33	Females that synchronized but were not impregnated on first service should return to heat between days 27 and 33.

Adapted with permission from American Breeders Service. 1983. A.I. Management Manual. W.R. Grace & Co., DeForest, WI.

CL. If ovarian palpation is not possible, females should be selected carefully, so that time and money are not wasted by injecting non-cycling females.

One-Injection System

This system is the most economical since only one injection is used. Table 6–13 outlines the schedule for this system.

Two-Injection System

Two injections of prostaglandin make it possible to have all females cycling within a 5-day period. Table 6–14 outlines the schedule for this system.

SYNCRO-MATE-B USE

All animals are treated on a single day without regard to their previous estrous cycle. When the implant is removed, a new estrous cycle begins (Table 6–15).

COMMON MISCONCEPTIONS

Cows must be cycling for prostaglandin products to be effective. Prostaglandin does not cause cows to come into heat sooner after calving. Females must be allowed a normal postpartum recovery period.

Calving season is not shortened unless strict culling procedures are used. Late calvers remain late calvers. A prostaglandin synchronization program, however, concentrates more cows (assuming they are cycling) at the beginning of the breeding season and therefore concentrates the number of cows calving during a given time period.

Heat detection is not required for the two-injection prostaglandin method and the Syncro-Mate-B method with timed in-

TABLE 6–14. Outline of Two-Injection System Using Prostaglandin

Day 1	All animals are injected with prostaglandin.
Day 11, 12, or 13	All animals are re-injected with prostaglandin.
Five days following second injection	All cycling females come into heat and can be heat detected and inseminated.
or:	
76 to 80 hr after second injection	Insemination is timed during a 4-hour period.
Days 32 to 38	Most repeat services occur. Additional AI calves may be obtained by insemination.

Adapted with permission from American Breeders Service. 1983. A.I. Management Manual. W.R. Grace & Co., DeForest, WI.

TABLE 6–15. Proper Use of Syncro-Mate-B

Day 1	Implant is inserted, and injection is administered.
Day 10	Implant is removed.
48 to 54 hr after removal	All heifers can be fixed-time inseminated within a 6-hr period.
or:	
Days 11 to 14	Treated heifers can be heat detected and inseminated during this time period.
Days 27 to 33	Breeder observes for estrus to detect open animals.

Adapted with permission from American Breeders Service. 1983. A.I. Management Manual. W.R. Grace & Co., DeForest, WI.

semination. With all other systems, heat detection is required. Table 6–16 summarizes the different heat synchronization systems.

EMBRYO TRANSFER

Embryo transfer is the removal of embryos from a donor female and the successful transfer to a recipient female. It is a management tool that can be used to upgrade the genetic pool when properly used; however, as with other such tools, it is not a cure-all for poor management.

Embryo transfer has progressed from a surgical technique that rendered many females infertile to a non-surgical recovery and transfer procedure. Today, it is possible to recover and transfer embryos on the farm.

Donor cows should be selected based upon their own individual performance,

TABLE 6–16. Summary of Heat Synchronization Systems

	No. of Injections	Heat Detection	"Go/No Go" Decision	No. Days AI	No. Times Cattle Handled	Drug Cost	Use Semen on Cycling Females Only
One-injection system	1	11 days	Yes	11	2	Minimal	Yes
Two-injection system with heat detection	2	5 days	No	5	3	Higher	Yes
Two-injection system without heat detection	2	None	No	1	3	Highest	No
Combination two-injection system	2	80 hr to 5 days	Yes	5	3	Moderate	No
Syncro-Mate-B with heat detection	1[a]	4 Days	No	4	3	High	Yes
Syncro-Mate-B	1[a]	None	No	1	3	High	No

[a] Plus implant insertion and removal.

Adapted with permission from the American Breeders Service. 1983. A.I. Management Manual. W.R. Grace & Co., DeForest, WI.

the performance of their offspring, and the popularity of their bloodlines. Good records are a definite requirement for selection of superior individuals. One must remember that ten pregnancies from a donor cow do not guarantee ten superior offspring. One should not expect the offspring from embryo transfer to be any more similar than any other group of full brothers or sisters. The only exception to this would occur in the case of clones, but one would not expect their performance to be similar. A breeder must consider cost versus return when contemplating the use of embryo transfer.

Cows are superovulated. One procedure for superovulation involves giving two daily injections of follicle-stimulating hormone (FSH) to a recipient female approximately 10 days after estrus. These injections are continued until the donor comes into heat. An injection of prostaglandin is given 48 to 72 hours after the first FSH injection. The donor usually comes into heat approximately 48 hours after prostaglandin administration. Multiple inseminations are recommended at 12, 24, and 36 hours after the onset of estrus. Embryos are then recovered 6 to 8 days later.

After recovery, the embryos are morphologically evaluated and quality-graded before transfer. The prospective recipient cow is palpated to determine the location of the corpus luteum. The embryo is then deposited in the proper uterine horn.

It has been reported that ten randomly selected donors normally average six transferable embryos per superovulation. Average pregnancy rate for embryos is highly variable, but should be somewhere between 55 and 60%.

The impact of embryo transfer or micromanipulation of embryos on beef cattle production may be unlimited. When sexing of embryos becomes commonplace, a breeder may be able to take his top animals to a transplant center and specify how many bulls or heifers he wants. Freezing of embryos may facilitate establishment of gene banks by preserving families of cattle. Embryo transfer calves do not suckle their genetic dams, however. Thus, weaning weight information of these calves cannot be compared accurately to the performance of other cattle in a breeder's herd. This is a disadvantage.

PREGNANCY TESTING

When the cow becomes pregnant, her estrous cycle ceases. Therefore, by close observation of estrus, one can get an indication of the herd's fertility, discover and treat reproductive disease, or detect and replace an infertile bull before a year's calf crop is lost.

Pregnancy is commonly determined by rectal palpation of the reproductive tract. An experienced technician using this method can detect pregnancy as early as 45 days after conception. The most practical time to perform a pregnancy test, however, is at weaning, when cows should be 3 to 5 months pregnant.

By "working the herd" (weaning, weighing, and vaccinating the herd) at this time, a number of other profitable management practices besides the pregnancy test are performed. Weaning is also an ideal time to cull open cows, and possibly, those that were bred late in the breeding season.

No producer can afford to maintain an open cow or heifer over the wintering period. A pregnancy test usually costs between $2 and $5 per cow, depending on the number examined and the handling facilities available. Such tests provide cheap "insurance" against the high cost of maintaining a non-producing female.

New technology, including electronic instruments and blood tests for hormones and proteins only occurring in pregnant females, will be used in the future. Electronic instrumentation and blood tests, however, will be used in the beef cattle industry only when they are cost-competitive with manual palpation.

REFERENCES

American Breeders Service. 1983. A.I. Management Manual. W.R. Grace & Co., DeForest, WI.

Arnett, D.N., G.L. Holland, and R. Totusek. 1971. Some effects of obesity in feed females. J. Anim. Sci. 33:1129.

Bearden, J.H., and J. Fuquay. 1980. Applied Animal Reproduction. Reston Pub. Co., Reston, VA.

Bellows, R.A., R.E. Short, D.C. Anderson, et al. 1971. Cause and effect relationships associated with calving difficulty and calf birth weight. J. Anim. Sci. 22:407.

Bond, J., and J.N. Wiltbank. 1970. Effects of energy and protein on estrus conception rate, growth and milk production of beef females. J. Anim. Sci. 30:438.

Brinks, J.S., J.E. Olson, and E.J. Carroll. 1973. Calving difficulty and its association with subsequent productivity in Herefords. J. Anim. Sci. 36:11.

Burfening, P.J. 1981. Genetic dilemma. Beef Digest. August. p. 35.

Dickerson, G.E., N. Kunzi, L.V. Cundiff, et al. 1974. Selection criteria for efficient beef production. J. Anim. Sci. 39:659.

Falk, D.G. 1981. The effects of prepartum dietary energy intake on cow reproductive performance. M.S. Thesis. University of Idaho. Moscow, ID.

Laster, D.B., H.A. Glump, L.V. Cundiff, and K.E. Gregory. 1973. Factors affecting dystocia and the effects of dystocia on subsequent reproduction in beef cattle. J. Anim. Sci. 36:695.

Moody, E.L., and D.K. Han. 1976. Effect of induced calving on beef production. J. Anim. Sci. 43:298 (Abstr.).

Patten, B.M. 1964. Foundations of Embryology. 2nd Ed. McGraw-Hill, New York.

Petty, R.R., Jr., and T.C. Cartwright. 1966. A summary of genetic and environmental statistics for growth and conformation traits of young beef cattle. Texas Agriculture Experiment Station Department, College Station, TX. Technical Report No. 5.

Stagmiller, R.B. 1980. Managing the beef cow herd to minimize calving difficulty. Proc., 2nd Ann. Western Beef Symposium. Boise, ID. October 28–29.

Stagmiller, R.B., and R.A. Bellows. 1981. Daytime calving. Proc., Range Beef Cow Symposium VII. Rapid City, SD. December 7–9.

Singleton, W.L., and D.C. Petritz. 1975. Management and economics of a beef cow herd AI program. Cooperative Extension Service, Purdue University. W. Lafayette, IN. ID-100.

Singleton, W.L., and L.A. Nelson. 1975. Assisting the beef cow at calving time. Cooperative Extension Service, Purdue University. W. Lafayette, IN. AS-405.

U.S. Meat Animal Research Center. April 1976. Progress Report No. 3. Clay Center, NE. ARS-NC-41.

Wiltbank, J.N., J. Bond, E.J. Warnick, et al. 1965. Influence of total feed and protein intake on reproductive performance of the beef female through second calving. USDA, Washington, DC. USDA Tech. Bull. No. 1314.

Selected Readings

Bellows, R.A. 1972. Problems at calving. Montana Stockman Magazine and Franklin Laboratories, Billings, MT.

Card, C.S., and E.P. Duren. 1978. Pegram Project—Beef Herd Health Program. Cooperative Extension Service, University of Idaho, Moscow, ID. CIS-430.

Eldsen, P. 1977. Non-surgical recovery of bovine ova. Charolais Bull-O-Gram. Kansas City, MO. June/July.

Hultine, V., G.H. Kiracofe, R.R. Schalles, et al. 1975. Induced calving in beef cattle. Cattlemen's Day Pub. 230. Kansas Agriculture Experiment Station, Kansas State University, Manhattan, KS.

Nelson, L.A., W.L. Singleton, W.H. Smith, and K.M. Weinland. 1974. Beef herd management calendar—Spring calving program. Cooperative Extension Service, Purdue University. W. Lafayette, IN. AS-414.

Nelson, C.F. 1982. The fast-growing speciality—A management and nutrition guide to embryo transfer. Anim. Nutrition and Health. January/February.

Rich, T.D. 1980. How to handle calving difficulties. In Cow-Calf Management Guide—Cattlemen's Library. Cooperative Extension Service, University of Idaho, Moscow, ID. CL-740.

Wilson, L.L., and W.M. Dillon. 1967. Structure and function of the cow's reproductive system. Cooperative Extension Service, Purdue University. W. Lafayette, IN. AS-359.

Wilson, L.L., and C.W. Floey. 1967. Estrous cycle and reproductive hormones of the cow. Cooperative Extension Service, Purdue University. W. Lafayette, IN. AS-361.

QUESTIONS FOR STUDY AND DISCUSSION

1. Name the reproductive organs of the cow, and list the major function(s) of each.
2. Define Graafian follicle.
3. What is the function of the corpus luteum?
4. When does ovulation normally occur in the cow?
5. Where are the following hormones produced, and what is their function?
   a. Estrogen
   b. Progesterone
   c. Follicle-stimulating hormone (FSH)
   d. Luteinizing hormone (LH)
   e. Oxytocin
6. Define puberty.
7. Describe and characterize the periods of a cow's estrous cycle.
8. What factors affect the onset of puberty in heifers?
9. What are some of the signs of estrus in the cow?

10. In what part of the reproductive tract does fertilization occur?
11. What function does the placenta serve?
12. Why are the nutrient requirements of the beef cow higher in late gestation (last 4 months) versus early gestation?
13. What is the function of prolactin?
14. What is meant by involution of the uterus, and when does it usually occur?
15. Define placentome.
16. What are the advantages and disadvantages of spring calving? Of fall calving?
17. How may time of calving be manipulated through feeding management?
18. Discuss how the following factors influence calving difficulty.
    a. Birth weight
    b. Age of dam
    c. Pelvic area
    d. Length of gestation
    e. Calf shape

19. Describe the stages of parturition.
20. When should calving assistance be given?
21. Outline the steps in giving calving assistance.
22. How would you correct the following:
    a. Hip lock
    b. Breech position
23. What are the disadvantages of induced calving?
24. Describe the benefits and limitations of artificial insemination.
25. List the visual signs of estrus.
26. How may heat, or estrus, be detected in beef cattle?
27. How do prostaglandins and Syncro-Mate-B synchronize estrus in beef cattle?
28. Outline procedures for synchronizing cattle with a prostaglandin and with Syncro-Mate-B.
29. Why should beef cattle producers test their cattle for pregnancy?

Chapter Seven

Female Selection

The development of a highly productive cow herd depends on the selection of females (1) that will produce a live salable calf every 12 months, (2) that will wean a calf whose weight at 7 months of age equals 60% of the cow's weight, (3) that will be pregnant at weaning time, and (4) that will bear and mother calves with little or no outside assistance. Female selection has many times been considered second in importance to sire selection; however, proficient beef cattle managers not only select outstanding sires but also practice strict cow culling and critical selection of replacement heifers.

FACTORS TO CONSIDER IN THE SELECTION PROCESS

Reproductive Efficiency

Economic returns from cow-calf enterprises are determined by the total pounds of calf weaned per cow exposed to the bull. Therefore, the goal of a cow-calf producer is a high percent calf crop. Net calf crop is calculated as follows:

Net % calf crop =

$$\frac{\text{Number of calves weaned}}{\text{Number of cows exposed to the bull}} \times 100$$

Table 7–1 shows the relative importance of factors affecting the net calf crop at the Livestock and Range Research Station at Miles City, Montana. The net calf crop that was weaned from 12,827 calvings was 71%.

The most important factor affecting the net calf crop was the failure of cows to become pregnant during the breeding season. Females not pregnant at the end of the breeding season accounted for 60% of the total reduction in net calf crop (17.4 ÷ 29 × 100).

Net calf crop gives an indication of the reproductive efficiency of the herd. Cows not calving or those failing to wean a calf cost almost as much to feed, and require the same amount of labor, as those that wean a calf. Table 7–2 illustrates the effect of conception rate and net calf crop on pounds of calf weaned per cow exposed to the bull.

A herd with a high net calf crop but a low average weaning weight could possibly produce more pounds of calf weaned per cow exposed to the bull than a herd with a heavier average weaning weight but poorer reproductive efficiency. Weaning weights and growth rates are important, but if a livestock producer loses 20 to 30% of the potential calf crop, he has lost valuable pounds of calf weight that would have increased his income.

One should have as many cows as possible breed early in the breeding season. Cows bred early in the season produce calves that are older at weaning and usually heavier than those born to cows bred late in the season.

Selecting and developing replacement heifers properly is the key to a high reproductive and calving rate. Few cattlemen

TABLE 7-1.   Factors Affecting Net Percentage of Calf Crop Weaned (14-Year Summary)[a]

Factors	No. Cows		Reproduction in Net Calf Crop (%)	
Females not pregnant at end of breeding season	2,232		17.4	
Perinatal deaths	821		6.4	
Calf deaths from birth to weaning	372		2.9	
Calf deaths during gestation	295		2.3	
Total potential calves lost		3,720		
Net calf crop weaned	Number	9,107	Percentage	71.0
Totals		12,827		100.0

[a] Includes females 14 months to 10 years old during breeding seasons lasting 45 or 60 days.
From R.A. Bellows, R.E. Short, and R.B. Stagmiller, Research areas in beef cattle reproduction, *from* Animal Reproduction, Beltsville Symposia in Agricultural Research: 3 (Montclair, MJ: Allanheld, Osmun, 1979), p. 5.

today can afford the luxury of breeding heifers to calve as 3-year-olds. Breeding heifers at 14 to 15 months of age to calve at 24 months of age reduces the overhead and increases income. Some beef cattle producers that have attempted to breed their heifers at 14 to 15 months of age have had problems with conception rates. Usually these problems are due to the fact that the heifers are not in heat because they are too young or light. Heifers must be fed properly after weaning so that they can be bred at 14 to 15 months of age. Nutritional regimes that cause slow growth rate delay the onset of puberty while a high level of nutrition, resulting in rapid growth, hastens the onset of puberty.

A study of the USDA Livestock and Range Research Station at Miles City, Montana was conducted to determine the effects of feeding on fertility and puberty of Angus-Hereford heifers (Table 7–3) (Short and Bellows, 1971). Heifers were fed during the winter to gain 0.6 (low), 1.0 (medium), and 1.5 (high) pounds daily for 153 days. Marked differences in weight gains during both winter and summer were related to winter nutrition. In October, at 1½ years of age, heifers that had been wintered as calves on a high plane of nutrition weighed 79 pounds more than heifers wintered on the low plane of nutrition. Twenty percent of the low-plane heifers failed to come into heat during the breeding sea-

TABLE 7-2.   Effect of Conception Rate and Percentage of Calf Crop on Weaning Weight per Cow Exposed to the Bull

Item	Performance Level		
	Excellent	Average	Poor
Conception rate (%)	95	85	75
Calves weaned (%)	95	90	80
Net calf crop (%)	90	77	64
Average weaning weight (lb)	500	425	350
Pounds of calf weaned per cow exposed to the bull	451	325	223

TABLE 7-3. Summary of Feed Effects on Heifer Reproduction

Data	Winter Gain Group		
	Low	Moderate	High
Winter gain (lb/day)	0.6	1.0	1.5
Summer gain (lb/day)	1.3	1.2	0.9
Body weight (lb)			
End of winter (5/6)	414	481	558
Onset of breeding (6/15)	458	527	584
October (10/15)	629	667	708
Puberty age (days)	434	412	388
Percentage in heat			
Before breeding season	7	31	83
During breeding season	73	66	17
After breeding season	20	3	0
Percentage bred and conceived			
First 20 days	30	62	60
Second 20 days	10	21	20
Third 20 days	10	3	7
Percentage not bred	20	3	0
October pregnancy (%)	50	86	87

Based on Short, R.E., and R.A. Bellows. 1971. Relationships among weight gains, age to puberty and reproductive performance in heifers. J. Anim. Sci. 32:127.

son. October pregnancies were 50%, 86%, and 87% for the three groups of heifers.

If a heifer is to be a good lifetime producer, she must be bred early. Only 30% of the heifers in the low-plane group were bred and conceived during the first 20 days of the breeding season. Those heifers were destined to be poor lifetime producers simply because of the feed level during their first winter as weaned calves. Heifers should weigh a minimum of approximately 66% of their mature body weight at breeding if they are bred at 14 to 15 months of age.

Interestingly, heifers wintered on the low plane of nutrition had smaller pelvic areas and more calving difficulty (Table 7–4). Inadequate nutrition delays development of the skeletal components as well as the muscle and fat components of the body.

The effect of early calving of heifers on their future reproductive performance and pounds of calf produced was studied at Colorado State University (Spitzer et al., 1975). In the study, 140 yearling Angus

heifers were assigned to two groups: "control" and "new management" (Table 7–5). Breeding of heifers started 20 days prior to the cow herd and estrus synchronization was used in the new management group. In addition, 70% more heifers were exposed than were needed as replacements. Replacement heifers returning to the new

TABLE 7-4. Effects of Rearing Nutrition on Reproduction, Pelvic Area, and Calving Difficulty in Heifers

Item	Winter Gain Group	
	Low	High
Number of heifers	30	59
Daily gain—winter (lb)	0.6	1.3
Daily gain—summer (lb)	1.3	1.0
October pregnancy (%)	50	86
Precalving pelvic area (cm^2)	240	252
Calving difficulty (%)	46	36

From Bellows, R.A. 1979. Heifer and cow management. Proc., Western Beef Symposium I, Live Calf—A Key to Pleasure. Boise, ID. October 22. p. 15.

TABLE 7-5. Effect of Management System on Estrus, Pregnancy, and Weaning Weights

| | Group | | | |
| | New Management | | Control | |
Item	Estrus (%)	Pregnancy (%)	Estrus (%)	Pregnancy (%)
Number of heifers	(51)		(32)	
First 25 days of breeding season	98	68	63	41
First 45 days of breeding season	100	83	81	63
By 90 days of breeding season			94	78
Number of cows	(195)		(199)	
First 25 days of breeding season	98	74	81	60
First 45 days of breeding season	100	87	93	78
By 90 days of breeding season			100	92
Calf weaning wt (lb)	444		411	

From Spitzer, J.E., et al. 1975. Increasing beef cow productivity by increasing reproductive performance. Colorado State University Experiment Station, Ft. Collins, CO. Gen. Series 949.

management herd were selected entirely on the basis of having conceived early in the breeding season. Based on an average calculated from four calf crops, more cows exhibited estrus and became pregnant early in the breeding season, which resulted in heavier weaning weight. Greater lifetime production can be derived from having heifers breed early in their first breeding season; however, it must be done economically.

Many heifers fail to rebreed, or they rebreed late during their second breeding period, because they take longer to come into heat following calving. This may be explained by comparing the body functions of a heifer to that of a mature cow. A heifer must perform all the body functions of a mature cow, such as body maintenance, lactation, and rebreeding. In addition, she is attempting to grow, and to further complicate matters, she is shedding incisor teeth. Therefore, the heifer must sacrifice some other activity, and this is usually rebreeding.

This problem may be overcome by breeding heifers 20 to 30 days earlier than the cow herd to assure early calving which allows more post-calving time during the second breeding season to return to heat. Also, early weaning of calves at 30 to 90 days of age removes the stress of lactation and allows the nutrients to be used for other body functions. Studies at the Texas Agricultural Experiment Station have shown pregnancy rates of 27% in Brahman cross heifers whose calves were weaned at 7 months of age in contrast to a pregnancy rate approaching 100% in groups whose calves were weaned at 60 days.

Proper nutrition of heifers and cows during gestation is also necessary to obtain the maximum calf crop. Inadequate nutrition during gestation can have a dramatic effect on the pregnancy rate after calving, as is illustrated in Table 7–6. Reproductive performance of cows exceeded that of heifers in every category. Cows returned to heat sooner after calving and had a higher pregnancy rate. Low feed levels prior to calving were more detrimental to heifers than cows.

Adequate nutrition from calving to rebreeding is necessary to maintain optimum reproductive performance. This period has the greatest nutritional demand. The cow is at her maximum milk flow, being suckled several times daily, recovering from the stress of parturition, and being expected to conceive during the breeding season. If the cow is not fed adequate energy, protein, or phosphorus, re-

TABLE 7-6. Effect of Gestation Feed Level on Reproduction in Heifers and Cows

Gestation Feed Level[a]	Dam	Postpartum Interval (days)	In Heat by Beginning of Breeding Season (%)	October Pregnancy (%)
Low	Heifer	100	17	50
	Cow	58	93	83
High	Heifer	77	47	78
	Cow	60	88	81

[a] Low = 8.0 lb TDN; High = 15.0 TDN fed last 90 days gestation.
From Bellows, R.A. 1979. Heifer and cow management. Proc., Western Beef Symposium I, Live Calf—A Key to Pleasure. Boise, ID. October 22. p. 15.

productive problems may develop. Proper post-calving nutrition is discussed in more detail in Chapter 8.

Other factors that may also reduce reproductive performance are calving difficulty (dystocia), reproductive diseases, and poor calving management.

PROCEDURES TO IMPROVE REPRODUCTION PERFORMANCE

1. Identify all animals, and keep reproductive information (percentage of calf crop; pregnancy rate; length of calving season; open cows).
2. Retain 50% more replacement heifers than needed.
3. Breed replacement heifers at 14 to 15 months of age.
4. Breed replacement heifers 20 days earlier than the cow herd; breed 40 days; cull open heifers.
5. Feed heifers after weaning so that they will weigh 66% of their mature body weight at breeding time.
6. Breed heifers to a bull that will sire calves with optimum birth weights.
7. Feed heifers and mature cows adequately during gestation and the post-calving period.
8. Cull all open cows.
9. Develop a sound immunization program for reproductive diseases.

Health

All selected cattle should be free of disease and parasites. Cattle purchases from an outside breeder should be subject to the animals being free of contagious diseases.

A health certificate should be furnished by a veterinarian when purchasing costly purebred animals. Every reasonable effort should be made to determine if cattle purchased are from herds free of infectious diseases. Newly acquired animals should be isolated for a minimum of 3 weeks and observed for disease before being turned in with the rest of the cow herd. Diseases to be concerned with when purchasing breeding animals are anaplasmosis, brucellosis, leptospirosis, tuberculosis, trichomoniasis, and vibriosis. A detailed description of each disease is found in Chapter 9.

Visual Appraisal

Desirable characteristics of a mature beef cow are shown in Figure 7–1. The heads of fertile females have a feminine appearance. Heifers should have a deep jaw, broad muzzle, and a large growthy appearance to their head. They should be clean and neat in the face, throat, and brisket. A good spring of rib is important in females to ensure that they have the capacity to consume adequate feed and support the growth and development of the fetus. Females should be long in the hip and wide at the pins. Width at the pins indicates the size of the pelvic opening (Fig. 7–2). In general, the smaller the pelvic area, the greater is the chance of calving difficulty. Correct feet and leg placement with adequate bone to support the skeleton is ex-

FIG. 7-1. Desirable characteristics of a mature beef cow.

tremely important. Females with structural abnormalities do not travel and forage as well as correctly structured females. The cow should be clean in the flank and trim along her underline. Feminine, high-producing females have long, clean, smooth muscling. Excessive muscling has been associated with calving difficulty.

A well-developed udder with squarely placed teats and strong udder attachment

is desirable. Cows with saggy, poorly attached udders or with large ballooning teats, which are difficult to nurse, should be culled.

The American Polled Hereford Association developed an udder scoring system to provide breeders with a standardized method for comparison and to draw attention to the significance of udder soundness. Teat size and udder suspension are included in this system since they are the most significant characteristics and the easiest to evaluate. Each characteristic should be scored separately using a numerical range from 1 to 50, with a higher score being more desirable. In evaluating teat size, smaller teats receive higher scores (Table 7–7; Fig. 7–3).

Scores for udder suspension indicate the degree of tightness with which the udder is held to the cow's body. The median suspensory ligament is the center support that ties the udder to the cow's body. It creates the characteristic cleft in the floor of an udder when viewed from the rear. When the ligament weakens, the udder floor drops and the teats begin to strut outward along the sides of the udder. This results in a broken-down, non-functional

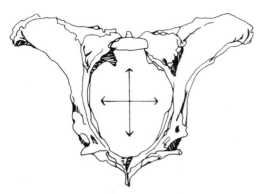

FIG. 7-2. Frontal view of pelvic area, through which the calf must pass at birth. The arrows indicate the critical dimensions that can be measured. (From Mankin, J.D. 1983. Pelvic size and calving difficulty. Cooperative Extension Service, Cattleman's Library, University of Idaho, Moscow ID. CL-742.)

TABLE 7-7. Teat Size and Udder Suspension Scoring System

Numerical Value	Teat Size	Suspension
50 45 40	Very small	Very tight
35 30	Small	Tight
25 20	Intermediate	Intermediate
15 10	Large	Pendulous
5 0	Very large	Very pendulous

Courtesy of the American Polled Hereford Association and Dr. Jim Gibb. Kansas City, MO. Polled Hereford World, November 1983.

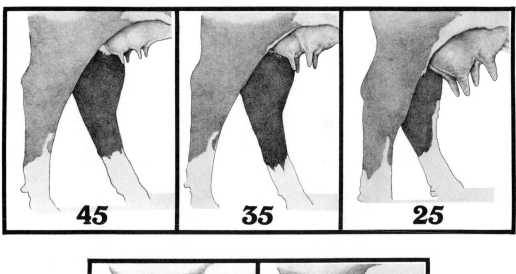

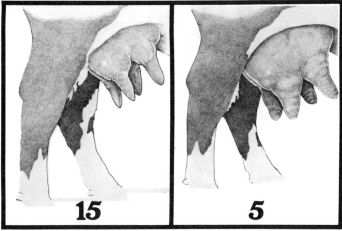

FIG. 7-3. Teat size evaluation. (Courtesy of the American Polled Hereford Association and Dr. Jim Gibb. From Polled Hereford World, November 1983.)

udder. Tighter udders should be given higher scores (Table 7–7; Fig. 7–4).

Udders should be scored immediately following calving; however, they can also be scored at the time when calves are weighed, tagged, and tatooed. Udder scores should then be used for cow culling, replacement heifer selection, sire selection, and donor dam selection.

Mature commercial cows should be "mouthed" each year to determine if they have their full complement of teeth. Cows without a full complement of teeth may have difficulty in masticating and digesting the fiber in forage; therefore, adequate nutrition is not derived from their diet, and body condition and production decline.

Age verification is an art, not a science. With normal development, mature cattle have 32 teeth, of which 8 are incisors, located at the front of the lower jaw (Fig. 7–5). The two front teeth (central incisors) are known as *pincers* (Fig. 7–6). The next two (on either side of the pincers) are *first intermediates*, the third pair (on either side of the intermediates) are *laterals*, and the

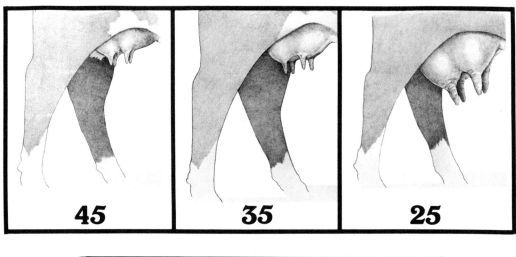

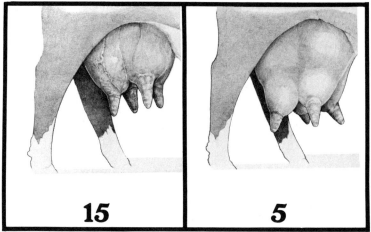

FIG. 7-4.   Udder suspension evaluation. (Courtesy of the American Polled Hereford Association and Dr. Jim Gibb. From Polled Hereford World, November 1983.)

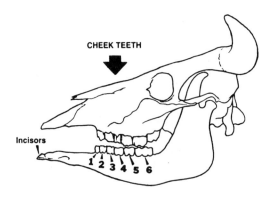

FIG. 7-5.   Upper and lower arcades (jaws) showing three pairs of premolars and three pairs of molars. Only the molars on the upper arcade are used in age determination. Premolars: 1, 2, and 3; molars: 4, 5, and 6. (Adapted from Bull-O-Gram, Douglas, AZ.)

outer pair are known as *corners*. Cattle and other ruminants have no upper incisor teeth; in their place is a dental pad.

The eruption of the incisor teeth is used to estimate the general age of cattle. From birth to one month of age, two or more temporary (deciduous) incisor teeth are present. Within the first month, the entire eight temporary (deciduous) incisors usually appear. Three pairs of pre-molars are present at birth or shortly afterward. The permanent incisors replace the temporary pair at 1½ to 2 years of age. The next pair, or middle permanent incisors, appear at 2 to 2½ years of age. The next pair of permanent incisors (laterals) are erupted at 3 years of age; and the outer pair (corners) at 3½ to 4 years. At 5 years, all incisors are in place, and the corners show slight wear. After 5 years, age determination is principally shown by wear; however, wear in cattle is an inaccurate means of calculating exact age, and even under the best conditions, provides merely an approximation.

A rule of thumb for estimating age of cattle of up to 5 years is as follows:

2 permanent teeth—2 years old
4 permanent teeth—3 years old
6 permanent teeth—4 years old
8 permanent teeth—5 years old

Production and Performance Records

Individual cow records for 1965 to 1978 from the Livestock and Range Research Station, Miles City, Montana show that 83.4% of those cows removed from the herd were removed because they were open (Table 7–8). Therefore, one of the primary herd performance records a beef cattle producer should keep is the number of open cows. Wintering an open cow costs almost as much as wintering a pregnant cow. Open cows have a tremendous economic impact if they are not identified at weaning and removed from the herd.

Beef cattle producers need to be aware of the *length of their calving season*. A long calving interval is one of the largest robbers of income in most beef operations.

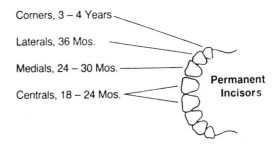

Corners, 3 – 4 Years
Laterals, 36 Mos.
Medials, 24 – 30 Mos.
Centrals, 18 – 24 Mos.
Permanent Incisors

FIG. 7-6. Permanent incisor development. (Adapted from Bull-O-Gram, Douglas, AZ.)

Table 7–9 summarizes the results of a survey for open cows conducted by the Cooperative Extension Service at the University of Idaho in 1982. Eleven percent of the 17,378 cows involved in the survey were open, and of those calving, 15% calved 80+ days into the calving season. Less than 50% of the cows calved in 40 days. Cattlemen should attempt to have as many females as possible breed early in the breeding season.

Weaning weights should be obtained for all calves. Adjusted weaning weights evaluate differences in mothering ability and in milk production of the dam, as well

TABLE 7-8. Cows Removed From Herd at Livestock and Range Research Station, Miles City, MT

Reason for Removal	Percentage of Herd	Percentage of Cows Removed
Death or Disappearance	1.5	7.9
Bad udder	0.2	0.9
Lump jaw	0.0	0.2
Cancer eye	0.4	1.9
Prolapse	0.2	1.2
Bad feet	0.0	0.3
Other injury or illness	0.2	0.9
Open cow	15.8	83.4
Performance	0.6	3.2
Total	18.9%	99.9%

From Greer, R.C., et al. 1979. Reasons for culling beef cows and estimates of the proportion culled for each reason. Montana Agriculture Experiment Station, Bozeman, MT. Bulletin 708.

TABLE 7–9. Survey to Identify Open Cows

Total cows and heifers	17,378
Open cows	11% (range: 0–35%)
Cows calving in first 40 days	46%
Cows calving in second 80 days	28%
Cows calving in 80+ days	15%

From Mankin, J.D. 1983. Growth, environment and production. Special Report—International Minerals and Chemical Corporation. Veterinary Products Division, Terre Haute, IN.

as in growth potential of the calf. Weaning weight ratios should then be calculated within sex groups. Individuals should be used in the selection process. In a commercial cattle operation, a heifer could possibly have a high 205-day adjusted weight ratio but be unacceptable because of her insufficient actual weight and the probability that she will not be heavy enough at 14 to 15 months of age to be cycling with older heifers.

Average birth weight, weaning weight ratios of all calves weaned, and most probable producing ability (MPPA) can provide valuable information for within herd comparisons. These values can be helpful in identifying both lowest-producing cows to be culled and consistently high-producing cows.

PROGRAM FOR SELECTING AND MANAGING REPLACEMENT HEIFERS*

Management—Weaning Until Breeding

1. Determine average weight of replacement heifers at weaning.
2. Determine number of days from weaning to expected breeding.

* Adapted from Beverly, J.R., and J.C. Spitzer. 1980. Management of replacement heifers for a high reproductive and calving rate. Texas Agricultural Extension Service, College Station, TX. The Texas A&M University System. B-1213.

3. Determine desired weight at breeding using 66% of mature body weight as a basis.
4. Calculate necessary rate of gain from weaning to breeding.

Example:

Average weight	=	450 lb
Days from weaning to breeding	=	210
Weight at breeding (1000-lb cow)	=	660 lb
Thus 660 lb − 450 lb	=	210 lb
200 lb ÷ 210 days	=	1 lb per day

5. Feed heifers to grow at calculated rate. Check development by periodic weighing of a sample group throughout the feeding period.

Management—At Breeding

1. Remove heifers that have failed to grow adequately.
2. Remove heifers showing noticeable unsoundness.
3. Initiate breeding 20 to 30 days prior to regular breeding season.
4. Breed for 45 to 60 days, and remove bulls.

Management—Soon After Breeding

1. Palpate for pregnancy 45 days after end of breeding season.
2. Market all open heifers.
3. Continue to grow heifers at 1.0 to 1.5 lb per day until calving.
4. Avoid fattening heifers, but feed for full growth capacity.

Management—At Calving

1. Move bred heifers into small, easily accessible pasture area.
2. Check all heifers at least three times daily during calving.
3. Provide assistance for heifers experiencing calving difficulty.

Management—Rebreeding

1. Keep heifers separated from mature cows.
2. Continue to grow heifers.
3. Breed for 60 days, and palpate for pregnancy.
4. Sell all open heifers.

CULLING THE BEEF HERD

A careful culling program is essential, owing to the high cost of maintaining a cow herd. Although most culling is done in

TABLE 7-10. Some Advantages and Disadvantages of Starting With
Different Types of Breeding Stock

1. Heifer calves without performance records.

   Advantages
   Investment per animal is low—about $2 to $3 per hundred lb weight above normal market price.
   Culled heifers can be marketed as feeders or finished in feedlot.

   Disadvantages
   First calf crop is not weaned for 2 years.
   No performance information on which to make initial selections is available.

2. Open heifers with performance records.

   Advantages
   Performance information on which to base selections is available.
   Breeder can choose breed and type of bull to mate with heifers.

   Disadvantages
   No early return on investment occurs.
   Nearly 2 years pass before their first calf crop is weaned.

3. Bred yearling heifers.

   Advantages
   Earlier return on investment occurs than with open heifers.
   If heifers are diagnosed as pregnant, there is less chance of having open heifers at calving.

   Disadvantages
   Purchase may be costly, especially for heifers bred to "exotic" breeds of bulls.
   Breeder may find some heifers not pregnant if he does not test them when purchased.

4. Mature cows.

   Advantages
   Mature cows offer quickest way to build a herd and quickest return on investment, especially if cows are nursing calves and are rebred.
   Breeder can sometimes purchase a bred cow about as cheaply as a heifer calf.

   Disadvantages
   Previous owner's problems may be inherited, unless cows of known health status and productivity are purchased from reputable producer.
   Productive years of the herd are fewer than with younger heifers.
   Proven mature cows are more costly than heifers.

From Nelson, L.A., et al. 1977. Selection, management, and nutrition of the cow herd. Cooperative Extension Service, Purdue University, W. Lafayette, IN. AS-396.

the fall when the calves are weaned, culling should be a year-round program. The following culling procedure has proven to be successful. It may require some adaptation to different ranching situations.

Procedure

1. Check for pregnancy and cull open cows.
2. Cull cows that would calve late in the calving season.
3. Remove cows with poor progeny records.
4. Remove cows that lack a full complement of teeth.
5. Remove cows showing noticeable unsoundness (lump jaw, cancer eye, poor udder attachment, balloon teats, etc.)
6. Cull cows that are highly nervous and act as a constant source of irritation to the entire herd. A good time to identify these cows is at calving.
7. Breed cows for 60 days, and remove bulls.

STARTING THE COW HERD*

For the established beef producer, the best source of replacement heifers is usually his own herd; however, one getting started in the cattle business must initially purchase his breeding stock. The following paragraphs describe four ways to start the

---

* Adapted from Nelson, L.A., et al. 1977. Selection, management and nutrition of the cow herd. Cooperative Extension Service, Purdue University, W. Lafayette, IN. AS-396.

cow herd. The advantages and disadvantages are summarized in Table 7–10.

Heifer Calves Without Performance Records

Good heifer calves are most numerous in the fall. Those selected should be individually weighed and identified at time of purchase, then developed on a high-roughage growing ration. Before the breeding season, heifers should be weighed again, and the larger ones that have gained well, but that are not excessively fat, should be selected for breeding.

Open Heifers With Performance Records

Heifers should be selected on the basis of their 205-day and 365-day weights and weight ratios. Only those that have performed in the top three fourths of the herd—that is, those having weight ratios above 90—should be considered. They should be purchased at 12 to 13 months of age so that they can be isolated 30 to 60 days before the breeding season begins (when they average 15 months of age).

Bred Yearling Heifers

About 25% more heifers than are needed in the herd should be purchased. It is advisable either to test them for pregnancy before paying for them, or to buy them subject to being guaranteed pregnant. The breeder should also consider the potential for calving problems resulting from the sire to which they were mated. The least desirable heifers should be culled prior to calving, or better yet, they should be allowed to calve, with the breeder selling those that lose their calves at or soon after parturition because of such problems as calving difficulty or failure to claim their calf.

Mature Cows

In general, purchase of mature cows is not advisable. The cows are most likely being culled from someone's herd because of poor reproductive performance, light

calves, old age, mastitis, or other health problems. If cows are nursing calves, however, the probability of infertility is less.

Before breeding stock is bought, the animals should be closely observed for potential health problems, such as internal and external parasites. They should be tested by a veterinarian for tuberculosis, brucellosis, and leptospirosis. Also, they should be isolated for 30 to 60 days after their arrival on the farm.

REFERENCES

Bellows, R.A. 1979. Heifer and cow management. Proc., Western Beef Symposium I, Live Calf—A Key to Pleasure. Boise, ID. October 22. p. 15.

Bellows, R.A., R.E. Short, and R.B. Stagmiller. 1979. Research areas in beef cattle reproduction. In Beltsville Symposium in Agriculture Research, 3 Animal Reproduction, Allanheld, Osman, Montclair, NJ, p. 5.

Beverly, J.R., and J.C. Spitzer. 1980. Management of replacement heifers for a high reproductive and calving rate. Texas Agricultural Extension Service. College Station, TX. The Texas A&M University System. B-1213.

Greer, R.C., R.W. Whitman, R.R. Woodward, and W.A. Yager. 1979. Reasons for culling beef cows and estimates of the proportion culled for each reason. Montana Agriculture Experiment Station. Bozeman, MT. Bulletin 708.

Mankin, J.D. 1983. Growth, environment and production. Special Report—International Minerals and Chemical Corporation. Veterinary Products Division, Terre Haute, IN.

Mankin, J.D. 1983. Pelvic size and calving difficulty. In Cow-Calf Management Guide. Cattleman's Library, Cooperative Extension Service, University of Idaho, Moscow, ID. CL-742.

Nelson, L.A., W.L. Singleton, and K.S. Hendrix. 1977. Selection, management and nutrition of the cow herd. Cooperative Extension Service. Purdue University, W. Lafayette, IN. AS-396.

Short, R.E., and R.A. Bellows. 1971. Relationships among weight gains, age to puberty and reproductive performance in heifers. J. Anim. Sci. 32:127.

Spitzer, J.E., J.N. Wiltbank, and D.G. LeFever. 1975. Increasing beef cow productivity by increasing reproductive performance. Colorado State University Experiment Station, Ft. Collins. CO. Gen. Series 949.

Selected Readings

Bone, J.F. 1979. Animal Anatomy and Physiology. Reston Publishing Co., Reston, VA.

Bull, R.C. 1982. Nutrition—Its effect on reproduction in cattle. Proc., Pacific N.W. Animal Nutrition Conf. 17:57.

Frischknecht, D.W., and J.D. Mankin. 1982. How to select, grow and manage replacement heifers.

In Cow-Calf Management Guide. Cattleman's Library. Cooperative Extension Service, University of Idaho, Moscow, ID. CL-745.

Mankin, J.D. 1982. Pelvic size and calving difficulty. *In* Cow-Calf Management Guide. Cattleman's Library. Cooperative Extension Service, University of Idaho, Moscow, ID. CL-742.

McReynolds, W.E. 1982. Cull the beef herd. *In* Cow-Calf Management Guide. Cattleman's Library. Cooperative Extension Service, University of Idaho, Moscow, ID. CL-735.

Minish, G.L., and D.G. Fox. 1979. Beef Production and Management. Reston Publishing Co., Reston, VA.

QUESTIONS FOR STUDY AND DISCUSSION

1. What is the most important factor affecting net calf crop?
2. Which herd in the following table would produce more pounds of calf weaned per cow exposed to the bull?

	Herd A	Herd B
Conception rate (%)	95	85
Calves weaned (%)	90	80
Net calf crop (%)	—	—
Average weaning weight (lb)	420	420

3. Why is reproductive efficiency one of the most important traits to be considered in the selection process?

4. What is the effect of nutrition on the age at puberty and on the fertility of heifers to be bred at 14 to 15 months of age?
5. Why do many heifers that are bred at 14 to 15 months of age either fail to breed or rebreed late during their second breeding season?
6. Outline a procedure to improve reproductive performance.
7. When visually appraising cows, why is it important to select:
   a. Those that have a good spring of rib?
   b. Those that are long and wide at the pins?
   c. Those that have long, smooth muscling?
   d. Those that have well developed udder with squarely placed teats.
8. Describe the American Polled Hereford Association's teat size and udder evaluation system.
9. Why should all cows be "mouthed" at weaning time?
10. How many teeth do cattle have?
11. Describe the development of teeth in the cow.
12. What performance records should a beef cattle producer keep? Why?
13. Which should be used in selecting heifers in a commercial cattle operation: 205-day adjusted weights or actual weaning weights? Why?
14. How much should heifers weigh at breeding if they are to be bred at 14 to 15 months of age?
15. Outline a procedure for selecting and managing replacement heifers.
16. Why should heifers be bred 10 to 30 days prior to the regular breeding season?
17. Outline a procedure to use when culling the beef herd.

Chapter Eight

Feeding Management of the Cow Herd

Beef cows are maintained to produce calves for slaughter. Beef cattle producers must maximize the pounds of calf weaned per cow in the most economical manner. The key factor influencing pounds of calf weaned per cow in the herd is reproductive efficiency. As emphasized in Chapter 6, a sound nutrition program is important to achieving the highest reproductive efficiency.

Nutritional needs of the beef cow are relatively simple, and this fact should be considered when building an economical feeding program. Beef cattle need the same nutrients as non-ruminants (poultry and swine), but as ruminants, they possess a complex stomach (four compartments: rumen, reticulum, omasum, abomasum) that can digest and utilize fibrous feed, convert non-protein nitrogen to true protein, and synthesize the B vitamins and vitamin K, which non-ruminants cannot synthesize.

Feed costs can constitute up to 65% of the total cost of producing calves. Cattle producers today must use available feedstuffs efficiently to satisfy the varying nutrient requirements of the beef cow during the annual reproductive cycle.

IMPORTANT NUTRIENTS FOR THE COW HEAD

Energy

Theoretically, energy is not a nutrient but is contained in several dietary components (carbohydrates, fats, and excess protein). Energy, which is often referred to as total digestible nutrients (TDN), digestible energy, metabolizable energy, or net energy, is important because it constitutes the largest portion of the total nutrient needs. The amount of energy required by heifers and cows vary with their size and physiologic state.

Good-quality forage (alfalfa hay, grass-legume hay, corn silage, grass silage) meets the energy requirement of the dry, pregnant cow. If low-quality roughage is fed, however, energy intake may not be adequate to maintain body weight and reproductive performance. This is often the case when feeding cornstalks, soybean residue, straw, and poor-quality hay. The high fiber content of low-quality roughage limits dry matter intake (Table 8–1). If low-quality roughage is fed, beef cows usually require energy and probably protein and mineral-vitamin supplementation.

Protein

Proteins are nitrogen-containing compounds that are essential to life. Proteins are involved in the formation and maintenance of muscle, organs, bone, milk, and other biochemical components in the body. In general, the amino acid composition or quality of proteins does not influence their utilization by cattle because microorganisms in the rumen convert feed protein to microbial proteins that are digested and absorbed by cattle.

TABLE 8–1. Dry Matter Capacity of Beef Cows

Ingredient	Expected Intake (Dry Matter Basis) (% Body Weight)
Pelleted hays	3.0–3.5
Grains	2.5–3.0
High-quality forage (legume hay, silage, green pasture)	2.5
Average-quality forage (non-legume hay)	2.0
Low-quality roughage (straw, dry grass, meadow hay)	1.5

The most common nutritional deficiency in cows grazing on dry, mature grass, or being fed low-quality roughage, is protein. A mature, pregnant beef cow weighing 1000 lb in the last trimester of pregnancy requires 1.6 lb of crude protein. When alfalfa or alfalfa-grass hay is fed, ample protein is provided; however, if low-quality roughage is fed, cows may not be able to consume enough dry matter to satisfy their protein requirement. For example, a 1000-lb cow needs to consume only 9.4 lb of alfalfa hay dry matter to satisfy her protein requirement, while the same cow would need to consume 57.1 lb of dry matter from corn cobs to provide the same amount of protein (Table 8–2). A 1000-lb beef cow does not have enough stomach capacity to consume 57.1 lb of dry matter from corn cobs.

Research at the University of Idaho has demonstrated the importance of feeding adequate protein to pregnant yearling beef heifers. Yearling Hereford heifers fed 0.7 lb of crude protein daily from 100 days prior to expected date of parturition until 45 days after parturition had a longer postpartum interval (86.4 versus 74.8 days) and a lower first service conception (33% versus 70% within 110 days post partum) than those fed 2.1 lb of crude protein daily (Sasser, personal communication). Also, in a field trial, heifers fed less than 2.0 lb of crude protein daily delivered more weak calves at birth than those fed greater than 2.0 lb of crude protein (Bull et al., 1974) (Table 8–3). These data re-emphasize the importance of feeding adequate levels of crude protein to cows during their reproductive cycle.

Minerals

Mineral elements of major importance in cattle feeding are calcium, phosphorus, salt (sodium and chlorine), magnesium, potassium, and selenium.

Calcium and phosphorus are important because they influence bone formation, reproduction, and nerve irritability. The ideal calcium-to-phosphorus ratio is 1.5 to 2 parts calcium for every 1 part phosphorus. Diets containing legumes (e.g., alfalfa or clover) as pasture, hay, or silage are high in calcium but may be deficient in phosphorus. The possibility of a phosphorus deficiency increases when cattle are grazed

TABLE 8–2. Meeting the Protein Requirement of Beef Cows

Feed	Protein Content (%)	Protein Required (lb)[a]	Amount Fed (lb)[b]	Dry Matter Intake (% body weight)[c]
Alfalfa hay, mid-bloom	17.1	1.6	9.4	.9
Barley straw	4.1	1.6	39.0	3.9
Ground corn cobs	2.8	1.6	57.1	5.7
Orchard grass hay	9.7	1.6	16.5	1.6

[a] From National Research Council. 1984. Nutrient requirements of domestic animals. Vol. 6. Nutrient Requirements of Beef Cattle. National Academy of Sciences, Washington, DC.
[b] Pounds of protein required. Protein content feed = $(1.6 \div 17.1 \times 100)$.
[c] Amount fed (lb) = Cow weight (lb) $\times$ $(9.4 \div 1000 \times 100)$.

TABLE 8-3. Relationship Between Protein Intake and the Incidence of
Weak Calf Syndrome in Beef Cattle Herds

Crude Protein Intake	No. Herds	Avg. Protein Intake per Cow (lb)	Weak Calf Syndrome (%)[a]
High (greater than 2.0 lb/day)	6	2.5	0.6
Medium (1.5 to 2.0 lb/day)	4	1.8	3.4
Low (less than 1.5 lb/day)	4	1.2	9.8

[a] Percentage of all herds.
From Bull, R.C., et al. 1974. Nutrition and weak calf syndrome in beef cattle. Current Information Series. University of Idaho, Moscow, ID. No. 246.

for long periods on mature forage in late summer or fall.

Legume forage contains more calcium than non-legume forage or low-quality roughage (e.g., corn silage, corn cobs, straw, or mature meadow hay). Cereal grain diets, because of their relatively high phosphorus content, require calcium supplementation. In general, as the proportion of legume forage in a diet decreases, the need for calcium supplementation increases. Common calcium and/or phosphorus mineral supplements are bone meal (32% Ca, 16% P), defluorinated rock phosphate (30% Ca, 12% P), dicalcium phosphate (22.5% Ca, 18.5% P), monoammonium or monosodium phosphate (0% Ca, 24% P), and limestone (35.9% Ca, 0% P).

Salt is universally deficient in all common feedstuffs used for animal feeding. Salt is often used as a tool to control grazing distribution of cattle on the range. All cattle rations must be fortified with salt either as a part of the diet or on a free-choice basis. Beef cows require about 1 to 2 oz of salt daily. Block salt should not be used as the only source of salt for cattle but can serve as a supplementary source. Intake of loose salt is usually higher than for block salt. Salt and mineral intake is low when mineral content of the drinking water is high. A consumption level equivalent to 0.25% of the feed dry matter intake is recommended. As a precaution against trace mineral deficiencies, trace mineralized salt may be substituted for plain salt in the mineral mixture.

Most diets fed to beef cattle contain enough magnesium to meet this nutrient's requirement in cattle; however, in spring, in early summer, or when cattle graze winter cereal grain pastures, forage may not contain enough magnesium, or magnesium may be unavailable, and grass tetany develops. Grass tetany may be prevented by incorporating approximately 20 to 25% magnesium oxide in the mineral supplement. Magnesium oxide is unpalatable, and a supplement containing 20 to 25% magnesium oxide must contain other ingredients that are more palatable, such as dehydrated molasses, trace mineralized salt, and oilseed meals, to ensure adequate intake.

Most beef cattle rations contain at least 0.6% potassium, which meets the requirements of beef cattle. Cattle require 0.1 part per million (ppm) selenium in their diet. A deficiency of selenium usually occurs when cattle graze or consume harvested vegetation from acid soils, but it may occur in other types of grazing areas. Signs of selenium deficiency are white muscle disease in baby calves and lowered fertility in cows. Selenium deficiency may be prevented by injecting cows and calves with a selenium-vitamin E mixture, by supplementing selenium in the salt mixture, or by using a combination of these two methods.

Vitamins

Vitamin A is the only vitamin likely to be deficient in the diets of beef cattle under most feeding programs. Serious vitamin A deficiencies do not occur often in spring-calving beef cow herds. A cow can store several months of vitamin A during the summer grazing season, when the carotene content of the forage is high. A lactating cow rapidly depletes her stores of vitamin A, however, and vitamin A deficiencies are possible, especially in fall-calving cows, in cows during dry years, and in young cows. Twenty thousand I.U. of vitamin A for pregnant cows, and 40,000 I.U. of vitamin A for lactating cows, meet the daily requirements. Supplemental vitamin A (1) can be added to the protein supplement, (2) can be injected intramuscularly (1,000,000 I.U. lasts about 3 months), or (3) can be added to the mineral mix, in which case a stabilized form of vitamin A must be used. The mineral mix must be in a covered feeder to protect it from moisture and sunlight, and fresh mix must be prepared each week.

LIFE CYCLE FEEDING

Successful beef production depends on proper management of the reproductive cycle of the cow. To attain production stability, cattlemen must understand and appreciate the cow's needs as she progresses through her reproductive cycle. Feeding programs must be managed to coincide with her reproductive needs.

The cow's reproductive cycle is fixed and well defined. Average gestation of the cow

TABLE 8–4. Reproductive Cycle of the Cow

Period	Duration in Days
First trimester of gestation	94
Second trimester of gestation	94
Third trimester of gestation	94
Post-calving period	83
	365
Pre-weaning period	(variable)

From Schoonover, C.O., and D.A. Yates. 1980. The biological cycle of the beef cow. *In* Cattleman's Library. Cooperative Extension Service. University of Idaho, Moscow, ID. CL-200.

is 282 days. The cow's reproductive cycle can be divided into four definite periods as shown in Table 8–4.

The first trimester begins on the day on which the cow is bred and conceives. The reproductive cycle remains constant, but the chronologic cycle varies according to the date on which the cow is bred. Table 8–5 illustrates how the reproductive and chronologic cycles work.

The biologic and chronologic cycles are important to beef cattle producers. The nutritional requirements of beef cows correspond to their reproductive cycles, and a thorough understanding of these cycles allows beef cattle producers to develop a sound feeding program.

First Trimester of Gestation (94 days)

During this period, nutrients are needed for maintenance and milk production, although milk production is declining and spring-born calves are eating pasture forage. Growth of the fetus commences with the fertilization of the female egg by a male sperm; however, rapid fetal growth does not occur until the third trimester of pregnancy. A cow should be gaining weight in preparation for the winter months.

Second Trimester of Gestation (94 days)

This period is referred to as *mid-gestation*. By the end of this second trimester, the fetus weighs about 15 lb. The calf is usually weaned near the beginning of the second trimester. Nutrient requirements of the beef cow are lowest after weaning to the last trimester of gestation. A cow is fed primarily to maintain her body weight, and feeding level depends on her condition. During this period, the cow can be fed minimally. If feed supplies are short, the cow can withstand nutritional stress at this time *if* she is fed properly during late gestation and during the postpartum period.

Spring-calving cows can utilize low-quality roughage, such as crop residues, poor-quality hays, and pasture aftermath. A dry (non-lactating) cow in good condition can usually meet her energy needs from forage alone if she is supplemented properly with a salt-mineral mixture, vitamin A, and possibly protein.

Third Trimester of Gestation (94 days)

The last trimester of pregnancy is a time of rapid fetal growth. Nutritional requirements of the cow increase rapidly, paralleling the rapid growth of the developing calf as it approaches parturition. Fetal weight increases from approximately 15 lb at the beginning of the third trimester of pregnancy to 70, 80, or more pounds at birth.

The consequences of inadequate nutrition during the last trimester of pregnancy include the following:

1. Lighter and weaker calves at birth
   a. Higher death loss
   b. Greater susceptibility to disease
2. Lighter weaning weights
   a. Lower milk production
   b. Less pounds of calf to sell
3. Cows returning to estrus more slowly
   a. Younger and lighter calves the following year
   b. Longer breeding season
   c. Longer calving season the following year
   d. More uterine infections and retained placentas

A study conducted at the University of Wyoming showed the importance of adequate nutrition during late gestation (Corah et al., 1975) (Table 8–6). One hundred days prior to anticipated calving date, cows were individually fed the same level of protein and energy up to 30 days precalving. Beginning 30 days prior to scheduled calving, cows of one group were given an increased energy level of 10.6 lb of TDN while cows of the other group were restricted to 4.6 lb of TDN. After calving, the proper level of energy and protein was fed to the cows as a group. Results demonstrated that a low level of energy 30 days prior to calving reduced birth weight, calf livability, milk production, and calf wean-

TABLE 8-5. Chronologic and Reproductive Cycles of the Cow

Biologic Cycle	Chronologic Cycle
Day 1	June 1—Breeds and conceives
Day 94	September 3—End of first trimester
Day 198	December 6—End of second trimester
Day 282	March 10—End of third trimester (birth of calf)
Day 365	May 31—End of post-calving period, beginning of next breeding period
Day 482	October 7—Weaning of calves

From Schoonover, C.O., and D.A. Yates. 1980. The biological cycle of the beef cow. *In* Cattleman's Library. Cooperative Extension Service. University of Idaho, Moscow, ID. CL-200.

ing weight while increasing the length of the postpartum interval.

TABLE 8-6. Effect of Energy Restriction on Cow and Calf Performance

Item	Continuous Low	Elevated Last 30 Days
Energy level (lb)		
TDN first 70 days	4.6	4.6
TDN last 30 days	4.6	10.6
Weight change (lb)		
First 70 days	−119.2	−114.6
Last 30 days	−23.0	−92.8
Birth weight (lb)	58.7	67.0
Calving livability (%)		
Birth	90.5	100.0
Weaning	71.4	100.0
Calves treated for scours (%)	52.0	33.4
Cow's milk production (lb)	9.1	12.0
Weaning weight (lb)	294.4	320.1
Percentage in estrus, 40 days post partum	37.5	47.6

Based on Corah, L.R., et al. 1975. Influence of prepartum nutrition on the reproductive performance of beef females and the performance of their progeny. J. Anim. Sci. 41:819.

Post-Calving Period

This is the most critical time of the reproductive cycle and the period of greatest nutritional needs. The cow should be gaining weight prior to and during the breeding season because she is lactating, and because she must be in good condition to conceive during the 83-day post-calving period if she is to calve every 365 days. At this time, cows should be fed forage of the best quality available. Nutrient requirements are approximately 50% higher during this period than during other periods of the reproductive cycle. Cows should not be fed low-quality roughage during this period unless energy and protein are supplemented. When supplemental feeds cannot be mixed with forage and must be fed separately, cows will tend to eat less forage and too much of the supplement.

Feeding Replacement Heifers*

Because the replacement heifer is still growing, she has the additional physiologic requirement of growth that a mature cow does not have. The heifer and mature cow have different nutritional requirements per unit of body weight, but their responses to the reproductive cycle remain about the same.

WEANING TO BREEDING

Weaning usually occurs when herd replacements are selected. At this time, the breeder must calculate the average weight gain necessary for the heifers to reach puberty sufficiently early so that they may be bred at 14 to 15 months of age. In general, for a heifer to calve at 2 years of age, she must reach puberty between 13 to 16 months of age. Breed, environment, nutrition, and other factors all influence age at puberty.

Most heifers, if fed properly, show estrus between 13 and 16 months of age. Exceptions include Brahman and Brahman crossbred heifers. Approximately 90% of Brahman crossbred heifers show estrus at 15 to 17 months, while 90% of purebred Brahman heifers may be close to 20 months of age before reaching puberty. Table 8–7 summarizes the weight necessary for 50%, 70%, and 90% of heifers to reach puberty. For example, 50% of Angus-Hereford heifers that are 13 to 16 months of age and weigh 575 lb could be expected to have reached puberty (575 lb is the average weight at puberty). The same heifers would need to weigh 675 pounds if the breeder wanted 90% of them to be in heat.

Ideally, selecting only the heaviest heifers at weaning for replacement would be most economically beneficial to beef production. Heavy heifers would not need to gain as much weight from weaning to breeding as would lighter heifers. Realistically, however, long, extended calving seasons, the need for large numbers of replacement heifers, and value as a registered animal, may cause retention of lightweight heifers. To show a satisfactory reproductive performance, these heifers must be fed to reach appropriate target weights.

Consider, for example, a producer who has 100 Angus-Hereford heifers with an average weaning weight of 450 lb on October 1. These heifers are to be maintained in drylot and fed a high forage diet until April 1, the start of the breeding season. Records indicate that the heifers will gain approximately 1.25 lb per day on this type of ration. These heifers would weigh 675 lb on April 1, which is an adequate weight for 90% or more to come into heat (Table 8–7). The average weight at weaning was 450 lb, however. The lightest heifer weighed 350 pounds, and the heaviest heifer weighed 550 pounds. If all heifers gained 1.25 pounds daily for 180 days, the lightest heifer would weigh 575, and the heaviest heifer 775 pounds. Therefore, some heifers would be too light and some too heavy.

---

* Much of the information in this section was adapted from Beverly, J.R., and J.C. Spitzer. 1980. Management of replacement heifers for a high reproductive and calving rate. Texas Agricultural Extension Service, Texas A&M University System. College Station, TX. B-1213.

TABLE 8–7. Estimates for Heifers Reaching Puberty at Various Weights (lb)

Breed or Cross	Heifers Reaching Puberty		
	50% in Heat	70% in Heat	90% in Heat
Angus	550	600	650
Brahman	675	725	750
Brangus	600	650	700
Charolais	700	750	775
Hereford	600	650	700
Santa Gertrudis	675	725	750
Shorthorn	500	550	600
Brahman x British	675	725	750
British x British	575	625	675
Charolais x British	675	725	775
Jersey x British	500	550	600
Limousin x British	650	700	775
Simmental x British	625	675	750
S. Devon x British	600	650	725

From Beverly, J.R., and J.C. Spitzer. 1980. Management of replacement heifers for a high reproductive and calving rate. Texas Agricultural Extension Service, Texas A & M University System. College Station, TX. B-1213.

Averages are thus inadequate for accomplishing this task.

If there is a wide range in weaning weights, the herd should be separated into two or possibly three groups, and each group should be fed properly to allow heifers to attain the desired weight for reaching puberty. Recent work, represented by Table 8–8, demonstrates an improvement in reproductive performance when heifers are sorted into light and heavy groups and are fed to reach target weights.

BREEDING TO CALVING

After replacement heifers are bred, they should continue to grow and gain weight. At calving, heifers should weigh approximately 85% of their mature weight. A 1000-lb mature cow would need to weigh 850 lb at calving. This requires a gain of .71 lb daily (850 lb − 650 lb ÷ 282 days) from breeding to calving. To ensure optimum reproductive performance after calving, however, bred heifers should be fed to

TABLE 8–8. Reproductive Performance of Light and Heavy Heifers When Fed Separately and as a Group

	Fed Together		Fed Separately	
	Light Heifers	Heavy Heifers	Light Heifers	Heavy Heifers
Number of heifers	10	10	19	20
Age at puberty (days)	423	404	405	389
Cycling at start of breeding (%)	60	90	79	90
Pregnant in 45-day breeding season (%)	60	80	79	90
	Combined 70%		Combined 85%	

From Varner, L.W., et al. 1977. A management system for wintering replacement heifers. J. Anim. Sci. 44:165.

TABLE 8–9. Nutrient Requirements of Medium-Frame Heifer Calves

Weight (lb)	Dry Matter Consumption (lb)	TDN		Crude Protein		Calcium (%DM)	Phosphorus (%DM)
		(lb/day)	(%DM)	(lb/day)	(%DM)		
Average Daily Gain—1.0 lb							
300	8.0	4.8	62.0	.91	11.4	.44	.22
400	9.9	6.1	62.0	1.01	10.2	.36	.20
500	11.8	7.3	62.0	1.11	9.9	.30	.21
600	13.5	8.4	62.0	1.19	8.8	.28	.20
700	15.1	9.4	62.0	1.28	8.9	.25	.19
Average Daily Gain—1.5 lb							
300	8.2	5.6	68.5	1.1	13.1	.59	.27
400	10.2	7.0	68.5	1.2	11.4	.45	.24
500	12.1	8.3	68.5	1.3	10.3	.38	.22
600	13.8	9.4	68.5	1.3	9.5	.32	.21
700	15.5	10.6	68.5	1.4	9.0	.28	.20

Adapted from National Research Council, 1984. Nutrient requirements of domestic animals. *In* Nutrient Requirements of Beef Cattle. Vol. 6. National Academy of Sciences, Washington, D.C.

gain 1.0 to 1.25 lb daily during the last trimester of gestation since fetal growth rate is approximately 1 lb per day.

CALVING TO BREEDING

After calving, the replacement heifer must regain calving weight loss, produce milk for her calf, ready herself for another pregnancy, and continue to grow. This is the most critical time period of a heifer's reproductive cycle. She requires special care and attention, especially in the area of nutrition, if she is going to rebreed. The daily nutrient requirements for dry matter,

TABLE 8–10. Nutrient Requirements of Pregnant Yearling Heifers in Last 3 to 4 Months of Pregnancy

Weight (lb)	Dry Matter Consumption (lb)	TDN		Crude Protein		Calcium (%DM)	Phosphorus (%DM)
		(lb/day)	(%DM)	(lb/day)	(%DM)		
Average Daily Gain—0.9 lb							
700	15.3	8.5	55.4	1.3	8.4	.27	.20
750	16.1	8.9	55.1	1.3	8.3	.27	.19
800	16.8	9.2	54.85	1.4	8.2	.28	.20
850	17.6	9.6	54.5	1.4	8.2	.26	.20
900	18.3	9.9	54.3	1.5	8.1	.26	.20
Average Daily Gain—1.4 lb							
700	15.8	9.6	60.3	1.4	9.0	.33	.21
750	16.6	10.0	59.9	1.5	8.9	.32	.21
800	17.4	10.4	59.6	1.5	8.8	.33	.21
850	18.2	10.8	59.3	1.6	8.6	.30	.21
900	19.0	11.3	59.1	1.6	8.5	.30	.21

Adapted from National Research Council. 1984. Nutrient requirements of domestic animals. *In* Nutrient Requirements of Beef Cattle. Vol. 6. National Academy of Sciences, Washington, DC.

TABLE 8-11. Nutrient Requirements of Dry, Pregnant Mature Cows

Weight (lb)	Minimum Dry Matter Consumption (lb)	Nutrient			
		TDN (lb/day)	Crude Protein (lb/day)	Calcium (%DM)	Phosphorus (%DM)
Middle Third of Pregnancy					
800	15.3	7.5	1.1	.17	.17
900	16.7	8.2	1.2	.18	.18
1,000	18.1	8.8	1.3	.18	.18
1,100	19.5	9.5	1.4	.19	.19
1,200	20.8	10.1	1.4	.18	.19
1,300	22.0	10.8	1.5	.20	.20
1,400	23.3	11.4	1.6	.20	.20
Last Third of Pregnancy, Average Daily Gain—0.9 lb					
800	16.8	9.2	1.4	.26	.20
900	18.2	9.8	1.5	.27	.21
1,000	21.0	11.2	1.6	.26	.21
1,200	22.3	11.8	1.7	.26	.21
1,300	23.6	12.5	1.8	.26	.21
1,400	24.9	13.1	1.9	.26	.21

Adapted from National Research Council. 1984. Nutrient requirements of domestic animals. *In* Nutrient Requirements of Beef Cattle. Vol. 6. National Academy of Sciences, Washington, DC.

TABLE 8-12. Nutrient Requirements of Cows Nursing Calves in
First 3 to 4 Months Post Partum

Weight (lb)	Minimum Dry Matter Consumption (lb)	Nutrient			
		TDN (lb/day)	Crude Protein (lb/day)	Calcium (%DM)	Phosphorus (%DM)
Average Milking Ability—10 lb Milk/Day					
800	17.3	10.1	1.8	.30	.22
900	18.8	10.8	1.9	.29	.22
1,000	20.2	11.5	2.0	.28	.22
1,100	21.6	12.1	2.0	.27	.22
1,200	23.0	12.8	2.0	.27	.22
1,300	24.3	13.4	2.1	.27	.22
1,400	25.6	17.0	2.3	.27	.22
Superior Milking Ability—20 lb Milk/Day					
800	15.7	12.1	2.2	.48	.31
900	18.7	13.1	2.4	.41	.28
1,000	20.6	13.8	2.5	.39	.27
1,100	22.3	19.5	2.6	.38	.27
1,200	23.8	15.2	2.7	.36	.26
1,300	25.3	16.9	2.8	.36	.26
1,400	26.7	16.5	2.9	.35	.26

Adapted from National Research Council. 1984. Nutrient requirements of domestic animals. *In* Nutrient Requirements of Beef Cattle. Vol. 6. National Academy of Sciences, Washington, DC.

crude protein, energy, calcium, and phosphorus are respectively 13%, 30%, 20%, 60%, and 54% higher for the lactating heifer than in the last trimester of pregnancy. Heifers should be fed as much good-quality forage or grazed forage as they can eat. A feed analysis should be conducted on home-grown forage to determine if supplemental protein is required.

FEED PLANNING GUIDE FOR WINTERING BEEF CATTLE

Adequate nutrition is necessary to maximize genetic potential and minimize disease and health problems in the beef herd.

A successful cattle feeding program requires the following:

1. Knowledge of the daily nutrient requirements of beef cattle according to weight and class.
2. Knowledge of the amount of roughage, feed grains, and pasture that is available.
3. Buying or providing feed that meets the animal's nutrient requirements at the least cost.
4. Achieving adequate consumption levels so that the daily nutrient requirements are met.

The following procedure should be used to put together a feeding program:

1. Determine the nutrient requirements of the cattle to be fed (Tables 8–9 through 8–12).
2. Complete a feed inventory (Table 8–13).
3. Obtain a feed analysis (Table 8–14).
4. Determine the total amount of feed required (Tables 8–14 and 8–15).
5. Compare current and projected feed prices (hay and grain, minerals, and protein supplement).

Nutrient Requirements

Tables 8–9 through 8–12 list the nutrient requirements for wintering some of the common classes of beef cattle according to the 1984 National Research Council (NRC).

Feed Inventories

Feed inventories indicate how much feed is on hand and how much needs to be purchased. One should start making the inventory before the first cutting of hay or grain and before any feed is purchased. It should be completed by the time the last cutting of hay is in the stack and grain harvest is finished. Table 8–13 can be used to complete the inventory.

Feed Requirements

Winter feeding periods vary from ranch to ranch, depending on location and season of calving. Fall-calving cows need more feed than those calved in the spring since they lactate during the winter months. The 60 days just before calving and the 60 days immediately after calving are the cow's most critical feeding periods. Table 8–15 can be used to estimate total feed required for a given number of animals on hand.

Feed Analysis

There is no substitute for a feed analysis to determine the nutritional value of a feedstuff. Feeds should be analyzed for dry

TABLE 8–13. Feed Inventory Listed in Tons or AUMs[a]

Item	Carry-Over in Storage	Harvested in Current Year	Total on Hand
Hay	————	————	————
Straw	————	————	————
Feed grain	————	————	————
Pasture (AUM)	————	————	————
Total tons	————	————	————

[a] AUM = Animal Unit Month, or the amount of feed and/or forage required to support an animal for 30 days. One 1000-lb cow is equal to 1 AUM.

TABLE 8–14. Average Composition of Common Feed Ingredients Fed to Beef Cattle (Dry Matter Basis)

Feedstuffs	Dry Matter (%)	Crude Protein (%)	Energy			Calcium (%)	Phosphorus (%)
			TDN (%)	NEm (Mcal/lb)	NEg (Mcal/lb)		
Low-Protein Concentrates							
Barley grain	89.0	13.0	83	.97	.64	.09	.47
Beet pulp	91.0	10.0	72	.73	.47	.75	.11
Beet pulp w/molasses	92.0	9.9	74	.92	.61	.61	.11
Beet molasses	77.0	8.7	89	.93	.62	.21	.04
Dent corn (#2)	89.0	10.0	91	1.04	.67	.02	.35
Ground ear corn	87.0	9.3	90	1.01	.63	.05	.41
Milo	89.0	12.4	80	.84	.56	.04	.33
Oats	89.9	13.2	76	.79	.52	.11	.39
Rye	89.0	13.4	85	.93	.62	.07	.38
Sugar cane molasses	75.0	4.3	72	.87	.54	1.19	.11
Wheat	89.0	14.3	88	.98	.64	.06	.41
Wheat bran	89.0	18.0	70	.70	.44	.16	1.32
Protein Supplements							
Cottonseed meal[a]	91.5	44.8	75	.77	.50	.17	1.31
Linseed meal[a]	91.0	38.6	76	.79	.52	.44	.91
Meat and bone meal	94.0	53.8	72	.73	.47	11.25	5.39
Rapeseed meal[a]	90.3	43.5	69	.69	.43	.44	1.00
Safflower meal[a]	91.8	23.3	55	.53	.22	.37	.92
Soybean meal[a]	89.0	51.5	81	.88	.59	.36	.75
Sunflower meal[a]	93.0	50.3	65	.64	.29	—	—
Urea	88.0	283.0	—	—	—	—	—
Dry Roughages							
Alfalfa hay							
Early-bloom	90.0	18.4	57	.55	.25	1.25	.23
Mid-bloom	89.2	17.1	55	.53	.22	1.35	.22
Full-bloom	87.7	15.9	53	.51	.18	1.28	.20
Dehydrated alfalfa meal (17% CP)	93.0	19.2	62	.60	.31	1.43	.26
Barley hay	87.3	8.9	57	.56	.26	.21	.30
Barley straw	88.2	4.1	41	.46	.06	.36	.09
Bromegrass hay	89.7	11.8	52	.50	.16	.28	.22
Clover hay							
Alsike	87.9	14.7	60	.59	.30	1.31	.25
Ladino	91.2	22.0	61	.60	.31	1.38	.40
Crested wheatgrass hay	93.0	12.4	64	.64	.35	.33	.20
Ground corn cobs	90.4	28.0	47	.48	.11	.12	.04
Oat hay	88.2	9.2	61	.60	.32	.26	.24
Orchard grass hay	88.3	9.7	57	.55	.25	.45	.37
Sorghum, Sudan grass hay	88.9	12.7	59	.57	.29	.56	.31
Wheat hay	85.9	7.5	66	.65	.039	.15	.20
Wheat straw	90.1	3.6	48	.47	.09	.77	.68

TABLE 8–14. Continued

Feedstuffs	Dry Matter (%)	Crude Protein (%)	Energy			Calcium (%)	Phosphorus (%)
			TDN (%)	NEm (Mcal/lb)	NEg (Mcal/lb)		
Silages							
Alfalfa							
Early-bloom	35.0	18.4	57	.61	.28	1.45	.23
Mid-bloom	35.0	16.0	57	.55	.25	1.28	.20
Full-bloom	35.0	13.6	55	.53	.21	1.25	.20
Corn (well-eared)	28.0	8.4	70	.71	.45	.28	.21
Oats	32.0	9.7	59	.58	.28	.37	.30
Milo	29.0	7.3	57	.55	.26	.25	.18
Sorghum, Sudan grass	23.0	10.2	59	.60	.29	.64	.22
Mineral Supplements							
Dicalcium phosphate	99.0	—	—	—	—	22.5	18.5
Limestone	99.9	—	—	—	—	35.9	—
Oyster shell	99.6	—	—	—	—	38.1	—
Steamed bone meal	100.0	—	—	—	—	32.0	16.0
Defluorinated rock phosphate	100.0	—	—	—	—	30.0	12.0
Diammonium phosphate	99.0	—	—	—	—	22.0	—
Monosodium phosphate	99.0	—	—	—	—	24.0	—
Phosphoric acid	100.0	—	—	—	—	31.6	—

[a] Solvent extracted.

Table values from National Research Council, 1984. Nutrient Requirements of Beef Cattle. No. 6. 6th Ed. National Academy of Sciences, Washington, DC. And from United States–Canadian Tables of Feed Composition. 1982. National Academy of Sciences, Washington, DC.

matter, energy, protein, calcium, and phosphorus. Used correctly, a feed analysis helps to prevent overfeeding or underfeeding important nutrients. Private laboratories and feed dealers offer feed analysis services. County extension agents, private consultants, and feed dealers may offer livestock nutrition recommendations.

Formulating a Ration

This step-by-step approach can be used to formulate a ration for beef cattle, using either a feed analysis or book values for dry matter, TDN (energy), calcium, phosphorus, and protein:

1. Multiply the percentage of DM of the hay by the amount of hay fed.

25 lb of hay × 90% DM = 22.5 lb DM

2. Multiply estimated percentage lost during feeding by amount of DM.

22.5 lb DM × 10%
= 2.25 lb feed lost during feeding

3. Subtract percentage lost during feeding from total DM to obtain DM actually available.

22.5 − 2.25 = 20.25 lb feed DM available

4. For TDN (energy), multiply percentage of TDN by the amount of DM available.

20.25 lb DM × 50% TDN
= 10.12 lb TDN (energy)

5. For crude protein (CP), multiply percentage of CP by the amount of DM available.

20.25 lb DM × 12% CP = 2.43 lb CP

TABLE 8–15. Estimated Feed Requirements for Wintering Beef Animals[a]

	Hay (tons)	Feed Grain (tons)	Pasture (tons)
1. Sixty days before calving (No. 1000-lb cows × 20 lb hay × 60 days)	_____	_____	_____
2. Sixty days after calving (No. 1000-lb cows × 27 lb hay × 60 days)	_____	_____	_____
3. Sixty days after calving (No. 1st calf heifers × 23 lb hay × 60 days)	_____	_____	_____
4. No. weaned calves × 12 lb hay × days No. weaned calves × _____ lb grain × days	_____	_____	
5. No. replacement heifers × 6 lb hay × days No. replacement heifers × _____ lb grain × days	_____	_____	
6. No. yearling calves × 9 lb hay × days No. yearling calves × _____ lb grain x days	_____	_____	
7. No. yearling bulls × _____ lb hay × days bulls × _____ lb grain x days	_____	_____	
8. No. mature bulls × 30 lb hay × days No. mature bulls × _____ lb grain × days	_____	_____	
9. Total feed required	_____	_____	_____

[a] The given requirements assume a hay analysis of 50% TDN, 12% CP, 90% DM, and a feed loss of 10% during feeding.

TABLE 8–16. Ration for 1000 lb Nursing Beef Cow of Average Milking Ability

	DM (lb)	TDN (lb)	CP (lb)	Ca (lb)	P (lb)
Animal requirements (from Table 8–12)	20.2	11.5	2.0	0.06	0.04
Amount fed (from calculations)	22.5	10.1	2.4	0.295	0.0445
Difference	+2.3	+0.6	+0.4	+0.235	−0.005

TABLE 8-17. Rations for Weaned Beef Heifers (Average Daily Gain—1.5 lb)

Item	Ration Number[a]							
	1	2	3	4	5	6	7	8
	lb/day, as-fed[b]							
Feed ingredient								
Alfalfa hay, mid-bloom	11.9	—	—	8.6	14.0	—	—	19.5
Alfalfa silage, mid-bloom	—	29.5	—	—	—	34.5	—	—
Corn silage, well eared	—	—	36.0	21.8	—	—	37.2	—
Dent corn (2)	3.3	3.3	—	—	3.3	3.3	—	—
Cottonseed meal	—	—	2.2	—	—	—	3.3	—
Total daily feed	15.2	32.8	38.2	30.4	17.3	37.8	40.5	19.5
Nutrient intake								
Dry matter	12.2	11.9	10.8	12.4	13.9	13.5	12.1	15.7
Crude protein	1.9	1.7	1.6	1.6	2.2	2.0	2.0	2.7
TDN	7.7	7.7	7.7	7.7	8.6	8.6	8.6	8.6

[a] Rations 1, 2, 3, and 4 are for heifers weighing 400 lbs, while rations 5, 6, 7, and 8 are for heifers weighing 500 lbs.
[b] Includes a 10% adjustment for feed wastage.

TABLE 8-18. Rations for Pregnant Yearling Heifers in Last 3 to 4 Months of Pregnancy (average weight—700 lb)

Item	Ration Number[a]							
	1	2	3	4	5	6	7[c]	8
	lb/, as-fed[b]							
Feed ingredient								
Alfalfa hay, mid-bloom	12.9	—	—	18.3	17.9	—	—	23.3
Alfalfa silage, mid-bloom	—	31.7	—	—	—	44.0	—	—
Corn silage, well eared	—	—	38.1	—	—	—	50.6	—
Dent corn (2)	3.3	3.3	—	—	3.3	3.3	—	—
Cottonseed meal	—	—	2.2	—	—	—	2.2	—
Total daily feed	16.2	35.0	40.3	18.3	21.2	47.3	52.8	23.3
Nutrient intake								
Dry matter	13.0	12.6	11.4	14.7	17.0	16.5	14.6	18.7
Crude protein	2.0	1.8	1.6	2.5	2.7	2.4	1.9	3.2
TDN	8.6	8.6	8.6	8.6	10.3	10.3	10.3	10.3

[a] Rations 1 through 4 are formulated to meet nutrient requirements of a 700-lb heifer gaining 0.9 lb daily, while rations 5 through 8 are formulated for a 700-lb heifer gaining 1.4 lb daily.
[b] Includes a 10% adjustment for feed wastage.
[c] Although ration 7 meets the nutrient requirements, it may be necessary to decrease the amount of corn silage fed and increase corn grain intake to ensure adequate dry matter consumption.

TABLE 8-19. Rations for Dry, Pregnant Mature Cows Weighing 1000 lb

Item	Ration Number[a]							
	1	2	3	4	5	6	7	8
	lb/day, as-fed[b]							
Feed ingredient								
Alfalfa hay, mid-bloom	19.9	9.9	—	8.6	22.6	11.3	5.7	—
Wheat straw	—	11.3	—	—	—	12.8	—	—
Alfalfa silage, mid-bloom	—	—	47.9	—	—	—	41.9	—
Corn silage, well eared	—	—	—	41.0	—	—	—	48.9
Cottonseed meal	—	—	—	2.2	—	—	—	2.2
Total daily feed	19.5	20.8	47.9	43.2	22.6	24.1	47.6	51.1
Nutrient intake								
Dry matter	16.0	17.2	15.1	12.1	18.2	19.5	17.7	14.2
Crude protein	2.7	1.7	2.4	1.7	3.1	1.9	2.8	1.9
TDN	8.8	8.8	8.6	8.6	10.0	10.0	10.0	10.0

[a] Rations 1, 2, 3, and 4 are suggested for cows in the middle third of pregnancy, while rations 4, 6, 7, and 8 are suggested for those in the last third of pregnancy. If wheat straw is fed, it is recommended that both the hay and straw be chopped.

[b] Includes a 10% adjustment for feed wastage.

TABLE 8-20. Rations for Cows Nursing Calves in First 3 to 4 Months Post Partum (average weight—1000 lb)

Item	Ration Number[a]							
	1	2	3	4	5	6	7	8
	lb/day, as-fed[b]							
Feed ingredient								
Alfalfa hay, mid-bloom	26.4	—	—	13.2	33.5	16.8	—	16.8
Alfalfa silage, mid-bloom	—	44.9	—	—	—	41.2	—	—
Corn silage, well eared	—	—	54.8	33.1	—	—	54.8	42.0
Dent corn (2)	—	5.0	—	—	—	—	4.2	—
Cottonseed meal	—	—	3.3	—	—	—	3.3	—
Total daily feed	26.4	49.9	57.1	46.3	33.5	58.0	62.3	58.8
Nutrient intake								
Dry matter	21.3	18.1	16.5	19.0	26.9	26.4	19.9	27.5
Crude protein	3.6	2.6	2.4	2.5	4.6	4.3	2.7	3.2
TDN	11.7	11.7	11.7	11.7	14.8	14.8	14.8	14.8

[a] Rations 1, 2, 3, and 4 are suggested for cows of average milking ability, while rations 5, 6, 7, and 8 are suggested for cows of superior milking ability.

[b] Includes a 10% adjustment for feed wastage.

TABLE 8–21.   Mineral Mixtures for the Cow Herd

Item	Mixture[a]		
	1	2	3
Feed ingredient			
Trace mineralized salt	20.0	25.0	33.3
Dicalcium phosphate	30.0	—	—
Monosodium phosphate	—	25.0	33.3
Magnesium oxide	25.0	25.0	—
Dry molasses	—	10.0	—
Ground grain or oilseed meal	25.0	15.0	33.4
	100.0	100.0	100.0
Nutrient composition (% DM)			
Calcium	6.8	—	—
Phosphorus	5.5	5.8	7.4
Magnesium	15.0	15.0	—

[a] Mixtures 1 and 2 are for grass tetany prevention. Mixtures 2 and 3 are for areas of the country where forages that are fed to cattle contain high levels of calcium. Mixture 1 is acceptable when corn silage is fed with legume hay or silage.

6. For calcium, multiply percentage of Ca by the amount of DM available.

    20.25 lb DM $\times$ 1.46% Ca = 0.295 lb Ca

7. For phosphorus, multiply percentage of P by the amount of DM available.

    20.25 lb DM $\times$ 0.22% P = .0045 lb P

These values should be entered on the "Amount fed" line in Table 8–16, and the amount fed should be subtracted from feed requirements. Then the amount of feed given is either increased or decreased, according to difference. Particular attention should be paid to the possible need for more TDN (energy), mineral, or protein in the ration.

GUIDELINE RATIONS FOR THE COW HERD

Common rations for beef cattle are shown in Tables 8–17 through 8–20. The rations include dry matter, crude protein, and TDN intakes. Feed consumption may differ from that shown in the tables. Loose salt and minerals should be provided as a free choice, with no limit on quantity, for all cattle (Table 8–21). Daily feed consumption is listed on an as-fed basis. A 10% feed wastage allowance has been applied to all

as-fed feed intake values. Level of mineral consumption should be monitored in grass tetany areas. Cows need to consume 2 to 3 oz of a high magnesium supplement to prevent tetany.

REFERENCES

Beverly, J.R., and J.C. Spitzer. 1980. Management of replacement heifers for a high reproductive and calving rate. Texas Agricultural Extension Service, Texas A&M University System. College Station, TX. B-1213.

Bull, R.C., R.R. Loucks, R.L. Edminston, et al. 1974. Nutrition and weak calf syndrome in beef cattle. Current Information Series. University of Idaho, Moscow, ID. No. 246.

Corah, L.R., T.G. Dunn, and C.C. Kaltenbach. 1975. Influence of prepartum nutrition on the reproductive performance of beef females and the performance of their progeny. J. Anim. Sci. 41:819.

National Research Council. 1984. Nutrient requirements of domestic animals. Nutrient Requirements of Beef Cattle. Vol. 6. National Academy of Sciences, Washington, DC.

Sasser, R.G. 1983. Personal communication. University of Idaho, Moscow, ID.

Schoonover, C.O., and D.A. Yates. 1980. The biological cycle of the beef cow. In Cattleman's Library. Cooperative Extension Service, University of Idaho, Moscow, ID. CL-200.

Varner, L.W., R.A. Bellows, and D.S. Christenson. 1977. A management system for wintering replacement heifers. J. Anim. Sci. 44:165.

Selected Readings

Bowden, D.M., R. Hironaka, P.J. Martin, and B.A. Young. 1981. Feeding beef cows and heifers. Communication Branch, Agriculture Canada, Ottawa. 1670 E.

Corah, L.R. 1979. Nutrition, health, and reproduction in beef cattle. Charolais Bull-O-Gram. August/September.

Frischknecht, W.D., and J.D. Mankin. 1980. How to select, grow and manage replacement heifers. In Cattleman's Library. Cooperative Extension Service, University of Idaho, Moscow, ID. CL-745.

Johns, J.T., and J.B. Neel. 1978. Grouping the commercial beef herd for winter feeding. In Southern Regional Beef Cow-Calf Handbook. University of Kentucky, Lexington, KY. SR 3006.

N.A.S. 1982. Nutritional data for United States and Canadian feeds. 3rd Ed. National Academy of Sciences, Washington, DC.

Ney, J.J., V.M. Thomas, and G.W. Gibson. 1979. A feed-planning guide for wintering beef cattle. Current Information Series. University of Idaho, Moscow, ID. No. 498.

Vatthauer, R.J., and D.M. Schaefer. 1982. Rations for beef cattle. University of Wisconsin, Madison, WI. A 2387.

Yates, D.A. 1980. Average composition of common feeds (dry basis). In Cattleman's Library. Cooperative Extension Service, University of Idaho, Moscow, ID. CL-315.

Yates, D.A., and C.O. Schoonover. 1980a. Developing the replacement heifer into a productive beef cow. In Cattleman's Library. Cooperative Extension Service, University of Idaho, Moscow, ID. CL-345.

Yates, D.A., and C.O. Schoonover. 1980b. Developing the replacement heifer into a productive beef cow. In Cattleman's Library. Cooperative Extension Service, University of Idaho, Moscow, ID. CL-350.

Yates, D.A., and C.O. Schoonover. 1980c. Nutrition management of the productive mature beef cow. In Cattleman's Library. Cooperative Extension Service, University of Idaho, Moscow, ID. CL-330.

QUESTIONS FOR STUDY AND DISCUSSION

1. List the four compartments of a beef animal's stomach.
2. What factor limits the amount of low-quality roughage that a beef cow can consume daily?
3. What is the dry matter capacity, expressed as a percentage of the body weight, of beef cows fed the following feed ingredients:
   a. Pelleted hay
   b. Cereal grains
   c. Corn silage
   d. Low-quality roughage
4. What is the ideal calcium-to-phosphorus ratio in rations for beef cattle?
5. Compare the calcium and phosphorus content of alfalfa hay, orchard grass hay, and barley grain.
6. How much salt do beef cows require daily?
7. What is grass tetany and how may it be prevented?
8. An Idaho study demonstrated the importance of feeding adequate protein to beef heifers during the last trimester of gestation. Briefly summarize the results of this research.
9. Why is a vitamin A deficiency uncommon in spring-calving cow herds?
10. Outline a beef cow's reproductive cycle.
11. List the consequences of inadequate nutrition during the last trimester of gestation.
12. Why is the post-calving period the most critical time of a cow's reproductive cycle?
13. How much would Angus, Hereford, and Charolais crossbred with British heifers need to weigh at 13 to 16 months of age for 90% of them to be in heat?
14. Why is it not realistic to select only the heaviest heifers for replacements at weaning?
15. Is there any advantage to grouping heifers according to their body weight and weaning and feeding each group separately? Why?
16. How much should a heifer that has an expected mature weight of 1300 lb weigh at calving?
17. List the components of a successful feeding program.
18. List and discuss the steps that should be followed to plan a winter feeding program for beef cattle.
19. a. Would 20 lb (as-fed) of mid-bloom alfalfa hay satisfy the TDN requirement of a 1100-lb cow with average milking ability?
    b. If your answer is no, how many additional pounds of alfalfa hay or dent corn #2 would need to be fed on an as-fed basis to satisfy her TDN requirement?

Chapter Nine

Cow-Calf Herd Health and Disease Management

The primary objective of a herd health and disease management program is to minimize economic loss from disease. Preventing disease and parasite infestation is cheaper than treating sick animals. Preventive rather than therapeutic medicine is the key.

All health programs should be developed in cooperation with a veterinarian. The veterinarian is trained in livestock disease control and is most aware of health problems in a given geographic area. No program can be successful without the cooperation of everyone involved.

PLANNING A HERD HEALTH PROGRAM

A herd health program should be designed for specific conditions on a ranch. The primary objective of a herd health program should be to maximize returns by applying proper management practices at the right time, in the right way, with benefits exceeding costs.

Current, accurate herd performance records and realistic goals for improving production are necessary to identify the current state of the herd and the directions in which the herd is developing. Maintaining optimum reproductive efficiency should be a primary objective of a herd health program.

Performance records should be kept on a herd basis and should cover at least the following categories.

1. Number of cows exposed to the bull
2. Number of cows pregnant
3. Number of calves born and weaned
4. Number of cows sold and replacement heifers kept

Records of weaning weights or weight per day of age should be kept to evaluate nutritional programs, genetic potential, and mothering and milking abilities of the cow.

A record of disease problems previously encountered and disease preventive activities should be maintained to help in diagnosing disease and establishing effective vaccination programs.

Production goals are necessary to improve herd efficiency; however, goals should be realistic and achievable. Unrealistically high goals can cause a sense of failure if they are not accomplished. Knowing the current condition of the herd and what it is capable of producing in the future allows progress to be made.

A herd health program should correct major health problems, be compatible with other management practices, and provide flexibility and opportunity for modification as necessary.

PRINCIPLES OF A HERD HEALTH PROGRAM

Sanitation is an integral component of a herd health program. Disease organisms can and do live outside the host. Use of the following guidelines can help to prevent exposure of animals to disease-producing organisms and situations.

1. Practice good sanitation and cleanliness. A disease cycle can be broken by cleaning and disinfecting. Provide clean, dry quarters with good ventilation and plenty of clean fresh water.
2. Isolate newly acquired animals for 10 days to 3 weeks.
3. Provide idle periods for facilities, such as cleaning barns to provide a break in the buildup of disease-causing organisms.
4. Dispose of dead animals quickly and properly. Large animals should be sent to a rendering plant when possible. Alternate disposal methods would be cremation or deep burial (minimum of 6 feet).

A level of resistance can be maintained in the animal population by:

1. Feeding animals according to their nutritional requirements at various stages of their life cycle.
2. Vaccinating animals using available immunization procedures. The frequency of administration and the type of vaccines used depend on previous disease problems in the herd, diseases prevalent in the area, whether new additions to the herd are routinely made, and the dependability and versatility of the vaccine product being used.
3. Retaining only those animals that are physically sound and free from disease.

Spread of disease can be prevented by taking these steps:

1. Isolate sick animals.
2. Secure an early diagnosis. Have a veterinarian perform a physical examination and/or necropsy.
3. Carefully observe all animals.
4. Treat animals based on the diagnosis.

DISEASE RESISTANCE AND CATTLE VACCINES*

The body's first line of defense against disease is the skin. The skin prevents penetration of numerous disease-producing microorganisms that are in the environ-

ment. Mucous membranes that line body openings such as the mouth, eyes, and reproductive and digestive tracts also act as a first line of defense. These membranes produce enzymes and control the acidity or alkalinity (pH) of specific body parts to counteract infection. Mucosal cells also produce antibodies, which produce a temporary local immunity in a specific area of the body. Microorganisms that normally line the mucosal surface help to counteract invasions of disease-producing organisms. The skin and the mucous membranes of the body are similar to the wall of a fort, protecting the body from disease-producing germs. If the wall is broken, the body must depend on other means of protection.

White blood cells (phagocytic cells) are located throughout the body. They serve as a second line of defense against disease. White blood cells are attracted to foreign objects, such as bacteria or viruses, and engulf them. If successful, they destroy the disease-producing organism. If the organism is not killed, it may be carried by the white blood cells to other parts of the body, which helps to spread the infection or disease.

Vaccines and Immunity

Antibodies are protein fractions in the fluid portion of the blood (serum) that are secreted by cells in the body upon stimulation by antigens. Antigens are usually protein or carbohydrate fractions of disease-producing organisms. Antibodies are highly specific and neutralize the disease-producing ability of antigens.

The quantity of protection or degree of immunity possessed by an animal is measured by antibody titer, which is the degree to which the blood serum can be diluted and still show protection toward a specific antigen.

Vaccines may be defined as biologic agents that are used to stimulate active or passive immunity and that do not cause disease. The following paragraphs offer brief descriptions of some common types of vaccines for beef cattle (Hall and Woodard, 1983).

Bacterins. Bacterins are composed of killed disease-producing bacteria and are safe to use on any animal without fear of causing the disease. Some examples of such vaccines are those for blackleg, redwater, leptospirosis, vibriosis, pasteurellosis, and hemophilus infections.

Live Bacterial Suspensions. Live bacterial suspensions produce a better immunity than bacterins but may cause disease unless they are especially modified or attenuated so that they do not produce the disease. Effectiveness of the vaccine depends on the number of live organisms injected. Brucellosis vaccine Strain 19 is one example of this type of vaccine.

Toxoids. Toxoids are prepared from bacterial metabolic by-products (toxins). They are used as antigens and produce an active immunity. Examples are *Clostridium perfringens* type D toxoid, *C. perfringens* type C toxoid, tetanus toxoid, and *C. botulinum* toxoid.

Modified Live Viruses. These viruses have lost their ability to produce disease (i.e., they are attenuated), but they still stimulate the production of antibodies. Examples are the vaccines for infectious bovine rhinotracheitis (IBR), bovine virus diarrhea (BVD), parainfluenza-3 virus (PI_3), and bluetongue.

Killed Viruses. In this type of vaccine, the live properties of the virus are destroyed, and there is no danger of a disease outbreak. Through the immune body reaction, it creates immunity to a killed virus particle that is not capable of growth and development.

Antitoxins. Antitoxins effectively neutralize toxins when they are injected parenterally into an affected animal. They produce a passive immunity that usually lasts a few weeks. Examples include *Clostridium perfringens* type B, C, and D antitoxins.

One must remember that no vaccine is 100% effective. Effectiveness depends on age of the animal, the passive immunity that the animal already has when vaccinated, the stress status of the animal, and other factors.

Immunity Types

SPECIES IMMUNITY

Species immunity is a natural type of immunity. Some phases of this form of immunity are the variations in susceptibility to disease observed in different species. An example is measles, to which humans are susceptible but not cattle.

INDIVIDUAL IMMUNITY

This type of immunity refers to the differences among individuals within a herd or flock in susceptibility to a specific disease. Differences between individuals may be due to age, maternal antibodies from the uterus or from colostrum, or unobserved infections followed by recovery.

PASSIVE IMMUNITY

Passive immunity results from antibodies that are borrowed from another animal and given to an animal that needs immediate protection. This procedure is referred to as administration of antiserum or antitoxin. It provides protection of shorter duration than an active immunity but is immediately available to the recipient. Another form of passive immunity is immunity acquired from antibodies obtained by the fetus through the placenta, or by the nursing animal through colostrum milk.

ACTIVE IMMUNITY

When an animal is vaccinated or recovers from disease, antibodies are usually produced. This is an active form of immunity.

IDENTIFYING THE SICK ANIMAL

Some signs of illness in cattle are given in the following list.

1. Nasal discharge or any sign of a respiratory problem (coughing) may indicate a potential problem.
2. Vomiting may also be a serious sign of illness, because it is uncommon in ruminants.
3. Spasms, convulsions, or drooling are common signs of poisoning.
4. Lack of appetite and scours are positive signs of digestive disturbances, parasitism, or gastrointestinal infection.
5. Elevated or depressed body temperature indicates a condition deviating from normal health. (See the section "Vital signs," in Chapter 10.)
6. Discolored urine is a sign of infection, and abnormal fecal odor may be a sign of a digestive disturbance.
7. Other signs of illness that cattlemen should look for are drooping of one or both ears, the head carried in an abnormal position, reluctance to rise and move, a stiff gait, dragging of the hind feet when walking, dehydration, and a rough hair coat.

COMMON DISEASES OF BEEF CATTLE

The following sections offer a brief description of some common diseases prevalent in beef cattle. Table 9–1 provides a more complete list of diseases affecting beef cattle.

Diseases Causing Abortion

LEPTOSPIROSIS

Leptospirosis is a common cause of abortion in cattle in many areas of the United States. It is caused by bacteria that belong to the genus *Leptospira*. At least 40 different strains of *Leptospira* have been identified throughout the world.

Animals become infected with *Leptospira* by coming in contact with contaminated water, feed, or urine. The organism gains entrance by penetrating the abraded skin of the feet and legs and by passing through the mucous membrane of the digestive and urinary tracts and eyes.

The acute form of leptospirosis is often seen in young animals, but rarely in older animals. Animals with acute leptospirosis usually exhibit a high temperature (103 to 107° F), failure to eat, depression, and bloody urine. Abortion in older animals usually occurs 2 to 3 weeks after infection. Leptospirosis is usually diagnosed by detection of antibodies to the organism in the blood of infected animals. Risk of abortion may be minimized by vaccination.

BRUCELLOSIS (Bang's disease)

Brucellosis is caused by the bacterium *Brucella abortus*. Many cows in newly infected herds abort after 5 months or more of gestation. The fetal membranes may be retained and metritis (inflammation of the uterus) may occur. The disease is usually acquired by ingesting contaminated material from uterine membranes and discharges from aborted fetuses. Control is possible by vaccination of heifers at 2 to 10 months of age. Nationwide eradication requires blood testing and slaughter followed by herd quarantine of unvaccinated animals. Cows that abort as a result of brucellosis are always considered carriers.

INFECTIOUS BOVINE RHINOTRACHEITIS (IBR; Rednose)

IBR is caused by a virus and commonly produces signs of respiratory distress. In very young calves, it may produce signs of enteric (intestinal) distress. Calves usually have an initial high fever (104 to 107° F) accompanied by a red nose and profuse nasal discharge. The disease also causes abortion in non-immune, pregnant cows. Abortions usually occur about 6 weeks following initial infection. IBR infection is spread primarily by airborne or contact transmission. Another method of transmission is congenital, by passage of the newborn through the infected vagina at the time of calving. Effective vaccines are available. Intranasal vaccines appear to protect against abortions that result from usual contact levels of the virus.

The IBR virus causes fetal death at any stage of gestation, and the fetus is subsequently resorbed or aborted. There may be a 7- to 14-day interval, however, from initial infection and sickness in the cow to abortion because the virus may take some time to pass through the membranes.

BOVINE VIRAL DIARRHEA (BVD)

BVD causes abortion in cows in up to 150 days of gestation. Aborted fetuses contain the virus, and virus isolation is used to make a diagnosis. The disease is transmitted by carrier animals and by contaminated objects such as boots or tires, and is contagious from cow to cow.

BOVINE VIBRIOSIS

Bovine vibriosis is an infectious disease of the genital tract of cattle, caused by the bacterium *Campylobacter fetus venerealis* (*Vibrio fetus*). The disease is characterized by infertility (repeat breeding) and occasional abortions. It is a venereal disease spread by breeding and is considered the most important cause of infertility in cattle.

Good vaccines for vibriosis are available, and afford good protection when used according to directions. In cattle, one or two injections of vaccine should be given, depending on the type of vaccine that was used the first year 30 to 60 days before breeding.

Other Diseases

BLACKLEG

Blackleg is disease caused by *Clostridium chauvoei* and primarily affects cattle under two years of age. It is usually seen in the better-growing calves. The organism is taken in by mouth. The first signs of the disease noted are lameness and depression. Swelling, caused by gas bubbles, can often be felt under the skin as a crackling sensation. A high temperature is present. Usually, sudden death occurs, with no illness observed.

The chances of survival are poor. If administered early, however, large doses of penicillin may save the life of the animal.

Prevention is readily accomplished by the use of blackleg bacterins. Vaccination

at less than 4 months of age does not produce a lasting immunity. Calves vaccinated at less than 4 months of age should be revaccinated at 5 to 6 months.

MALIGNANT EDEMA

Malignant edema is a disease of cattle of any age caused by *Clostridium septicum*. It occurs as a wound infection. *C. septicum* is found in the feces of most domestic animals, and large numbers of the organism are found in the soil where livestock populations are high. The organism gains entrance to the body in deep wounds and can even be introduced into deep vaginal or uterine wounds in cows following difficult calving.

Signs of the disease are primarily depression, loss of appetite, and a wet doughy swelling around the wound; the swelling often moves to lower portions of the body. Temperatures of 106° F or more are associated with the infection, and often, death occurs in 24 to 28 hours.

The disease can be prevented by the use of *C. septicum* bacterins, which are usually produced in combination with other bacterins.

CLOSTRIDIUM NOVYI INFECTIONS

Infections caused by *Clostridium novyi*, sometimes called "black disease," occur sporadically in cow-calf operations. They are usually seen under feedlot conditions. The routes of infection and transmission are not known; however, the organism is thought to gain entrance into the body by a wound infection, or possibly by ingestion. Only sudden deaths are thought to occur, with virtually no recognition of sick cattle.

Diagnosis is based on the history of sudden death, significant postmortem lesions, and positive laboratory confirmation of the organism on fresh tissue. No treatment is available, owing to the sudden death aspect of the disease.

REDWATER DISEASE

Clostridium hemolyticum causes an infection commonly known as redwater dis-

TABLE 9-1. Cattle Disease Identification Guide

Disease	Organism/Cause	Time of Occurrence	Signs of Disease	Vaccination/Prevention	Remarks
Anaplasmosis	Red blood cell parasite	Most common during insect season	Loss of appetite, weight loss, difficult breathing, dehydration, anemia.	Two injections of killed vaccine 4 to 6 weeks apart, followed by yearly injection just before insect season (does not completely prevent disease).	Spread by biting insects, ticks, and unsanitary conditions (e.g., bleeding needles).
Anthrax (splenic fever)	*Bacillus anthracis*	Summer months	Staggering, labored breathing, convulsions, leading to death.	Noncapsulated sterne-strain vaccine.	Occurs in lowlands, swamps, and high alkaline and calcareous soils.
Blackleg	*Clostridium chauvoei*	Anytime	Lameness, depression, swelling under skin. Usually, sudden death occurs with no illness observed.	Bacterins; calves vaccinated less than 4 month of age should be vaccinated at 5 to 6 months of age.	Primarily affects cattle under 2 years of age and calves that are growing at a rapid rate.
Blue tongue	Virus	Summer	Fever, depression, reluctance to move, profuse slobbering, crusty and burned appearance of muzzle.	Vaccination available; vaccinate only if disease is diagnosed in the herd.	Consult closely with veterinarian.
Bovine virus diarrhea (BVD)	BVD virus	Anytime	Fever, diarrhea, excessive salivation, depression.	Vaccination of cows 30 days before breeding and calves between 5 and 8 months of age.	Transferred in feces and discharges from eyes and mouth. Affects mostly young stocker cattle.
Brucellosis	*Brucella abortus*	Anytime	Abortion in last half of gestation; increased retained placenta.	Vaccination of females 2 to 10 months of age.	Spread by infected carriers through normal body discharge.
Cattle grubs	Heel fly	Spring; early summer	Warbles on back.	Pour-on insecticide September to November.	Disrupts heat detection programs (AI). Insecticides less effective when grubs are on animal's back (Dec. through May).
Enterotoxemia (overeating)	*Clostridium perfringens* type C	1 week after birth	Acute abdominal pain and kicking at stomach, leading to death within 12 hours.	Type C antitoxin immediately after birth, or vaccination of cows with type C toxoid in doses during pregnancy and yearly booster thereafter.	Usually in calves 1 week old or less.

Disease	Cause	When occurs	Symptoms	Prevention/Treatment	Remarks
Foot rot	Perhaps several causes	Anytime ground tends to cause foot wounds	Lameness, foot swelling, broken skin around foot.	Good hoof health care; good nutrition program with all needed vitamins and minerals (iodized salt).	Cattle in infected areas should be watched closely, and immediate treatment should be administered.
Foothill abortion	Psittacoid virus	Anytime	Abortion in 6th to 9th month of gestation.	No vaccine available; spray cattle to prevent tick vector.	Immunity seems present after first abortion.
Grass tetany	Magnesium deficiency	Late spring, when cattle are placed on pasture	Staggering, nervousness, muscle twitching.	No vaccine available.	Keep plenty of magnesium mineral available.
Infectious bovine rhinotracheitis (IBR, rednose)	IBR virus	Anytime	Difficult breathing, profuse watery nasal discharge, abortion, vulvo-vaginitis.	Vaccination at 4 to 6 weeks of age or older.	More prevalent in feedlots. May show up as a form of pink eye.
Leptospirosis	*Leptospira* spp.	Anytime	Fever, prostration, jaundice, bloody urine, and anemia, leading to death; abortion in approx. the 7th month.	Vaccination is effective for 9 to 12 months.	Can be stopped within 7 to 10 days after outbreak with vaccination.
Malignant edema	*Clostridium septicum*	Anytime	Depression, loss of appetite, wet, doughy swelling around wound.	Bacterins.	Occurs as a wound infection.
Metritis	Various bacteria	Shortly after calving	Discharge of pus from vagina; repeat breeding.	Sanitation at calving; antibiotics effective as treatment.	Use clean calving equipment. Retained placenta mode of spread; sulfa boluses placed in uterus effective.
Redwater (bacillary hemoglobinuria)	*Clostridium haemolyticum*	Rainy, summer months	Depression, anemia, bloody diarrhea, red urine, high temperature.	Bacterin must be given every 6 months to prevent outbreak; treatment usually ineffective.	Primarily found in marshy lowlands; associated with liver fluke infestations.
Shipping fever (hemorrhagic septicemia)	Multiple infections with bacteria and/or virus and stress	Anytime animal is exposed to undue stress (e.g., marketing time)	Fever, hacking cough, discharges, diarrhea.	Several modified and inactivated vaccines; avoidance of undue stress.	Calves most susceptible.
Tetanus	*Clostridium tetani*	Anytime	Muscle spasms brought on by sudden sounds or touch.	Antitoxin following surgery or wounding.	Uncommon in cattle.
Trichomoniasis	Protozoa	Anytime	Infertility in cows.	No vaccine available.	Carried by bulls.
Vibriosis	*Vibrio fetus*	Anytime	Drop in conception rate, especially in heifers (herd history important).	Vaccinate annually 30 to 40 days prior to breeding.	Causes infertility with delayed conception of 2 to 8 weeks. Spread by bulls.

Adapted with permission from Extension Bulletin 1165. 1982. Cattle disease identification. Washington State University, Pullman, WA. October.

ease, which affects cattle in areas of poorly drained soil and lowland pastures. The organism is ingested by the animal and is frequently associated with liver fluke infection. Liver tissue damage caused by the flukes allows the bacteria to proliferate, grow, and produce powerful toxins that destroy red blood cells, spilling the released red hemoglobin into the urine (hence, the name "redwater" disease).

Signs of the illness are depression, anemia, bloody diarrhea, red-stained urine, high temperature, and collapse, leading to death in 1 to 3 days. Postmortem examination reveals an extremely jaundiced animal, with red-stained urine in the bladder, thin watery blood, and usually a large necrotic area in the liver.

Treatment is usually ineffective unless it is begun early. Large doses of penicillin may help. A bacterin is available for use in areas where the disease is prevalent, but it must be given every six months. In heavily infected areas, more frequent vaccination may be necessary.

ENTEROTOXEMIA

This disease condition is caused by *Clostridium perfringens*. This organism is found throughout the world in the lower intestinal tract of man and animals. The disease entity seen most frequently in the cow-calf operation is hemorrhagic enterotoxemia caused by *C. perfringens* type C.

There does seem to be somewhat of a geographic limitation to the condition, as it is seen most frequently in the mountain states and in the western parts of Kansas, South Dakota, and North Dakota. It is seen sporadically, however, throughout the entire Great Plains area.

Clostridium perfringens is a normal inhabitant of almost all mammals, and the following set of circumstances must exist for the disease to be present in the animal. (1) The type C strain of the bacteria must be present in the intestinal tract. (2) For the bacteria to attack, it must have an abundance of nutrients, especially carbo-

hydrates, as is present in milk, for instance. (3) There must be at least a partial slowdown or stoppage of intestinal tract movement brought about by ingesting a particularly large amount of feed, which allows the toxins to accumulate and be absorbed in the gut.

These conditions could be met in the case of a vigorous one-week-old calf who after exercise develops a real hunger and drinks more than its normal amount of milk from a good milking dam, overloading its digestive tract.

The disease is usually seen in calves one week of age or younger. Although riders may find only dead calves, the signs usually observed are those of acute abdominal pain, as evidenced by kicking at the stomach and straining. Later, the calves go down, frequently exhibiting "paddling"-type convulsions. Death usually occurs within 12 hours after signs of the disease are noted. Infrequently, bloody diarrhea develops prior to death.

There is no treatment of value as the animals almost always die soon after signs of the disease are noted. The disease can be prevented by injecting the calf with *Clostridium perfringens* types C and D antitoxin (antiserum) as soon as possible after birth. Single preventive injections seem to protect almost all calves through the dangerous early period of life.

If there is a history of a problem with the disease on the premises, however, a more efficient method of protection is to vaccinate the cows with *Clostridium perfringens* type C toxoid. Two doses are given during late pregnancy, and a yearly booster is given thereafter. This method allows the cow to produce her own antitoxin in the colostrum, and therefore protects the calf after nursing.

Common names for some clostridial disease conditions are recounted in the following list:

Clostridium chauvoei—blackleg
Clostridium septicum—malignant edema
Clostridium novyi—black disease

Clostridium perfringens C and D—enterotoxemia, or overeating disease

The clostridial diseases as a group present unique problems in control and diagnosis. The cattlemen should work closely with a local veterinarian to evaluate the prevalence of these agents in the area. As has been noted in the discussion, prompt postmortem examinations and tissue collection for laboratory testing are essential for an accurate diagnosis.

PI₃ (Parainfluenza-3 virus)

PI_3, IBR, and BVD are viruses that are together referred to as the *respiratory disease complex;* they are frequently associated with shipping fever. PI_3 is widespread in the cattle population, but it is usually manifested only when calves become stressed. The infection seldom causes death on its own, but often leaves the respiratory membranes vulnerable to bacterial infection. Calves should be vaccinated according to the recommendation of a local veterinarian.

PASTEURELLOSIS

True infections of *Pasteurella* spp. are rarely seen. The terms "shipping fever" and "hemorrhagic septicemia" have been used in the past to describe this condition, but are losing favor.

Most cases of pasteurellosis occur in younger animals and are associated with viral or chlamydial respiratory pathogens, together with stress conditions.

Pasteurella bacteria can be found in the respiratory passages of normal calves, and usually cause no disease unless the animal is stressed, either by other pathogenic microorganisms or by adverse environmental conditions.

When pasteurellosis is a problem, stress should be minimized during critical stages such as weaning and transportation. Calves should be immunized well in advance of anticipated stress situations with suitable viral vaccines.

The use of *Pasteurella* bacterins is under serious debate, and unless the immunologic features of the organism can be improved upon, the indiscriminate use of this bacterin cannot be relied upon to produce a satisfactory long-lasting immunity.

HEMOPHILUS SOMNUS

Hemophilus somnus has been identified as being associated with respiratory diseases in young stock, such as shipping fever, diphtheria, tracheitis, and thromboembolic meningoencephalitis (TEME) of feedlot cattle. TEME is a condition of small brain abscesses and results in a central nervous system disturbance. Animals affected by this condition die suddenly.

The organism is transmitted through respiratory tract discharges and secretins. Cattle that recover continue to carry the organism, and serious problems have developed when new feeder cattle have been mixed with carriers or penned next to them.

Vaccines are commercially available to help in preventing or reducing the occurrence of those conditions associated with *Hemophilus* spp. previously described. A *Hemophilus* vaccine should be part of a preconditioning program for weaned calves.

EXTERNAL AND INTERNAL PARASITE CONTROL

Parasites deprive the animal of nutrients required for maintenance and production, and cost livestock producers millions of dollars annually. Economic losses can be broken down into the following sources:

1. Reduced daily gain, feed conversion, and milk production.
2. Wasted labor and animal holding space.
3. Reduced reproductive performance.
4. Increased susceptibility to other diseases.

Prevention and control of parasitism should be based on available knowledge of factors that affect the life cycle of parasites. A better understanding of these factors enables cattlemen to treat parasite problems more effectively.

Internal Parasites

The nematodes, or roundworms, constitute the largest class of internal parasites. All have direct life cycles, which do not require an intermediate host; all with the exception of *Trichuris*, the whipworm, have both a free-living and a parasitic phase to their cycle (Brauer and Corwin, 1984).

Roundworms with a free-living phase develop from the eggs to the infective larvae in the dung pat; the larvae then migrate out of the pat and wait on the adjacent herbage to infect the prospective host. *Trichuris* develops to the infective stage within the egg, and the egg itself is infective for the host. Most roundworms gain entry to the host by ingestion, but *Strongyloides*, the threadworm, and *Bunostomum*, the hookworm, can also penetrate the host's skin.

The trematodes, or flukes, and the cestodes, or tapeworms, have indirect life cycles, which require intermediate hosts. Fluke larvae leave the egg in moist surroundings and penetrate snails. Within the snail, they develop to an intermediate stage. The intermediate stages swim free of the snail and attach to forage plants, where they develop to an infective stage. They are then ingested along with the forage.

A tapeworm egg contains an embryo that is ingested by a non-parasitic soil mite. In the mite, the embryo develops to an infective stage. The mite, and the infective-stage embryo it contains, are ingested along with forage as the host animal grazes.

The sporozoa, or coccidia, are complex single-celled organisms. Each begins life not as an egg but as an oocyst. Within the oocyst, the organism multiplies by simple division to form eight cells, and the oocyst is then infective. When the infective oocyst is ingested by the host, each of these cells penetrates a host mucosal cell and begins dividing again. The organisms subsequently leave the original host cells and penetrate other cells to undergo further division, and eventually, to form new oocysts.

The life cycle of all of these parasites can be divided into two periods: the developmental period, during which the parasite develops from egg or oocyst to the infective-stage embryo, and the prepatent period, which is the time interval from infection to the appearance of eggs or oocysts in the feces of the host animal.

In the case of the coccidia, the developmental period can be as short as 24 to 48 hours under optimum conditions. The developmental period for the nematodes is of intermediate duration, at 7 to 21 days, and trematodes and cestodes take the greatest amount of time, 1 to 2 months. Many factors affect the length of this period, however. Temperature is particularly important; low temperature can extend the developmental period of many nematodes to as long as 15 weeks, while *Eimeria* oocysts may not sporulate at all below 32° F. The presence of an intermediate host is important to the trematodes and cestodes, and their transmission is effectively limited to those times at which the intermediate host is present.

These same factors also determine how long the parasite can survive outside the host after it reaches the infective stage. As a rule, when developmental periods are at their shortest, so are survival periods.

The prepatent periods of most cattle parasites are more or less fixed. Coccidian oocysts appear in the feces 15 to 20 days after infection. Depending upon the species, nematode eggs appear in 14 to 60 days, cestode eggs in 6 weeks, and trematode eggs in 13 to 19 weeks. Some parasites can prolong their prepatent period, however, by arresting their own development. Common internal parasites of cattle are listed in Table 9–2.

Some of the signs of gastrointestinal parasitism are listed below:

1. Anemia
2. Submandibular edema (bottle jaw)
3. Digestive disturbances

4. Persistent diarrhea
5. Progressive weight loss
6. Slow weight gain
7. Weakness
8. Rough coat
9. Poor appetite
10. Dehydration
11. Emaciation
12. Unthriftiness

Beef cows should be treated after calves are born and again at weaning time. Fall-born calves (November, December) should be treated as they start grazing the new growth of spring pasture; treatment may need to be repeated as often as every 90 days depending on the degree of exposure to worm larvae. It is usually desirable to worm feeder cattle when they enter the feedlot.

Common dewormers on the market are described in Table 9–3.

Liver Flukes

Liver flukes are common in the northwestern and Gulf states and become a problem especially where heavy rainfall or irrigation floods the low-lying pasture land. The liver fluke depends on the freshwater snail to complete its life cycle. In the snail, the immature fluke multiplies, and the offspring emerge and swim to blades of grass. Cattle eating the grass or hay contaminated with fluke cysts can develop liver damage. Signs of fluke infestation include weakness, emaciation, unthriftiness, and anemia. Diagnosis of liver flukes requires microscopic examination of fresh cattle feces for fluke eggs.

External Parasites*

Beef cattle pests can increase stress and reduce animal production. Pests causing injury to cattle are horn flies, face flies, stable flies, houseflies, grubs, mosquitoes,

---

* This section is adapted with permission from Stoltz, R.L., and H.W. Homan, 1983. Control of biting flies attacking cattle. Cooperative Extension Service, University of Idaho, Moscow, ID. CIS-515.

lice, and ticks. Effective control is necessary to maximize animal production.

FLIES

Several kinds of flies are pests of beef cattle. Most cause problems during the warmer months of the year. They have four life stages: egg, larva, pupa, and adult. Most flies have either sponging or piercing-sucking mouthparts. Houseflies and face flies use sponging mouthparts to soak up liquid foods. Horn flies, stable flies, horseflies, deerflies, and mosquitoes have piercing-sucking mouthparts, which they use to pierce the animal's skin and suck its blood. Some flies, such as adult cattle grub flies, do not have mouthparts and cannot feed at all. Flies breed in a variety of habitats, ranging from manure and other fermenting organic matter to living tissue of a host animal. Face flies and horn flies are found mostly in pasture situations, whereas stable flies and houseflies are found around barns, feedlots, and other concentrated cattle-raising operations.

Efforts to control flies should combine sanitation with the use of insecticide.

LICE

Lice are most serious during the winter months, when cattle have heavier coats and less oily skin. Cattle are the only hosts of cattle lice. Usually, 1 to 2% of the cattle in a herd are carriers that harbor high numbers of lice throughout the entire year. Lice spread by contact from carriers to other animals in the herd. Cattle lice spend their entire lifetimes on the animal and live only a few days if they are removed from the host.

Cattle should be examined for lice whenever they are handled. The breeder should part the hair with his fingertips and examine the animal in several places, including the neck, withers, brisket, shoulders, mid-back, tailhead, and behind the rounds. Many chemicals are available for lice control in the form of dusts, dips, sprays, and pour-on products.

TABLE 9-2.   Common Internal Parasites of Cattle

Parasite	Class	Site	Damage	Signs
Haemonchus placei (barber-pole worm)	Nematode	Abomasum	Both larvae and adults suck blood; early larvae also invade tissue.	Anemia, submandibular edema ("bottle jaw"), constipation.
Ostertagia ostertagi (medium brown stomach worm)	Nematode	Abomasum	Both larvae and adults damage the gastric mucosa, disrupting digestive function and causing mucosal hypertrophy ("Morocco leather").	Anemia, submandibular edema, diarrhea, emaciation, leading to death.
Trichostrongylus axei (small stomach worm)	Nematode	Abomasum or small stomach	Adults suck tissue fluids from the mucosa, causing necrotic patches with a "frosted" appearance.	Anorexia, diarrhea, depressed growth.
Trichostrongylus colubriformis (bankrupt worm)	Nematode	Small intestine	Adults cause minor hemorrhage, localized necrosis, and cratered lesions.	Diarrhea, depressed growth.
Strongyloides papillosus (intestinal threadworm)	Nematode	Small intestine	Larvae penetrate the skin and migrate through the lungs, causing dermatitis and pneumonia. Adults cause small hemorrhages in the intestinal mucosa. Mostly affects young calves.	Dermatitis, coughing, diarrhea, depressed growth.
Cooperia pectinata, C. punctata, C. oncophora (Cooper's worms)	Nematode	Small intestine	Adults suck blood and disrupt digestive function in the intestinal mucosa.	Anorexia, diarrhea, depressed growth.
Bunostomum phlebotomum (hookworm)	Nematode	Small intestine	Larvae penetrate the skin and migrate through the lungs, causing dermatitis and pneumonia. Adults suck blood and cause hemorrhage in the intestine. Most severe in young calves.	Severe dermatitis, coughing, anemia, submandibular edema, diarrhea, weight loss.
Moniezia benedeni (tapeworm)	Cestode	Small intestine	Adults cause focal inflammation of the mucosa at the site of attachment and occasional impaction of the intestine.	Depressed growth.
Eimeria bovis, E. zurnii (coccidia)	Sporozoa	Small intestine	Most pathologic development occurs in the colon.	
Trichuris discolor (whipworm)	Nematode	Cecum and colon	Adults deeply penetrate the mucosa and suck blood.	Anemia, depressed growth.
Oesophagostomum radiatum (nodular worm)	Nematode	Cecum and colon	Larvae produce nodules in gut wall that become small abscesses.	Anorexia, diarrhea, depressed growth.
Eimeria bovis, E. zurnii (coccidia)	Sporozoa	Cecum and colon	Coccidia multiply exponentially inside the mucosal cells, causing erosions and hemorrhage. Mostly affects weaning calves.	Bloody diarrhea, weight loss.

Dictyocaulus viviparus (lungworm)	Nematode	Bronchioles and lung parenchyma	Adults traumatize the bronchioles, causing excess mucus secretion, bronchitis, and emphysema. *D. viviparus* also predisposes to secondary bacterial infections.	Dyspnea, coughing ("husk"), weight loss.
Fasciola hepatica (liver fluke)	Trematode	Bile duct	Adults cause traumatic hepatitis, resulting in cirrhosis and fibrosis. *F. hepatica* also predisposes to infections with *Clostridium novyi* (black disease) and *C. hemolyticum*.	Anemia, abdominal pain, weakness, weight loss, leading to death. (Death due to black disease is rapid, often without other clinical signs.)
Fascioloides magna (giant deer fluke)	Trematode	Liver parenchyma	Cattle are abnormal hosts. The fluke is unable to develop completely, which causes thick fibrous cysts that lead to condemnation of the liver at slaughter.	

Adapted with permission from Brauer, M.A., and R.M. Corwin. 1984. The control of internal parasites of cattle. American Hoechst Corporation, Animal Health Division, Somerville, NJ 08876.

CATTLE GRUBS*

Adult cattle grubs are known as *heel flies* or *warble flies*. They appear in late spring or early summer and can be easily detected by the breeder through his observation of gadding cattle. Two species are involved. The northern cattle grub lays its eggs singly on hairs on the lower part of the animal's body. The common cattle grub attaches her eggs in the same areas, but may attach as many as a dozen eggs to a single hair. Adult flies do not feed, and consequently, they live only 3 to 10 days.

The eggs hatch in about 4 days, and the tiny larvae crawl down the hair to the skin, which they attempt to penetrate using their mouthparts and enzymatic secretions (Fig. 9–1). Where several larvae have entered the skin at the same place, a reaction called *hypodermic rash* develops on the animal. Affected animals constantly lick and scratch these spots and do not feed normally. The grubs suffer high natural mortality during their development because of the host's natural resistance. Cattle of all breeds are the normal hosts, but occasionally, horses are attacked. Calves and yearlings tend to have more grubs than older stock.

Little is known about the larva after they enter the animal's body, but in 1 to 2 months, the first common cattle grubs have migrated to just beneath the lining of the esophagus. Northern cattle grubs appear in the spinal canal at approximately the same time interval. The grubs live in these locations for about 6 months and then migrate to the topline of the animal. When the larvae first arrive in the animal's back, they cut a small breathing hole in the hide, then molt in pouches (warbles) between the inner and outer layers of the skin. After growing rapidly for 6 to 8 weeks, the fully grown larvae cut holes, exit the warbles,

---

* Adapted with permission from Stoltz, R.L., and H.W. Homan, 1979. Cattle grubs and their control. Cooperative Extension Service, University of Idaho. Moscow, ID. CIS-512.

TABLE 9–3. A Guide to Antiparasitic Drugs for Cattle[a]

Class	Generic Name	Trade Name(s)	Manufacturer	Dosage Form(s)	Spectrum of Activity
Amprolium	Amprolium	Amprol 25% Corid	Merck Animal Health	Feed premix, solution, soluble powder	*Eimeria bovis, E. zurnii*
Benzimidazoles	Albendazole	Panacur	SmithKline	Oral suspension	*Fasciola*
	Fenbendazole	Safe-Guard	American Hoechst	Oral suspension, paste	*Haemonchus, Ostertagia, Trichostrongylus, Cooperia, Nematodirus, Bunostomum, Oesophagostomum, Dictyocaulus*
	Thiabendazole	Omnizole TBZ Thiabenzole	Merck Animal Health	Bolus, paste, oral suspension, feed, premix, feed top-dressing	*Haemonchus, Ostertagia, Trichostrongylus, Cooperia, Nematodirus, Oesophagostomum*
Imidazothiazoles	Levamisole hydrochloride	Levasole	Pitman-Moore	Bolus, drench, injectable solution	*Haemonchus, Ostertagia, Trichostrongylus, Cooperia, Bunostomum, Nematodirus, Oesophagostomum, Dictyocaulus*
		Ripercol-L Tramisol	American Cyanamid	Bolus, drench, feed premix, injectable solution, gel	
Avermectins	Ivermectin	Ivomec	Merck Animal Health	Injectable solution	*Haemonchus, Ostertagia, Trichostrongylus, Cooperia, Nematodirus, Dictyocaulus, Hypoderma, Linognathus, Haematopinus, Psoroptes, Sarcoptes*
Organophosphates	Coumaphos	Baymix	Haver-Lockhart	Feed premix, feed top-dressing	*Haemonchus, Ostertagia, Trichostrongylus, Cooperia, Nematodirus*
	Haloxon	Halox Loxon	Wellcome/Jen-Sal	Bolus, drench	*Haemonchus, Ostertagia, Trichostrongylus, Cooperia*
Phenothiazine	Phenothiazine	Various commerical names	Tefenco West-Agro	Feed premix for either therapeutic or low-level continuous prophylactic feeding	*Haemonchus, Ostertagia, Trichostrongylus, Oesophagostomum, Bunostomum*
Quinolines	Decoquinate	Deccox	Rhone-Poulenc	Feed premix	*Eimeria bovis, E. zurnii*
Sulfonamides	Sulfaquinoxaline	Bovo-Cox	International Multifoods/Osborn	Feed premix, bolus	*Eimeria bovis, E. zurnii*
Tetrahydropyrimidines	Morantel tartrate	Rumatel Nematel	Pfizer	Bolus, feed premix	*Haemonchus, Ostertagia, Trichostrongylus, Cooperia, Nematodirus, Oesophagostomum*

[a] Label instructions should be followed when using those products relative to warning/cautions and withdrawal times. Adapted with permission from Brauer, M.A., and R.M. Corwin. 1984. The control of internal parasites of cattle. American Hoechst Corporation, Animal Health Division, Somerville, NJ 08876.

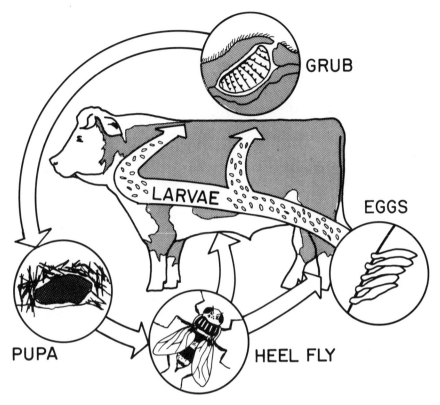

FIG. 9–1. Life cycle of the cattle grub. (Courtesy of Kathy Dawes Graphics, Moscow, ID.)

and fall to the ground. There they crawl under some object and transform into pupae. The pupal stage lasts 2 weeks to 60 days, depending on the weather. The new adults then emerge, mate, find hosts, and repeat the annual cycle.

Area-wide grub control programs are much more effective than individual efforts. Where every breeder treats his cattle for grubs, grub populations become low enough to prevent gadding the following year, and eventually the populations become so low that treatment is seldom necessary, unless grubby cattle are imported.

Control of grubs can be achieved easily and effectively using systemic insecticides at the proper time. Application should be timed to kill the grubs during the 30- to 60-day early migration period after heel fly activity has stopped in the fall. Such treat-ment kills the grubs before they reach the gullet or spinal column. Killing the grubs at this stage is desirable, because cattle that are heavily infested with large first-stage grubs when they are treated can be injured, and sometimes killed, by the reaction of their bodies to the dead grubs. To minimize this problem, cattle should be treated immediately after heel fly activity ends in the breeder's geographic area, which is usually after the first killing frost of the fall.

Reaction to the common cattle grub can cause the gullet to swell until it is closed, causing the animal to bloat. Reaction to the northern cattle grub can cause paralysis of the hindquarters. Cattle should be observed for side reactions for 48 hours after treatment. If such reactions persist, a veterinarian should be consulted. Signs to look for are stiffening of hindquarters, bloating, and excessive salivation.

Dipping and pour-on products are the most convenient and practical methods of application. Spraying works effectively, but it is not convenient to apply. Treated mineral block or treated ration is effective only when animals can receive regular dosages.

A HERD HEALTH MANAGEMENT PROGRAM

Calves

The following procedures should be undertaken at calving time:

1. Dip the naval cords of all newborn calves in a 7% tincture of iodine solution to prevent infection.
2. Ensure that calves nurse and receive colostrum (cow's first milk) within 2 hours of birth. Keep some frozen colostrum on hand for feeding weak or orphaned calves.
3. Identify calves soon after birth with an eartag.
4. Record birth weight, sex, and numbers of calf and cow.
5. If white muscle is a problem in the herd, inject the calf with a selenium-vitamin E solution.

These procedures should be followed for a calf of 1 to 3 months of age:

1. Brand, castrate, and dehorn the calf. These operations are easier and cause fewer problems when performed early.
2. Implant an approved growth promotant in calves that are not intended for herd replacement.
3. Vaccinate the calf for clostridial diseases that are prevalent in the area, and for scours if necessary. Consult a local veterinarian about the use of these vaccinations.
4. Check with the veterinarian about the need for other vaccinations.
5. Vaccinate replacement heifers for brucellosis (Bang's disease). Usually, a single vaccination is given between 2 and 10 months of age. Consult a veterinarian regarding specific state regulations.

At 3 to 4 weeks prior to weaning, the calf should be treated as follows.

1. Vaccinate heifers for brucellosis if such vaccination has not been given previously.
2. Administer a clostridial booster shot.
3. Administer two injections, 2 to 4 weeks apart, for pasteurellosis and *Hemophilus somnus* infection.
4. Treat the calf for internal and external parasites.
5. Implant or reimplant the calf with an approved growth-promoting agent. (Do not treat calves intended for herd replacement.)

WEANING

All vaccinations should be completed prior to weaning to avoid stress. At weaning, the following should be performed:

1. Treat the calf for internal and external parasites if such treatment has not been given earlier.
2. Reimplant growth-promoting agent in steers and heifers not intended to be used as herd replacements if reimplantment has not been performed earlier.
3. Review vaccination program and change as necessary.
4. Provide a fresh water supply.
5. Feed the calf a good-quality hay (grass or a grass-legume, but not a straight legume hay). Calves should be consuming feed in the amount of 2 to 3% of their body weight daily.
6. Check the calf for sickness two to three times daily.

Replacement Heifers and Bulls

At 10 to 15 months of age, replacement cattle of unknown origin (excluding pregnant heifers) should be vaccinated for the following conditions:

1. IBR and PI$_3$
2. Vibriosis
3. Leptospirosis
4. Clostridial diseases
5. *Hemophilus* and *Pasteurella* infections

In addition, these cattle should be treated for internal and external parasites, and should receive a vitamin A injection if their body condition is poor.

Second working cattle should receive the following injections.

1. BVD
2. *Hemophilus* booster

The following treatment should be given to replacement cattle of unknown origin that were immunized as calves and at weaning:

1. Give booster vaccinations as required (IBR; BVD; leptospirosis; vibriosis; clostridial infections).
2. Administer treatment for internal and external parasites.

Mature Cows and Bulls

At not less than 30 days prior to breeding, open cows should be treated for lice and be vaccinated for the following:

1. Vibriosis
2. Leptospirosis
3. Clostridial diseases
4. IBR, BVD, and PI_3

Pregnancy tests should be performed on all females after the breeding season, and open cows should be culled. Pregnant cows should be treated as follows:

1. Begin control of internal and external parasites as close to the end of the heel fly season as possible.
2. Use a killed product or an intranasal vaccine to provide immunization from IBR.

Bulls should be treated as follows:

1. Examine for breeding soundness at least 45 days prior to the beginning of the breeding season.
2. Vaccinate for clostridial infection and leptospirosis prior to breeding, if required.
3. Follow internal and external parasite control procedures developed for the herd by the local veterinarian.
4. Consult with the veterinarian to see whether virgin bulls require vaccination for IBR, PI_3, and BVD prior to breeding.

REFERENCES

Brauer, M.A., and R.M. Corwin. 1984. The Control of Internal Parasites of Cattle. American Hoechst Corporation, Animal Health Division, Somerville, NJ 08876.

Extension Bulletin 1165. 1982. Cattle disease identification. Washington State University, Pullman, WA. October.

Hall, Richard F., and Lynn F. Woodard. 1983. Vaccines and their use. *In* Cow-calf Management Guide. Cattleman's Library. Cooperative Extension Service, University of Idaho, Moscow, ID. CL-605.

Schipper, I.A. 1980. Disease resistance in livestock. *In* Cow-Calf Management Guide. Cattleman's Library. Cooperative Extension Service, University of Idaho, Moscow, ID. CL-600.

Stoltz, R.L., and H.W. Homan. 1979. Cattle grubs and their control. Cooperative Extension Service, University of Idaho. Moscow, ID. CIS-512.

Stoltz, R.L., and H.W. Homan. 1983. Control of biting flies attacking cattle. Cooperative Extension Service, University of Idaho, Moscow, ID. CIS-515.

Selected Readings

Blakely, J., and D.D. King. 1978. The Brass Tacks of Animal Health. Doane Agricultural Service, St. Louis.

Card, C.S., and E.P. Duren. 1978. Beef herd health program. Cooperative Extension Service, University of Idaho, Moscow, ID. CIS-430.

Christensen, C.M., F.W. Knapp, and R.A. Scheibner. 1979. Beef cattle pests. Cooperative Extension Service, University of Kentucky. Lexington, KY. Entomology 39.

Gates, N.L. 1981. Beef herd health program. Cooperative Extension Service, Washington State University. Pullman, WA. E13-0962.

Meyerholz, G.W., and R.E. Bradley. 1983. Internal parasitism of cattle. *In* Cow-Calf Management Guide. Cattleman's Library. Cooperative Extension Service, University of Idaho, Moscow, ID. CL-690.

Morter, R.L. 1979. Cow-calf herd health and disease management. Cooperative Extension Service, Purdue University. W. Lafayette, IN. VY-47.

NCA. 1981. Control of bovine respiratory disease in the cow-calf herd. Management guidelines prepared by the Animal Health Committee of the National Cattlemen's Association and the American Association of Bovine Practitioners. Beef Business Bulletin, Englewood, CO. July 17.

QUESTION FOR STUDY AND DISCUSSION

1. What records would be helpful to a veterinarian who is diagnosing a disease problem within a beef cattle herd? Why?
2. How may the following be achieved within a beef cattle herd?
   a. Prevention of exposure of cattle to disease-producing organisms
   b. Maintenance of a high level of disease resistance in the herd
   c. Prevention of the spread of disease
3. What is the body's first line of defense against disease?

4. Define the following terms:
   a. Antibody
   b. Antigen
   c. Vaccine
   d. Bacterin
   e. Toxoid
   f. Modified live virus vaccine
   g. Passive immunity
   h. Active immunity
5. List five signs of a sick animal.
6. List the causative agent, signs, method of transmission, and preventive measures for the following diseases:
   a. Leptospirosis
   b. Brucellosis
   c. IBR
   d. BVD
   e. Vibriosis
   f. Blackleg
   g. Malignant edema
   h. Black disease
   i. Redwater
   j. Enterotoxemia
   k. Parainfluenza
   l. Pasteurella
7. List five signs of gastrointestinal parasitism and how to prevent or correct an intestinal parasite problem.
8. Why are liver flukes a greater problem in areas where heavy rainfall or irrigation floods pasture land?
9. Describe the difference(s) between the northern and common cattle grubs.
10. Outline an effective grub control program.
11. Describe a herd health program for a calf immediately after birth to 3 months of age.
12. When should replacement heifers be vaccinated for brucellosis?

Chapter Ten

Calf Management—Birth to Weaning

The primary management objective relative to the calf crop should be to manage calves for maximum health and growth. Poor nutrition of the dam, difficult calving, contaminated surroundings, and bad weather are predisposing causes of calf mortality. Proper management of the calf from birth to weaning dramatically affects the profits of the beef cow-calf enterprise. Beef cattle producers must minimize calf mortality and maximize growth. This chapter outlines management practices that maximize liveability and growth of the calf crop.

MANAGEMENT AT BIRTH

Basic management objectives to ensure low calf mortality and promote health in the calf are outlined in the following list:

1. Provide clean, well-ventilated barns or sheds as calving areas. Scours and respiratory outbreaks are most often attributed to fecal contamination of the calf's surroundings. Fecal material from sick calves spreads infective bacteria and viruses to healthy calves. Make the calving area as large as possible.
2. Provide clean, fresh drinking water. Check that the water supply is not contaminated with fecal material. Ensure that lots are well drained.
3. Feed from bunks in confined areas to prevent calves from chewing hay or feed that is contaminated with fecal material. Feed bunks should be portable unless they are equipped with a hard surface apron that can be cleaned to prevent mud and manure accumulation.
4. Do not use pens that have contained calves with scours for other calves.
5. Remove mucus from the calf's nose and mouth as soon as possible after birth.
6. Disinfect the navel cord with iodine (7% tincture) to guard against navel infection. An easy method is to pour a small amount of iodine into a wide-mouthed bottle. Place the mouth of the bottle over the cord, and shake to dispense the iodine.
7. A newborn calf should nurse as soon as possible after birth. If the cow does not attend her calf, pen both in an isolated area, and leave them undisturbed for a few hours. Sometimes, it may be necessary to help the calf nurse. A weak or orphaned calf, or one whose mother displays problems in nursing, should be force-fed colostrum (first milk) with a calf feeding tube. Colostrum milk contains a high concentration of antibodies, which protect the young calf against calfhood diseases. Within 6 hours after birth, the calf has been shown to lose 50% of its ability to absorb antibodies. Keep some frozen colostrum available. Thaw colostrum slowly to prevent antibodies from being destroyed. Calves should receive 2 to 3 quarts of colostrum daily, divided into 2 or 3 individual feedings.
8. Identify each calf using individually numbered plastic tags plus a permanent ear tattoo of the same number. Record the calf's birth date, weight, and sex, as well as identification of the dam.

SCOURS OUTBREAK

In a disease outbreak, a diagnosis should be made as soon as possible so that corrective measures and proper treatment can be initiated. Vaccines may be used in the treatment of scours if the condition is of viral origin.

Calves with scours should be kept warm and dry. An attempt must be made to replace the calves' lost water and electrolytes and a readily available source of energy must be provided. Liquid antibodies may also be used to prevent secondary disease. If calves are less than 8% dehydrated, fluids may be given orally; however, once calves are more than 8% dehydrated, fluids should be given intravenously (Table 10–1).

Many homemade solutions for treating scours in calves have been developed and used successfully. One such solution contains 1 teaspoon of sodium chloride, ½ teaspoon of sodium bicarbonate, 4 oz of corn syrup, and 4 pints of water. This solution should replace all of the milk and be fed at least twice daily to overcome dehydration in the calf. Total fluid intake should never be less than 8 pints per day or 1 pint for every 10 lbs of body weight. After two or three days of treatment, or as soon as the scours cease, introduce acidophilus milk. Freeze-dried acidophilus milk cultures can be obtained from veterinarians or veterinary supply houses. Acidophilus milk is loaded with helpful bacteria that take the place of harmful organisms in the gut. Prolonged use of antibiotics should be avoided since they interfere with normal digestive processes.

Also, several commercial preparations are available that can be mixed with water and given to the calf to replace fluids.

VITAL SIGNS

Not only do beef cattle producers need to recognize changes in the condition and appearance of their animals, but they need to learn to what extent an abnormality may be reflected in the vital signs of the body. An understanding of these vital signs helps the beef cattle producer to describe clinical signs of illness to the veterinarian and may also help him to treating minor infection and disease on his own (Table 10–2).

The procedure for taking an animal's temperature consists of shaking the mercury column into the bulb end of the thermometer, lubricating the tube, inserting the full length of the tube into the rectum, and leaving it in place for three minutes. Temperature varies considerably even in healthy animals; therefore, the temperature must be taken at regular intervals to classify a fever. Elevated temperature in cattle is only one tool that can be used in diagnosing disease. Such other signs as

TABLE 10–1. Degree of Dehydration and Clinical Signs of Calves With Scours

Dehydration Level (% Body Weight)	Clinical Signs
0–5%	1. Mild depression
	2. Decreased urine output
6–8%	1. Sunken eyes
	2. Tight skin
	3. Depression, but calf in upright position
	4. Dry mouth and nose
	5. Further urine reduction
9–11%	1. Increase in above signs
	2. Cold extremities
	3. Recumbency
12–14%	1. Shock
	2. Death

TABLE 10–2. Vital Signs of Beef Cattle

Temperature	
Normal	100.4–103.1°F
Mild fever	103.1–104.6°F
Moderate fever	104.6–105.8°F
High fever	105.8–107.0°F
Very high fever	107.0–110.0°F
Pulse rate	40–70 heartbeats per minute
Respiration rate	10–30 inspirations per minute

From LeViness, E.A. 1982. Vital signs in animals—What stockmen should know about them. In Cow-Calf Management Guide. Cattleman's Library. Cooperative Extension Service. University of Idaho, Moscow, ID. CL-610.

chill, pulse, respiration rate, appetite, and digestion should also be used.

DEHORNING AND CASTRATION*

Calves are worth more at weaning if they are properly dehorned and castrated. Horned cattle use more feedbunk space, prevent other cattle from feeding properly, and may cause bruises, which lower carcass values. Dehorned cattle look more uniform, feed better, and can be sold at a higher market price. Dehorning and castration of males should be performed at the same time.

Castration is recommended for all bull calves destined to be sold as feeders or finished in the feedlot. They should be castrated before they are 2 months old (preferably before fly season) to minimize the shock effects of the operation.

Bloodless Dehorning

DEHORNING LIQUID

Dehorning liquid with collodion base dries and forms a rubber-like covering that is not easily rubbed or washed off. It is applied with a brush or swab. Calves of up to 10 days of age can be dehorned with dehorning liquid.

DEHORNING PASTE

Dehorning paste is placed on the horn "button" with a small wooden paddle. The paste should be prevented from coming in contact with the skin of either the calf or the operator. Several commercial dehorning pastes are on the market.

CAUSTIC SODA

Caustic soda is also called "caustic stick." The hair is clipped from around the small, undeveloped horns or buttons of the

*Adapted from Bagley, C.V., and N.J. Stenquist. 1981. Dehorning, castration, and branding. In Cow-Calf Management Guide. Cattleman's Library. Cooperative Extension Service, University of Idaho. Moscow, ID. CL-750. And from Strum, G.E., and I.E. Schipper. 1975. Tips for better selling calves. Great Plains Beef Cow Calf Handbook. Oklahoma State University, Stillwater, OK. GPE-3051.

calf, and petroleum jelly is applied to prevent the caustic from coming into contact with the skin. The end of the caustic to be held in the hand should be wrapped in paper or cotton. The other end is moistened, and the moist end of the caustic is rubbed on the undeveloped horn. Two or three applications are necessary. After each application, the caustic should be allowed to dry. Drying takes only a few minutes. If applications are thorough, no further horn growth will occur. Calves treated with caustic should be protected from rain for a few days following the treatment.

ELECTRIC DEHORNER

Most electric dehorners have a cupped attachment. The horn tissue is burned by the cup, which is placed over the horn buttons. This method is bloodless, but must be done when calves are young. It is not an entirely satisfactory method, as a regrowth of horn tissue sometimes occurs unless the burning application is liberally applied to destroy all potential new horn tissue.

TUBE DEHORNER

The tube dehorner is used on calves of up to 4 months of age. Tubes are available in various sizes; the one used should fit the base of the horn. The horn is gouged out by a turning action. This is an excellent method for removing horn buttons (Fig. 10–1).

SPOON DEHORNER

This tool is used on small calves to cut or gouge out horn buttons. Some ranchers use a heavy knife to cut off the horn buttons, then follow by circling the edges with a caustic stick.

Mechanical Dehorners

CALVES

Mechanical dehorners for calves are widely used by ranchers and farmers. These instruments have cutting blades that

FIG. 10–1.   Using a tube dehorner.

can remove horns on calves from 2 months to 8 or 10 months of age. A ring of skin is taken off with the horn so that new skin can grow over the horn base, which prevents further horn growth.

MORE MATURE CATTLE

Mechanical dehorners are available to dehorn more mature cattle. These are designed for speed in operation. A cut is made at a point ¼ to ½ inch below the junction of the horn with the skin or hide. Skin must be removed around the entire horn to prevent further horn growth.

DEHORNING SAW

When only a few cattle are to be dehorned, a dehorning saw can be used. Dehorning saws have blades especially designed for cutting horns. A cut is made about ¼ to ½ inch below the junction of the horn with the skin or hide to prevent further horn growth. A fine-toothed, stiff carpenter's saw can also be used for dehorning.

Disinfection Precautions

The instruments used in castration and dehorning must be disinfected thoroughly between each animal. Such disinfection helps to prevent infections and the spread of diseases such as anthrax, blackleg, warts, and anaplasmosis. Disinfection of the operator's hand also helps prevent the spread of disease from animal to animal and protects the operator himself from infection.

Many good disinfectants are available, such as Lysol, various quaternary preparations (Roccal, dairy utensil cleaners), and chlorine preparations. Iodine is an ideal skin disinfectant, but is corrosive to instruments. Kerosene has no disinfectant qualities. A veterinarian should be consulted regarding disinfectants and fly repellents.

Wound Treatment

If dehorning or castration is done in cool weather when there are no flies, no wound treatment is needed. If flies are present, there is the danger of maggot infestation.

Wounds should be checked and painted weekly until they heal. Pine tar is the favorite wound dressing of many stockmen. Aerosol preparations and smears can be obtained from the veterinarian.

Castration

Bull calves from a few weeks to 8 months of age may be castrated without serious harm. Older animals usually bleed more, so greater care must be used in castrating mature bulls. If a bull calf is not castrated before 8 months of age, he may become "staggy," which is objectionable in the feeder and market steer. The four methods most often used in castrating calves are described in the following paragraphs.

The lower end of the scrotum is grasped and stretched out tightly, then the lower third is cut off. This method exposes the ends of both testicles. The testicles are removed one at a time. The testicle is pulled out of the scrotum and cut to allow 3 to 4 inches of the cord to remain on the testicle. The cord is severed by scraping and simultaneously exerting tension by drawing on the testicle. Cutting with a sharp instrument may cause bleeding.

The second method consists of slitting each side of the scrotum. The incision is made on one side, and the testicle is removed from that side before the incision is made on the other side. The incision is made over the center of the testicle, from about the top third to the lower end. It is essential to extend the slit well toward the lower end of the scrotum to allow for proper drainage.

A third method used on young calves is to grasp the lower end of the scrotum, stretch it out tightly, stick the knife through the scrotum about midway, and cut the sack open. Each testicle is then pulled out with the long cord attached. This method is fast and clean, and it allows a full cord to develop.

The all-in-one lamb castrator is a handy instrument for calf castration. In addition to having a scissors-like cutter for removing the end of the scrotum, it provides an excellent claw arrangement for grasping the testes and extracting them from the calf.

When castrating older cattle, some cattlemen prefer to draw the cord tightly over the index finger of the left hand and cut it by scraping with a knife. Another satisfactory method is to place the cord under tension and sever it with an emasculator. Both of these methods of cutting the cords on older animals tend to check the flow of blood. The operation must be performed with clean instruments and under sanitary conditions.

BLOODLESS CASTRATION

The method known as "bloodless castration" calls for a special type of pincers, pliers, or clamps, which crush each cord separately an inch or two above the testicle. The method is a satisfactory means of castration when performed properly, but if the operation is performed too hastily, the cord may not be completely crushed and the steer is likely to develop stagginess later on.

The cords are pinched one at a time. Care should be taken to see that the cord is placed between the jaws of the pincers before they are closed. As there is no break in the skin of the scrotum, there is no external bleeding. This is an advantage in areas in which screwworms are troublesome. Steers castrated in this manner usually develop larger and fuller cords by the time they are ready for market. Some cattlemen consider this trait desirable for well-finished steers.

Another type of bloodless castration is *elastration*, which uses rubber bands. (Fig. 10–2). Disadvantages of this method are the possibility of tetanus (lockjaw) and the lack of cord development as steers become fat.

CHEMICAL CASTRATION

Recently, a chemical castration product has been developed. This product (Chem Cast) was developed to avoid the calf

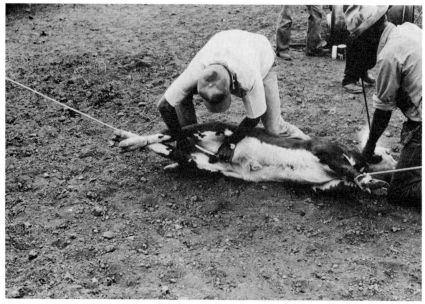

FIG. 10–2. Using an elastrator to castrate and a hot iron to dehorn calves.

stress, pain, and hemorrhage inherent in surgical and clamping methods of castration. With proper administration, it painlessly destroys the testes and testicular tunics within 10 days after castration. It is available by veterinary prescription and is approved for bull calves of up to 150 lb. The product is injected into the testes with a small needle.

CREEP FEEDING

Creep feeding is the practice of providing supplemental feed to nursing calves in a facility that prohibits the brood cow from having access to the feed (Fig. 10–3). Creep feeding is not a replacement for rapid growth potential, good milk production in cows, or improved pastures, all of which are essential in producing the quality calves sought by cattle feeders. Creep feeding consistently results in heavier calves at weaning for both spring- and fall-born calves. Creep feeding may add 24 to 50 lb to a calf's weaning weight, but those extra pounds require 10 or more lb of feed. Therefore, one must determine the cost-benefit ratio when contemplating creep feeding.

Creep feeding is likely to be profitable when the following conditions are met:

1. Dams are first- or second-calf heifers.
2. Calves are born in the fall, when grazing is not possible.
3. Pastures begin to decline in quality or quantity.
4. Calf prices are high in relation to feed prices.
5. Cows and calves are kept in confinement.

Creep feeding is not likely to be beneficial under the following conditions:

1. Price of feed is high in relation to calf prices.
2. Pasture quality and quantity remains good throughout the grazing season.
3. Dams are good milkers.
4. Calves are to be wintered on a high-roughage, growing ration.
5. Heifers are to be kept for herd replacements.

Importance of Milk Production

For the first 90 to 100 days of lactation, milk production is adequate for maximizing calf growth rate. After this period, a beef cow's milk production usually begins to decline. This decline in milk production

FIG. 10-3. Example of a creep-feeder for calves. (Courtesy of International Minerals and Chemical Corporation, Terre Haute, IN.)

plus the calf's increased nutrient requirements makes the calf more dependent on other feed ingredients. Usually, this is the time period in which creep feed should be fed to suckling calves.

Restricting Gain From Creep Feeding

Creep feeding may harm the lifetime productivity of a heifer calf if it becomes too fat. Fat is deposited in the heifer's udder and inhibits formation of milk-secreting tissue. Also, excessive weight gain may cause fat to be deposited in the reproductive tract of heifers, which may increase calving difficulty. Calves from dams that were themselves creep fed often weigh less at weaning than calves from dams that were not creep fed.

Unrestricted creep feeding may also cause overfattening of small-framed calves. In general, overfattened calves bring less profit per pound as feeder cattle, gain more slowly the first 2 to 3 months in the feedlot, and finish at lighter weights than desired. For this reason, cattle feeders may prefer calves that have not been creep fed.

Cost Considerations

Table 10-3 shows what creep feeding gains cost at various feed costs and feed conversion rates. For example, if it takes 8 lb of feed per pound of gain with feed that is priced at $5.00 per 100 lb, that extra pound of gain costs $.40, not including added equipment or labor.

A rule of thumb regarding the profitability of creep feeding is that ten times the cost of creep should be less than the price of calves because about 10 pounds of feed per pound of gain is required. If feed costs $.06 per pound, the price of calves must be greater than $.60 per pound for creep feeding to be profitable.

Creep Rations

Three major considerations must be considered when selecting creep rations:

1. Cost
2. Palatability
3. Quality of feed or nutrient content

Commercial feed companies offer excellent prepared creep feeds (often pelleted)

TABLE 10-3. Cost Per Pound of Extra Gain Achieved by Creep Feeding

Feed Needed per Pound of Extra Gain	Cost per Pound of Gain When Cost of Feed per cwt (100 lb) is:				
	$4.00	$5.00	$6.00	$7.00	$8.00
8 lb	$ .32	$ .40	$ .48	$ .56	$ .64
10 lb	.40	.50	.60	.70	.80
12 lb	.48	.60	.72	.84	.96

Adapted from Nelson, L.A. 1974. Creep feeding of calves. Cooperative Extension Service, Purdue University. W. Lafayette, IN. AS-415.

that are both palatable and properly formulated in nutrient content; however, the cost of these feeds should be compared with farm-designed rations.

Creep feeding grains should be processed. For optimum digestion and utilization by calves, shelled corn and wheat should be cracked, and barley should be processed by either rolling or flaxing. Crimping or rolling oats aids digestion but is not critical. Addition of cane molasses to creep rations at the rate of 5 lb for each 100 lb of creep decreases dust problems and increases a creep ration's palatability. Examples of creep rations are shown in Table 10–4.

Where to Place Creep Feeders

Select a site where cows lie and browse a few hours every day. These sites usually are shaded, have water available nearby, are open to prevailing winds on hot summer days, and are large enough for the entire herd to congregate. If the pasture is large, more than one site should be selected.

GROWTH-PROMOTING IMPLANTS

Growth stimulants increase the rate of gain and weaning weight of calves. Growth-promoting implants are one of the cheapest management tools used to increase the pounds of calf weaned per cow exposed to the bull.

Three implants are available for suckling calves under 400 lb. Ralgro, the trade name for zeranol, a resorcyclic acid lactone, is applied at the 36-mg level in three 12-mg ear implant pellets. The second implant is called Compudose, which contains a natural steroid, estradiol-17β. Each Compu-

TABLE 10-4. Composition of Seven Creep Rations for Beef Calves Using Normal Moisture Contents for Ingredients

Feedstuff	Ration						
	1	2	3	4	5	6	7
Cracked shelled corn	85%	—	65%	—	90%	—	38%
Oats	—	100%	35%	70%	—	—	30%
Processed barley	—	—	—	30%	—	40%	—
Protein supplement[a]	10%	—	—	—	10%	—	—
Dehydrated alfalfa pellets	—	—	—	—	—	60%	—
Soybean meal	—	—	—	—	—	—	20%
Cane molasses	5%	—	—	—	—	—	10%
Dicalcium phosphate	—	—	—	—	—	—	1%
Trace-mineralized salt	—	—	—	—	—	—	1%

[a] This refers to an all-natural protein supplement. Crude protein content is 30 to 35%. Soybean meal at 75% the amount can be substituted.

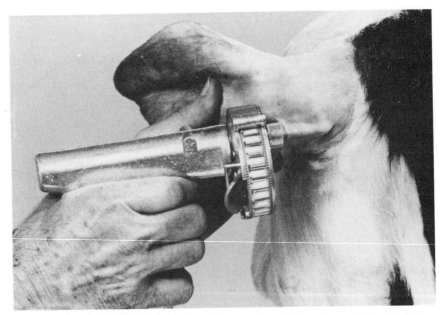

FIG. 10–4. Implanting with Ralgro. (Courtesy of International Minerals and Chemical Corporation, Terre Haute, IN.)

dose implant provides an effective daily dose of estradiol for 200 days. The most recent implant approved for suckling calves is Synovex-C. It contains 10 mg of estradiol benzoate and 100 mg of progesterone. It is approved for steer and heifer calves.

Ralgro, if implanted properly, improves the growth rate of suckling calves. (Fig. 10–4). Implanting Ralgro in cattle every 60 to 100 days has been shown to maximize growth to slaughter weight. The product is not approved for heifers or bulls retained for breeding purposes. Bull calves implanted with Ralgro have been shown to have smaller testicles at one year of age than calves not implanted with Ralgro. Calves must be implanted with Ralgro at least 65 days prior to slaughter.

Each Compudose silicone rubber implant contains 24 mg of estradiol. Figure 10–5 shows the proper implant position for suckling calves, and Figure 10–6 shows the proper position for weaned or larger calves. Implanting calves with Compudose every 200 days is recommended. Compudose is approved for suckling steers, pas-

ture-grown steers, and finishing steers. Compudose has been shown to improve average daily gain and feed efficiency of cattle. Compudose has no mandatory withdrawal time.

A summary of seven field trials conducted in five different geographic loca-

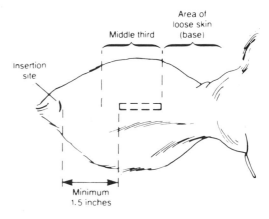

FIG. 10–5. Correct Compudose implant site for suckling calves. (Courtesy of Elanco Products Company, a division of Eli Lilly and Company, Indianapolis, IN.)

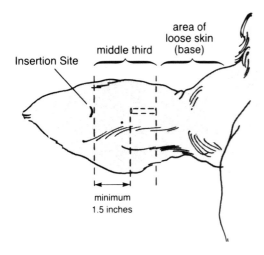

FIG. 10–6. Correct Compudose implant site for weaned calves or larger. (Courtesy of Elanco Products Company, a division of Eli Lilly and Company, Indianapolis, IN.)

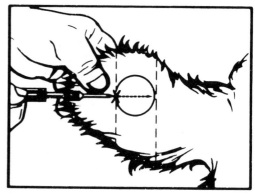

FIG. 10–7. Correct implant site for Synovex. (Courtesy of Syntex Agribusiness, Inc., Animal Health Division, Des Moines, IA.)

tions showed that calves (male and female) implanted with Synovex-C gained 8.3% more weight than non-implanted controls during a 110- to 120-day period. Synovex-C is not approved for use in veal calves, calves intended for reproduction, or calves of less than 45 days old. No withdrawal is required prior to slaughter of calves.

Other growth-promoting implants available for calves weighing 400 lb or more are Steer-oid and Synovex-S and -H. Steer-oid and Synovex-S contain a combination of the naturally occurring hormones estradiol benzoate and progesterone. Both products are for steers only. Synovex-H is a combination of estradiol benzoate and testosterone and is for heifers only. Cattle should be reimplanted every 90 to 100 days. Proper implant site for Synovex is shown in Figure 10–7.

All growth implants described in this section are intended to be inserted in the ear of the targeted beef animal. The recommended site of an implant varies with the implant used, however. The latest information concerning Food and Drug Administration (FDA) clearance of the implants just described should be obtained before one uses these products.

WEANING MANAGEMENT

Calves should be weaned at 6 to 9 months of age. Weaning at this time coincides with the weight requirements of most performance testing programs, places spring-born calves at about the right age and weight to be sold in the fall as feeder calves, and allows cows to graze forage regrowth of stalk or stubble fields so that they can regain some weight in the late fall and early winter.

Calves should be removed from the dams and kept out of sight and sound of one another. They should be provided with fresh, clean water, medium-quality hay, and 2 to 3 lb of grain per head daily. If calves have been creep fed, they should continue to be creep fed during the weaning period.

Early Weaning

Early weaning may be considered to be weaning at any time earlier than the normal weaning age. Several conditions that may make early weaning feasible are given in the following.

1. *Forage quantity or quality is inadequate to promote optimum weight gain in calves.* On many

Western ranges, calf weight gains during May and June may be as high as 2 lb per day; during July, they may be about 1.5 lb per day, and during August, 1 lb per day. Calf weight gains are low on most western ranges during September. Therefore, when supply of both feed and water becomes extremely short, early weaning should be considered. Cows gain weight when they are not nursing calves; as a result, they begin the winter in better shape. Dry, non-lactating cows need less water than those producing milk, so that dry cows range farther away from water. Calves can be weaned and fed in drylot while cows are left on open range.

2. *A higher reproductive rate for cows is desired.* Suckling delays the onset of estrus in beef cows. Several studies have shown that weaning calves at 40 to 50 days of age increases the number of cows cycling and induces estrus among cows that are too thin to breed while suckling calves.
3. *Induced twinning is desired.* At present, several researchers are working on producing multiple births, which potentially could increase the yield of beef per acre. It has been demonstrated, however, that cows with twins do not rebreed even after 100 days. Many beef cows do not produce enough milk to raise two calves and obtain maximum growth; therefore, early weaning of one or both calves may be necessary to maintain an acceptable growth rate in the calf and reproductive rate in the cow.
4. *Accelerated rebreeding of first- and second-calf heifers is desired.* Rebreeding heifers after the birth of their first calf is sometimes difficult. After calving, heifers require feed for continued growth, milk production, and maintenance and repair of the reproductive tract. When calves are weaned early, the onset of estrus is no longer delayed by the suckling reflex, and the consumed nutrients that would otherwise be used for milk production can be used for reproductive functions.
5. *Dams are old and have poor teeth or broken-down udders.* These cows may continue to produce for a number of years if calves are weaned early. The weight gain in calves produced may more than offset the cost of early weaning, which could be especially useful in the case of an outstanding purebred cow.
6. *Cows are kept in confinement.* In some areas where cattle are kept in confinement and their feed is hauled to them, it may be advantageous to wean the calves early and give a lower-quality feed to the cows.

7. *Calving takes place in the fall without winter pasture.* Permanent pastures are not adequate for cows during the winter months, when the temperature is ideal for breeding cows. Dead grass with protein supplement, however, can maintain these cows if they do not nurse calves.

Situations in which early weaning may not be feasible include the following:

1. *Costs of feed are high.* When grain prices are high and cattle prices are cheap, it probably is not profitable to wean calves early.
2. *Labor is not available to wean and feed very young calves in drylot.*
3. *Preventive medicine program is too poorly defined to control disease with early weaning.* Younger calves are less resistant to disease than other calves.
4. *Feed is plentiful, and cows are milking well.*

A weaned calf usually consumes about 3% of its body weight in feed each day. By the time the calf weighs 300 lb, it eats approximately 8 to 9 lb per day of a ration that is 50% roughage. Calves weaned early are fed in a manner that is opposite to that of feedlot cattle: they must first be given high concentrate, then later be switched to higher levels of roughage as their rumen capacity increases with size.

PRECONDITIONING

Preconditioning is the preparation of a calf that has been nursing its mother and that is destined for the feedlot to withstand the stresses associated with shipping and with adapting to a feedlot environment. It involves a complete health management program.

The following are requirements for a preconditioning program:

1. Calves should be weaned at least 30 days prior to sale.
2. All calves must be castrated and dehorned no less than 3 to 4 weeks prior to sale. The earlier these operations take place, the less stress is placed on the calves.
3. All calves must be accustomed to drinking from troughs and eating from feed bunks 30 days prior to sale or shipment, which ensures adaptation to feedlot rations and environment, and results in heavier calves and less shrinkage. If calves are to eat all at

once, 18 to 22 inches of bunk space should be provided.

4. The preconditioned calf must be vaccinated against several diseases. Most preconditioning programs require calves to be vaccinated for the following diseases 3 weeks prior to shipping:
   a. IBR (Rednose)
   b. PI_3 (Parainfluenza-3)
   c. BVD (Bovine viral diarrhea)
   d. Clostridial infections

   Vaccination programs vary somewhat from state to state. Other diseases for which vaccines can be incorporated into a preconditioning vaccination program are *Hemophilus somnus* infection, pasteurellosis, and leptospirosis.

5. Preconditioned calves must receive treatment for grubs and lice. Worming is not mandatory in most preconditioning programs.

6. All calves must be in the owner's possession for at least 60 days prior to sale.

7. Most states require individual identification of each preconditioned calf and written certification of the practices involved in preconditioning. An example of a preconditioning certificate is shown in Figure 10–8.

Feeding

Calves should be offered either rations to which they are already accustomed or rations that they can rapidly learn to eat. High-quality grass or grass-legume hay, or corn silage, should be offered along with a small portion of grain and supplement. Complete mixed rations that are specifically designed for preconditioning are available in meal or pelleted forms. The object is not to feed cattle a high level of grain, but to nourish them with energy and protein as quickly as possible to reduce stress. High levels of antibiotics or antibiotic sulfa-drug combinations (as prescribed by the feed dealer or veterinarian) may help reduce sickness. (Commercial stress supplements containing these drugs are available. Some cattlemen prefer to add these to the water if tanks or metering devices are available. *Gradual changes* from one feeding program to another reduce stress and digestive problems.

Table 10–5 contains concentrate mixtures that can be handfed with the indicated roughage. Concentrate should be fed at the rate of 1 lb per 100 lb of body weight. When calves are placed on feed in this manner, care should be taken to see that all calves have adequate bunk space. Also, the daily amount of concentrate per head should be restricted to 2 to 3 lb for the first few days until one is assured that all calves are eating. The silage or hay portions should be fed ad libitum. Silage that has been in the bunk for 24 hours should be removed. Feeding the calves small

TABLE 10-5. Composition of Concentrate Mixtures for Hand Feeding With Different Forages

Ingredient[a]	Pounds per Ton of Concentrate		
	Corn Silage	Mixed Hay	Alfalfa Hay
Corn, No. 2 shelled	1,640	1,740	1,992
Soybean meal	325	150	—
Molasses	—	100	—
Dicalcium phosphate	10	10	—
Sodium tripolyphosphate	—	—	8
Ground limestone	25	—	—
Vitamin A	4 million I.U./ton	4 million I.U./ton	4 million I.U./ton

[a] Salt should be provided on a free-choice basis.

Expected daily roughage intake for a 400-lb calf: 15 to 18 lb corn silage, 5 to 6 lb mixed hay, and 5 to 7 lb alfalfa hay.

THE IOWA VETERINARY MEDICAL ASSOCIATION

PRECONDITIONING CERTIFICATE

Date Of Sale _____

Owner _____

Address _____

Buyer _____

Address _____

Date Issued _____

_____ breed _____ age _____ # steers _____ # heifers _____

have been preconditioned as follows:

	Date	Product Name or Mfg.	SERIAL NUMBER (vaccine)
CLOSTRIDIAL GROUP-CSNS, C&D	_____	_____	_____
IBR · PI₃	_____	_____	_____
HAEMOPHILUS SOMNUS BACTERIN	_____	_____	_____
Second Injection (Optional but recommended)	_____	_____	_____
BVD	_____	_____	_____
Second Injection (Optional but recommended)	_____	_____	_____

MANDATORY Must be 4 Months of Age

MANDATORY · 3 WEEKS PRIOR TO SHIPMENT

		By Whom	Method Used
CASTRATED	_____		
DEHORNED	_____	_____	_____
GRUB TREATED · Aug. 1 · Nov. 15	_____	_____	_____
After Nov. 15, Lice Treated	_____	_____	_____

The animals identified by IVMA PC Tag # _____ _____ _____

MANDATORY FOR ISSUANCE OF CERTIFICATE

WEANED (DATE) _____

Ration During Preconditioning Period _____

I CERTIFY: Weaning 30 Days
Ownership of 60 Days   _____
                                        Owner Signature

	Date	Product Name or Mfg.	SERIAL NUMBER (vaccine)
PASTEURELLA	_____	_____	_____
LEPTO	_____	_____	_____
BOVINE RESP. SYNCYTIAL VIRUS	_____	_____	_____
WORM TREATMENT	_____	_____	
IMPLANT OF CHOICE Steers ☐ Heifers ☐	_____	_____	Method _____
OTHER	_____	_____	_____

OPTIONAL

To the best of my knowledge this is an accurate statement of the procedures performed on these calves. This is not to be construed as a health certificate.

CERTIFIED BY _____

Recommend that local veterinarian examine certificate and cattle at time of arrival.

FIG. 10–8.   Preconditioning certificate. (Courtesy of Iowa Veterinary Medical Association.)

amounts several times a day is desirable, for it keeps the feed fresh and encourages the calves to eat.

Economics

Most surveys report that preconditioned calves bring $3 to $7 per 100 lb more than normal calves. A cattleman can more easily receive a premium for preconditioned calves if he sells them through a sale specifically for preconditioned calves. Preconditioning programs must be tailored to the buyers. Before selecting a preconditioning program, one must determine the costs versus returns. Preconditioning costs usually range from $20 to $30 per calf, with about 50% of the costs attributable to feed and 15 to 20% attributable to labor. The costs of veterinary products and processing prices range from $5 to $7 per head.

REFERENCES

Bagley, C.V., and N.J. Stenquist. 1981. Dehorning, castration, and branding. In Cow-Calf Management Guide. Cattleman's Library. Cooperative Extension Service, University of Idaho, Moscow, ID. CL-750.

LeViness, E.A. 1982. Vital signs in animals—What stockmen should know about them. In Cow-Calf Management Guide. Cattleman's Library. Cooperative Extension Service, University of Idaho, Moscow. ID. CL-610.

Nelson, L.A. 1974. Creep feeding of calves. Cooperative Extension Service, Purdue University. W. Lafayette, IN. AS-415.

Strum, G.E., and I.A. Schipper. 1975. Tips for better selling calves. Great Plains Beef Cow-Calf Handbook. Oklahoma State University, Stillwater, OK. GPE-3051.

Selected Readings

Absher, C., and N. Gay. 1979. Conditioning cattle for growing or finishing. Cooperative Extension Service, University of Kentucky. Lexington, KY. ASC-22.

Burris, R. 1982. Growth stimulating implants for beef cattle. Cooperative Extension Service, University of Kentucky. Lexington, KY. ASC-25.

Hall, R.F., P.J. Shouth, and D.G. Waldhalm. 1982. Management in reducing newborn calf disease. In Cow-Calf Management Guide. Cattleman's Library. Cooperative Extension Service, University of Idaho, Moscow, ID. CL-646.

Lincoln, S.D., and D.D. Hinman. 1981. Preconditioning of calves. Cooperative Extension Service, University of Idaho. Moscow, ID. CIS-573.

Lusby, K.S., and R.P. Wettemen. 1980. Effect of early weaning calves from first calf heifers on calf and heifer performance. Animal Science Res. Report, Oklahoma State University. Stillwater, OK.

Mankin, J.D., and W.D. Frischknecht. 1982. Creep Feeding. In Cow-Calf Management Guide. Cattleman's Library. Cooperative Extension Service, University of Idaho, Moscow, ID. CL-335.

Minyard, J.A., J.H. Bainley, and F.W. Crandall. 1982. Preconditioning feeder calves. South Dakota State University, Cow-Calf Day, Dec. 8, 1982. Chamberlain, SD.

Nelson, L.A., W.L. Singleton, K.S. Hendrix, and R.L. Morter. 1982. Management of the beef calf crop. Cooperative Extension Service, Purdue University. W. Lafayette, IN. AS-397.

Strohbehn, D.R. 1976. Creep feeding, profit or paradox? In Creep Feeding Special Report. Shield Pub. Co., Lindsbury, KS.

QUESTIONS FOR STUDY AND DISCUSSION

1. Outline five management procedures to reduce calf mortality.
2. Define colostrum. Why must calves consume colostrum as soon as possible after birth?
3. Describe how calves with scours should be handled to minimize death loss.
4. What is the normal temperature range for beef cattle?
5. Discuss the advantages and disadvantages of various methods of castration and dehorning.
6. Define creep feeding.
7. When may it be profitable to creep feed calves?
8. What are some of the potential problems that can develop with creep feeding?
9. What factors should be considered when selecting creep rations?
10. List two growth-promoting implants that are approved for use in beef cattle. What is the active compound in each that promotes increased growth?
11. At what age are most calves weaned?
12. Define early weaning. Describe several conditions in which early weaning may be feasible.
13. Outline the requirements for a preconditioning program.

Chapter Eleven

Pasture and Range Management

Forage refers to the vegetative portions of plants that are consumed by animals. Forages include pasturelands, rangelands, forest range, and forage crops. Forage provides greater than 80% of all feed units consumed by cattle (Table 11–1). More than one billion acres of land area in the United States, consisting of grasslands, shrublands, and open forest are grazed by livestock (Fig. 11–1). Forage provides a relatively inexpensive source of energy, protein, vitamins, and minerals. The demand for forage is expected to increase in the future. Much of the forage land, however, is currently producing less than its potential, and therefore, stricter cultural and management practices are needed to increase production.

NEW PASTURE AND/OR RANGE SEEDINGS

Prior to revegetation, management practices must be critically reviewed to correct the misuse that made reseeding necessary. Only after this examination has been completed and new management techniques have been acquired should seeding be planned. Planning should first consider the profit versus cost to protect against loss of, or insubstantial profit from, dollars in-

Appreciation is expressed to Dr. M. K. Petersen, who while a graduate student at the University of Idaho made a significant contribution to the preparation of this chapter. Dr. Petersen is currently Assistant Professor of Animal-Range sciences at Montana State University.

vested. In some situations, reseeding may not be an acceptable alternative to increasing forage productivity. Not all range or pasture can be seeded successfully.

Usually, most seedings consist of a grass-legume blend or a single species. If a grass-legume blend is considered, the compatibility of components should be reviewed with the following in mind:

1. Similar maturity dates (especially important for harvested forages)
2. Equal height
3. Equal competitiveness
4. Same palatability.

A grass-legume mixture can be advantageous over a pure grass or legume stand for the following reasons:

1. It eliminates the need for nitrogen fertilizer because the legume in the mixture provides nitrogen for itself and for grass growth.
2. It lengthens the life of the pasture or hayland because the grass remains after the legume stand is reduced. A legume can be reintroduced by pasture renovation, if desired.
3. It reduces the problem of legume "heaving." This is the process whereby legumes are raised from the soil surface by freeze-thaw action in the late winter and early spring and plant damage occurs.
4. It reduces soil erosion on steep slopes. Grasses have a more massive root system and are better for soil conservation purposes than pure legume stands.
5. It improves livestock performance. Animal gain and beef breeding performance can be more improved by a grass-legume mixture than by a pure grass stand, especially when

TABLE 11-1.   Use of Feedstuffs by Livestock and Poultry in the United States (1978)

Animal Type	Proportion of Ration	
	Concentrate (%)	Forage (%)
All livestock and poultry	35.7	65.2
All dairy cattle	38.8	61.2
Beef cattle on feed	72.4	27.6
Other beef cattle	4.2	95.6
All beef cattle	17.0	83.0
Sheep and goats	8.9	91.1
Hens and pullets	100.0	0.0
Turkeys	100.0	0.0
Broilers	100.0	0.0
Hogs	85.3	14.7
Horses and mules	27.8	72.2

Adapted from Council for Agricultural Science and Technology. 1980. Food From Animals: Quantity, Quality and Safety. Ames, IA. Report No. 82.

the grass is tall fescue. The mixture can also reduce animal performance problems associated with bloat, grass tetany, and fescue toxicosis.

Only certified seed should be used. Although other seeds may cost less, certified seed guarantees purity, germination, qual-

FIG. 11-1.   The beef cow—a harvester of forage.

ity, and freedom from unwanted seeds. A cattleman should consult his county extension agent for those varieties recommended for his location.

Spring seeding is usually recommended, but in those areas that receive less than 12 inches of rainfall, a late fall (dormant) seeding may be best. The seeding must be done late enough, however, so that germination does not occur until spring. In irrigated regions that have a long growing season, seed may need to be planted as late as mid-August or early September so that seedlings become established before the winter-kill.

Legume seeds should always be inoculated with the proper strain of nitrogen-fixing bacteria. Different strains are available for different legumes. For best results, the seed should be inoculated immediately before planting, and direct sunlight on the inoculum should be avoided.

Prerequisites for successful forage establishment:

1. Selection of varieties adapted to local climatic and soil conditions
2. Use of high-quality seed
3. Proper season selected for planting
4. Uniform seed distribution, placement, and rate of planting
5. Firm seed bed
   a. Good soil contact
   b. Avoidance of moisture loss
6. Adequate soil nutrient and soil amendments
7. Adequate soil moisture
8. Minimum competition from companion crops or weeds
9. Protection from diseases and insects
10. Careful management during first growing year

PRINCIPLES OF ESTABLISHED PASTURE MANAGEMENT

The production of forage, either pasture or range, is controlled by the environment (temperature, moisture, light, etc.). These factors seldom are similar for any 2 years. Therefore, forage growth varies from year to year. Knowledge of the forage-environment response is essential for the rancher to determine management of grazing plants; seldom can such management be based on a calendar date.

Legumes and grasses store energy as readily available carbohydrate in plant crowns and roots. These energy reserves are utilized for initial spring growth and regrowth after forage removal (grazing or cutting). In addition, carbohydrates are required to resist cold and heat, to support life during dormancy (drought or freeze), and to promote flower and seed formation.

Maintenance of a required minimum energy reserve is essential for plant life. Forage plants go through periods of carbohydrate storage and usage. This cyclic pattern revolves around early growth and maturity. The cyclic pattern of a typical forage plant is shown in Figure 11–2.

At the beginning of spring growth (also after cutting or grazing), the carbohydrate reserves are utilized for growth, and energy storage begins after 23 weeks with legumes, or after stem elongation (48 weeks for some species) in grass species.

Cutting or grazing when plant energy reserves are at a low level may leave the forage with insufficient energy for regrowth. Thus, grazing when energy levels are low decreases the vigor and productivity of the stand, and continued removal exhausts and eventually kills the plant. Also, plants weakened by too early, too heavy, or continuous removal cannot withstand drought, heat, cold, insects, or invading plants.

As a plant matures, its ability to maintain productivity and vigor after removal increases. Allowing forage to mature, however, is not compatible with maximum nutrition for livestock. As maturity progresses, fiber increases and protein decreases. Therefore, a compromise must be made between the quality of the feed and the energy level required by the plant for rapid regrowth and stand vigor.

The majority of research indicates that plant damage is greatest when plants are grazed during the early heading stage. One should remember that cattle regraze the same plant approximately every 7 to 9 days; that most forage species require 30 to 40 days to recover from defoliation; that the recovery period may be longer depending on the amount of leaf area removed; and that the best management technique is time-control grazing, to avoid the period of greatest susceptibility or to control the interval between bites.

To maintain a vigorous forage, most grasses should be allowed to grow until early boot stage to ensure plant energy re-

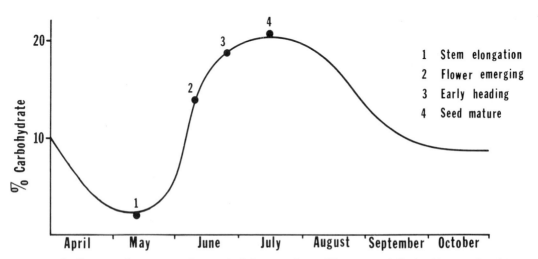

FIG. 11-2. Cyclic growth pattern of a typical forage plant. (Courtesy of Kathy Dawes Graphics, Moscow, ID.)

TABLE 11-2. Characteristics of Forage Species

Species	Longevity	Palatability	Bloat Hazard	Drought Tolerance	Winter Hardiness	Cool (C) or Warm (W) Season	Use and Comments
Legumes							
Alfalfa	Perennial	VH	Yes	G	G	C	Well-drained, mod. deep to deep fertile soils; hay, silage, pasture.
Alsike clover	Perennial	H	Yes	F	G	C	Withstands wet, acid, poorly drained soils, pasture, hay.
Ladino or white clover	Perennial	VH	Yes	P	F	C	Well-drained soils of acid to weak salinity, season-long moisture; pasture only.
Hairy vetch	Winter annual	H	Yes	F	F–P	C	Included with small grains for forage; can be used as a winter cover crop in some areas.
Birdsfoot trefoil	Perennial	VH	No	F	G	C	Moist lowland, fertile, mod.- to fine-textured soils; slow establishment.
Red clover	Perennial	H	Yes	F	G	C	Fertile, well-drained, shallow soils; rapid establishment; excellent pasture renovation crop.
Sweet clover	Annual or biennial	M	Yes	G	G	C	Tolerant to frost; mod. shallow to deep soils; imperfectly drained to non-fertile soil; hay, pasture, silage.
Lespedeza, striate	Summer annual	H	No	F	—	W	Late summer pasture or hay; must reseed itself annually to persist.
Grasses							
Kentucky bluegrass	Perennial	VH	—	P	G	C	Well-drained fertile soils; rapid regrowth; slow growth when hot; excellent quality but low-yield pasture.
Smooth bromegrass	Perennial	VH	—	G	G	C	Well-drained fertile soils; tolerates wet soils; slow recovery after grazing; pasture, hay.
Bermuda grass	Perennial	M	—	G	P	W	Moderate, well-drained soils with adequate moisture; pasture, hay.
Big bluestem	Perennial	H	—	G	G	W	Moist, well-drained soils, high fertility; pasture, hay.
Reed canary grass	Perennial	L	—	G	G	C	Wet, weakly salty, fertile, med. to deep soils; pasture (immature stages), hay.

Species							Characteristics
Tall fescue	Perennial	M	—	F	F	C	Tolerates low to mod. wet soils of mod. salinity, weak acidity.
Orchard grass	Perennial	M–H	—	F	F	C	Well-drained, salt-free, fertile soils; fast recovery; pasture, hay.
Timothy	Perennial	H	—	P	G	C	Adapted to cool, humid climates; used in pasture mixtures with legumes; pure stands for pasture.
Perennial ryegrass	Perennial	H	—	P	F	C	Pasture (quick establishment); requires soils of med. to high fertility; predominately W. of Cascades and Sierra-Nevada mountains.
Crested wheatgrass	Perennial	M	—	G	G	C	Good seedling vigor; med. salt tolerance; good early season forage; predominately on W. range.

VH = very high
H = high
M = medium
L = low

G = good
F = fair
P = poor

serves and adequate nutrition. In addition, grass should be allowed to build up energy reserves in the fall before the first killing frost (so that the dead top growth may be grazed after the freeze). This practice ensures adequate energy for rapid spring growth.

FORAGE SELECTION GUIDELINES

Selecting the appropriate forage for silage, hay, or pasture is an important decision that needs to be made before the crop is sown. This decision is as critical as selecting the appropriate variety within a forage species.

Soil type, crop use, and moisture requirements influence the forage selection. Legumes are usually better adapted to an area with medium-textured soils where annual precipitation averages 14 inches or more. Grasses have a wider adaptation to soils and precipitation. Some grasses (Kentucky bluegrass, orchard grass, tall fescue) require more than 22 inches of annual precipitation or season-long irrigation for good performance. Other grasses such as crested and bluebunch wheatgrass persist and produce economic forage yields in cooler areas (less than 140 frost-free days, 32° F) with 9 inches of annual precipitation. Table 11–2 gives the characteristics of some of the more common forage species.

GRAZING SYSTEMS

Maintenance or enhancement of the target plant community is the goal of any pasture or range management plan. The accomplishment of this goal rests on control of the frequency of "bite" on each plant. Most systems, such as short duration, rest rotation, or deferred rotation, cannot accomplish this control. Instead, they are designed to spread out utilization over as much of the plant community as possible and then provide as much rest as possible during successive growing seasons. This lower management input has certain drawbacks or costs: it results in forage that feeds fewer animals, and that has a lower nutritive quality and limited flexibility.

Time-control grazing concentrates animals in small paddocks to achieve uniform forage use, then bases the length of the grazing period on plant growth rate. This method allows considerable flexibility, high-quality forage, and maximum control of bite frequency.

Continuous Grazing

Under this system, animals are allowed to graze one area for the entire grazing season. This system must include regulation of animal numbers to maintain an adequate forage stand. Kentucky bluegrass and white clover are two forage species that are best adapted to withstand continuous grazing without a decline in pasture vigor.

In many cases, total potential production per acre is less with continuous grazing than with other systems. Stocking rate is often based on forage availability at or near the point of the plants' lowest growth rate. This often results in undergrazing and wasted forage resources during times of rapid forage growth. If excess forage is not harvested as hay when grazing is insufficient, forage quality declines.

Rotational Grazing

A rotational grazing system allows pasture to be grazed during optimum yield and quality while forage regrowth is occurring in a previously grazed pasture or paddock. This system works best in areas receiving greater than 20 inches of annual precipitation, or in irrigated pasture in drier regions. A method for determining the number of equal area paddocks needed for a particular operation is illustrated as follows:

$$\text{Pasture units} = \frac{\text{Time for grass regrowth (days)}}{\text{Length of grazing period (days)}} + 1$$

Livestock should first be turned out when the grass is 8 to 12 inches high (depending on variety) and rotated when the grass is 4 to 5 inches high if the animal population is dense. Cattle grazing under this system are forced to spread over the entire area and use all available forage uniformly. Trampling is reduced because feed is relatively abundant in a small area. The number and size of the paddocks determine the number of cattle grazed and the regrowth period for each paddock.

In rotational systems (especially in irrigated pasture), the lush, highly nutritional forage should be utilized in the spring of each year. A producer may prefer to open a paddock and utilize this forage during this period. Once the rotation has been established, a system called "top grazing" can be utilized. This method allows yearling steers, first-calf heifers, or calves to graze the best forage selectively during the first half of a paddock grazing period (days 1, 2, 3, and 4), followed by grazing by the cow herd for the last half of the grazing period (days 4, 5, 6, and 7). This system allows animals with the highest nutritional requirements to consume the most nutritious feed. Figure 11–3 is an example of a rotational grazing system.

Deferred Rotation

Under this system, grazing on one portion of the range or pasture is deferred until seeds are allowed to mature. Usually, a paddock is deferred for 1 or 2 years in areas of the United States that have limited precipitation. In the first year, the seed matures, and in the second year, it germinates and is protected from grazing until the seedlings are established. An example of a simple deferred rotation grazing system is illustrated in Figure 11–4. A regular rotation can be operated among non-deferred paddocks if the operator desires.

Rest Rotation

Rest rotation is similar to deferred rotation, except that one paddock is left ungrazed for the entire year. Thus, the paddocks actively grazed are used more heavily, and the unused portion has a

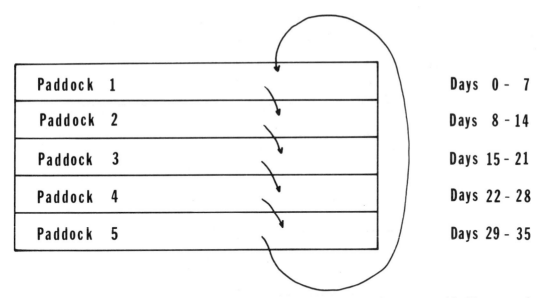

Paddock 1	Days 0 - 7
Paddock 2	Days 8 - 14
Paddock 3	Days 15 - 21
Paddock 4	Days 22 - 28
Paddock 5	Days 29 - 35

FIG. 11–3.   Rotational grazing system (5 paddocks, 7 days grazing: 30 days regrowth). (Courtesy of Department of Animal Sciences, University of Idaho, Moscow, ID.)

longer rest period. Figure 11–5 illustrates a rest rotation system.

The primary rest rotation system used in the Western United States is known as the "Hormay system," which uses five pastures in a 5-year sequence. Each paddock is systematically grazed and rested so that it provides for production of livestock and other resources of value simultaneously. Maintenance and improvement of the resources are accomplished almost entirely by timely resting of the range from use. The purposes for resting the range after use are:

1. To allow the plant the opportunity to make and store energy (recover vigor).
2. To allow seed to mature.
3. To allow seedling to become established.
4. To allow litter to accumulate between plants.

An example of the use of the Hormay system on one paddock over a 5-year period is shown in Figure 11–6.

Short-Duration, High-Intensity Grazing

This system uses a high stocking rate for a short period, interspersed with short rest periods. The pastures in this system are arranged in grazing cells. A "cell" is any set of paddocks being managed for the same purpose. Paddock numbers are based on achieving 30 days rest during rapid plant growth and 60 to 90 days rest during slow growth.

A layout that has been used successfully contains a central point that consists of the watering sites, corral system, and all handling facilities, with the pastures radiating out from this center in a kind of a "wagonwheel" fashion. Additional fences or paddocks can be made with no need to change or improve handling centers.

Usually, as the number of paddocks increases, paddock size gets smaller, stock density increases, grazing periods become shorter, and dung and hoof distribution and hoof action become more uniform. Also, intervals per paddock become shorter, and plant production improves.

Movement of cattle to a new paddock is determined by nutrient and feed requirements of the livestock involved. Stock can be moved to new pastures more frequently during periods of high nutrient requirement, but less often during periods of low

PADDOCK A   **PADDOCK B**

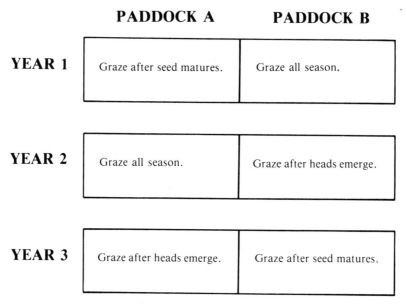

FIG. 11-4. Deferred rotation system. (Courtesy of J. Popa, Bozeman, MT.)

nutrient requirement; thus, available forage is not wasted. This system requires a
high degree of both grassland management and livestock management skills.

Strip Grazing

Strip grazing is an intensive form of rotational or short-duration high-intensity

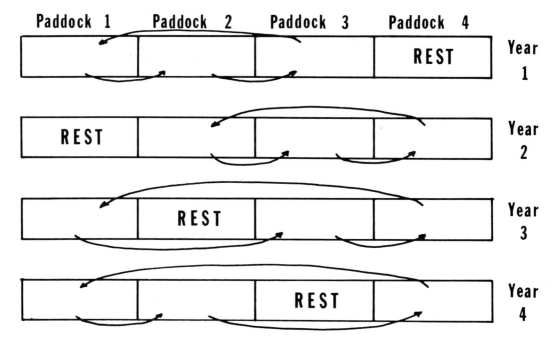

FIG. 11-5. Rest rotation system. (Courtesy of Department of Animal Sciences, University of Idaho,
Moscow, ID.)

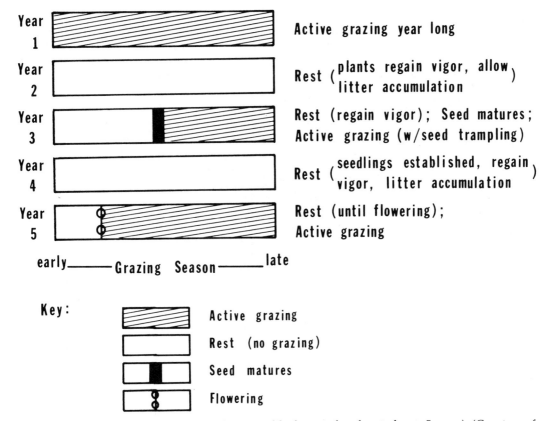

FIG. 11-6. Hormay rest rotation system (same paddock grazed and rested over 5 years). (Courtesy of Department of Animal Sciences, University of Idaho, Moscow, ID.)

grazing in which forage on a limited area is grazed down rapidly and cattle are moved to new areas every one to two days. This system is similar in principle to short-duration, high-intensity grazing, but is usually restricted to irrigated pastures or high-quality pastures in areas where annual precipitation is higher than on native range.

A portable electric fence is usually used to section off an area containing forage for one to two days. More labor and management are required, but total production can be increased through an increase in stocking rate and more complete utilization of forage. Figure 11–7 illustrates one type of strip grazing.

Grazing System Selection

The choice of grazing system for a particular operation depends upon the following:

1. Climate and length of growing season
2. Type of livestock
3. Number of livestock
4. Forage species
5. Intensity of forage management
6. Health and vigor of forage plants
7. Alternate sources of forage at either end of grazing season
8. Previous use of pasture

Before choosing a system, one should be sure that it fits the situation. Vegetation management objectives should be at least

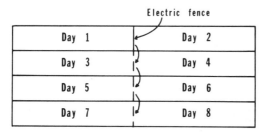

FIG. 11–7.   Strip grazing. (Courtesy of Department of Animal Sciences, University of Idaho, Moscow, ID.)

equal in priority to the livestock objectives. The costs of the system for development and maintenance should be considered. Watering areas should be developed, and salt should be used, to encourage use of lightly grazed areas. The system cannot be expected to do everything, however. Good management is needed to solve problems.

When rotational and continuous grazing are compared under conditions of equal grazing pressure and forage maturity, one is likely to find that continuous grazing results in the best animal performance. This is attributable to the greater amount of selective grazing allowed. Selective grazing produces areas of overused forage adjacent to areas with unused forage. Pasture condition declines, and nutrient intake is reduced. A rotational system forces use of unused portions of the forage base with a corresponding depression in animal productivity. In the long run, however, one frequently observes that a higher output per land area is realized, owing to forage management (better stand vigor, better health, etc.). Thus, with moderate to heavy grazing and good management, time-controlled grazing may possibly carry a greater number of animals.

NUTRITIVE INTAKE OF GRAZING ANIMALS

The effective management of forage and animal resources requires an understanding of the nutrients available in forage in relation to the animals' nutrient requirements. Once the nutrient requirements

have been established for a specific level of production, the nutrients in forage must be determined.

The nutritional value of forage is influenced by forage species and type, season of the year, soil fertility, stage of growth, method of sampling for analysis, selective grazing, grazing management, and dry matter intake. An understanding of the problems associated with determining the nutritive value of forage helps in making a reasonable estimate.

Stage of growth is the most important factor affecting nutrient content of grass. Cool season grasses (e.g., Kentucky bluegrass and orchard grass) are typically high in protein during the peak growing season in the spring, but are only moderately high in energy. Moisture content of the grass is high during early growth. Protein content and energy content drop rapidly with increasing maturity, and growth during the summer months is slow. Cool season grasses again grow as fall approaches with adequate moisture. Protein and energy levels improve but not to the levels attained in the spring.

Spring growth occurs later with warm season grasses (e.g., Bermuda grass and big bluestem) than with cool season grasses—in June and July versus April and May). In general, warm season grasses are not as high in protein content during their peak growth as are cool season grasses. The protein level falls progressively through the fall, and in the dormant state may contain only 3 to 4% protein. Total digestible nutrient content, which peaks during May and June, falls as low as 40% in the dormant state.

In general, as forages mature, nutrient content decreases, and fiber and lignin increase (Fig. 11–8).

The decline in nutrient content that occurs with increasing maturity may be explained as follows (Beaty et al., 1982). Forages are made up of green leaves, dead leaves, and stems. During early growth, the grass plant is almost entirely composed of green leaf, which is high in protein,

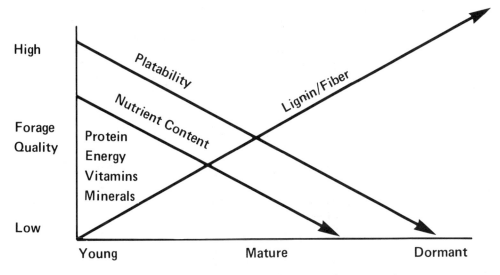

FIG. 11-8. Stage of plant growth relating forage quality to nutrient content.

energy, calcium, and phosphorus. The green leaves may be called "nutrient accumulators" and have a digestibility of about 70%. After the plant puts out its seedhead, the leaves become "nutrient dissipators," with mature leaves being sacrificed to provide nutrients for stem and seedhead growth. Although top leaves may still be green, stems and dead leaves accumulate rapidly, and forage digestibility declines. The mature stems are approximately 35% digestible. The dead leaves become a fibrous residue with a digestibility of less than 45%. Table 11–3 lists the nutrient composition of some common forage plants at three stages of growth.

Forage quality has a dramatic effect on *dry matter intake*. In general, as maturity increases, the grazing animal is limited in the amount of forage it can consume, owing to distension of the rumen (Fig. 11–9). As a forage becomes more mature, fiber content increases, which increases the bulk of the diet and physically impairs the ability of the animal to consume more feed. A feedstuff of low energy per unit volume such as barley straw would fall on the extreme left-hand side of the chart while a feedstuff of high energy content would fall on the extreme right-hand side. As feedstuffs increase in nutrient digestibility, more pounds of dry matter are consumed until the animal meets its energy needs. The resulting increase in energy consumption and digestibility produces an increase in average daily gain. Table 11–4 gives approximate values for forage dry matter intake by beef cattle.

The information in Table 11–4 can be used to estimate the amount of forage and TDN that an animal can consume daily. This estimate is important when balancing fall and winter rations with supplemental feed for increased animal performance. Chapter 12 discusses specific supplementation programs for backgrounding and stocker programs.

HEALTH PROBLEMS ASSOCIATED WITH GRAZING PASTURE AND RANGE

Toxic Compounds

ALKALOIDS

Five to ten percent of forage plants contain one or more alkaloids. Larkspur, lupine, locoweed and death camas contain

TABLE 11-3.　Nutrient Composition of Some Common Forage Grasses

Grass	Stage of Growth[a]	Nutrient Content (%) in Dry Matter			
		Crude Protein	Crude Fiber	Calcium	Phosphorus
Bluestem	Young	11.0	28.9	0.63	0.17
(Andropogon spp.)	Mature	4.5	34.0	0.40	0.11
	Dormant	3.1	35.6	0.30	0.08
Crested wheatgrass	Young	23.6	22.2	0.46	0.35
	Mature	9.8	30.3	0.39	0.28
	Dormant	2.5	32.2	0.22	0.06
Intermediate	Young	12.0	28.0	0.22	0.13
wheatgrass	Mature	4.2	33.0	0.285	0.05
	Dormant	3.2	39.0	0.29	0.03
Kentucky	Young	17.3	25.1	0.56	0.47
bluegrass	Mature	9.5	32.2	0.19	0.27
	Dormant	3.3	42.0	—	—
Reed canary	Young	19.1	28.9	—	—
grass	Mature	7.8	33.1	0.27	0.20
	Dormant	—	—	—	—
Side oats	Young	10.76	—	0.59	0.16
grama	Mature	6.02	—	0.58	0.08
	Dormant	5.45	—	0.51	0.06
Switchgrass	Young	10.96	—	0.34	—
	Mature	6.26	—	0.25	—
	Dormant	—	—	—	—

[a] Young—Period between ⅓ and ⅔ of growth before heading out; immature stage. Mature—Stage at which the plant would normally be harvested for seed. Dormant—Stage at which plants have been cured on the stem, the seeds have been cast, and weathering has taken place.

toxic alkaloids. Alkaloids usually affect the nervous system, and animals respond with altered respiration, excitation, incoordination, and possibly death from blockage of the central nervous system. Alkaloid content is not much affected by environmental conditions.

GLYCOSIDES

Glycosides are compounds that combine harmless sugars with some toxic compounds. Cyanogenic glycosides, which contain cyanide, are present in arrow grass, sorghums, chokecherry, and sometimes mountain mahogany. The toxicity of these plants is influenced by environmental effects. When sorghum is stressed (frosted or wilted from drought), prussic acid (hydrocyanic acid) is liberated in the plant tissue. Prussic acid motivates an enzyme response for linking atmospheric oxygen with animal cells. Consequently, when cattle consume stressed sorghum forage, no oxygen is transported to the cell, and the animal suffocates at the cellular level. Signs of prussic acid poisoning are nervousness, abnormal breathing, muscle trembling, blue discoloration of the lining of the mouth, and spasms or convulsions continuing at short intervals until death.

Mineral Poisoning

NITRATES

Nitrates tend to be high in some invader species such as pigweeds, lamb's-quarters, Russian thistle, and kochia. High nitrate levels may also occur in annual forage species when stressed (hail-damaged corn silage; drought-stricken oat forage). Plants containing 1.5% of potassium nitrate

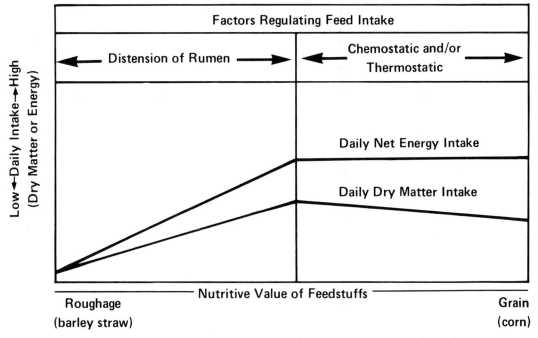

FIG. 11–9.   Regulation of feed intake as affected by the dietary nutritive value. (From Montgomery, M. J., and B. R. Baumgardt. 1965. Regulation of food intake in ruminants. II: Rations varying in energy concentration and physical form. J. Dairy Sci. *48:* 1623.)

(KNO_3) may be lethal to livestock. Nitrates are converted to nitrites by microorganisms in the rumen. Nitrites inhibit the ability of red blood cells to carry oxygen and result in asphyxiation. Nitrates cause red oxygen hemoglobin to be converted to a chocolate-brown methemoglobin, a color that indicates nitrate poisoning.

SELENIUM

Selenium is a mineral found in excessive amounts in certain soils derived from cretaceous marine shales. Most plants growing in selenium soils passively take up some selenium and deposit it in plant parts grazed by livestock. Certain plants are known to accumulate high amounts of selenium in their tissues (*Astragalus*, milk vetch; *Xylorhiza*, woody aster; *Oonopsis*, goldenweed; *Stanleya*, prince's plum).

Acute selenium poisoning occurs when livestock graze plants that have accumulated 250 or more parts per million (ppm) of selenium. Death may occur within an hour

to several days following ingestion of a lethal dose. Progressive signs are uneasiness, watery diarrhea, abundant urinary excretion, unconsciousness, and death.

One type of chronic poisoning is often called *blind staggers* and results from the continued intake of plants containing less than 200 ppm selenium. Clinical signs include reduced animal condition, impaired eyesight, wandering, loss of appetite, and paralysis.

A second type of chronic poisoning is called *alkali disease*. It results from contin-

TABLE 11–4.   Estimates of Forage Dry Matter Intake and TDN Content

Forage Type	Percentage of Body Weight (%)	TDN (%)
High quality	2.5	55–60
Medium quality	2.0	50–55
Low quality	1.5	45–50

ued intake of range grasses, cereal crops, or hays with low amounts of selenium (10 to 30 ppm) over several weeks or months. Signs of the disease are dullness, emaciation, lameness and loss of long hair, and partial sloughing of hooves.

Recommendations for managing livestock in areas of chronic selenium poisoning include the following:

1. Change to a steer program where feasible.
2. Check bulls frequently for lameness.
3. Use earlier breeding program for cows where feasible.
4. Graze selenium range in winter and fall, when plant content of selenium is lowest.
5. Rotate livestock between seleniferous and selenium-free range to lower selenium accumulation in plant tissues.
6. Stock lightly or moderately to allow selective grazing.

Bloat

Bloat in ruminants is a complex problem and has been observed in a variety of circumstances. Bloat is most commonly seen in association with the feeding of succulent legumes. This condition occurs when gas produced in the rumen (carbon dioxide and methane) cannot be released by the normal eructation process due to its being trapped in a stable foam (similar to rising bread). The left side of the rumen wall becomes distended, owing to the gas accumulation, and eventually, the animal dies if relief is not given. Although it has not yet been established, suffocation is currently believed to be the actual cause of death.

Usually, several hours pass between the beginning of pasture grazing and the death of the animal. As bloat progresses, breathing becomes labored, and there may be profuse salivation. Rumen motility can usually be detected until the animal's condition is critical. Beyond this point, the animal may begin to stagger, vomit, and eventually collapse. Death follows soon after collapse.

Treatment of a bloated animal requires quick reduction in ruminal pressure, which can be accomplished by insertion of a stomach tube (which is not always successful) or trocarization of the rumen. The trocar should be left in the animal until all danger is past. Defoaming agents can also be administered. At least 1 pint of readily available and useful defoaming agents such as vegetable oil (peanut, corn, soybean, etc.) should be administered. Cresol, turpentine, and formaldehyde have been recommended; however, they may cause off-flavors in milk and in lactating cows.

Recommendations to prevent bloat are as follows:

1. Maintain pastures that do not exceed 50% legumes.
2. Practice strip grazing, which induces whole plant consumption.
3. Feed at least 10 lb/head of a dry fibrous hay before grazing legume-rich pastures.
4. Administer 100 to 200 mg oral antibiotics (procaine penicillin) every 24 to 48 hours. (This is not often recommended, however.)
5. Administer polyoxypropylene (poloxalene), which can be added to water, supplied as a top dressing, or fed in mineral blocks.

Grass Tetany

Grass tetany is a metabolic disorder caused by low blood magnesium levels. It occurs frequently in the spring, and less frequently in the fall, when lactating cows are grazing lush green pastures. Although tetany is more prevalent in lactating cows, it can occur in dry cows.

CLINICAL SIGNS

Quite frequently, clinical signs are not observed prior to death. Affected animals may become excitable, displaying a wild stare with erect ears and appearing to be blind. They are uncoordinated and tend to lean backward or stumble.

An animal affected often has trembling muscles and grinding teeth, followed by violent convulsions, deep coma, and death. Cows often resemble milk fever cows and may have low blood calcium as well as low magnesium.

Positive diagnosis is a problem. If tetany is suspected, it is desirable to obtain blood samples from several animals within the herd. A normal blood magnesium level is 2.25 mg/100 ml of serum. The level in affected animals is usually less than 1.0 mg/100 ml.

PREDISPOSING CONDITIONS FOR OCCURRENCE

Grass tetany is most likely to occur in April and May, although it does occur in the fall and winter. Cows nursing calves under 2 months of age are most vulnerable to the disease, which is more prevalent in beef animals than dairy animals. It occurs most often in cloudy, rainy, windy weather with daytime temperatures between 40 and 60°F. The animals' low energy intake due to the high moisture content of lush green pastures has been shown to increase the incidence of tetany in herds. In addition, high nitrate levels in forages and high levels of ammonia in the rumen have been shown to interfere with magnesium utilization in the body.

PREVENTION

Cattlemen should keep plenty of magnesium mineral available from October to May and should see that mineral boxes are filled and scattered at several locations.

Legume or legume-grass pastures should be grazed first. Tetany is seldom a problem when cows graze legumes. Less susceptible animals (heifers, dry cows, and stocker cattle) should be grazed on high-risk pastures.

If cows develop tetany one year, they tend to be more prone to developing it the following year and should therefore be culled from the herd.

If soil magnesium is low, a long-term program to increase this nutrient should be developed. Dolomitic limestone is a good source of magnesium.

Magnesium oxide (MgO) mineral should be kept available from October to May. Supplemental legume or grass legume hay, or a concentrate with MgO, should be fed during the early part of the spring grazing season or following calving. Daily intake of magnesium is important since little or no storage of magnesium occurs in older animals.

An effective mineral mix should contain at least 10% magnesium, and a cow should receive approximately 2 oz of MgO in high-risk areas.

Pulmonary Emphysema

Pulmonary emphysema occurs naturally in cattle after abrupt change from sparse, poor-quality forage to lush, green pasture, resulting in respiratory distress and death. In Britain, the disease is referred to as "fog fever," reflecting its common occurrence after cattle are given access to lush regrowth (foggage) pasture.

Current evidence indicates that pulmonary emphysema results from the conversion of the amino acid tryptophan in lush forage to the compound 3-methylindole (3MI). This compound is absorbed through the rumen and is metabolized to a highly reactive compound that interacts directly with lung cells, causing toxicity. Accumulation of fluid and air and other changes result in respiratory distress, and death may occur.

The most promising means of prevention involves a combination of management schemes and antibiotic supplementation to inhibit the formation of ruminal 3MI. Management procedures should emphasize a gradual transition in both the amount and composition of forage consumed. Such antibiotics as monensin (Rumensin) and lasalocid (Bovatec) are effective in lowering ruminal 3MI concentrations and preventing pulmonary emphysema (Carlson and Breeze, 1983).

Fescue Toxicosis

Hemken and associates (1984) recently reviewed the toxic factors in tall fescue. Tall fescue (*Festuca arundinacea* Schreb), a well-adapted perennial grass grown in the transition zone of the United States, frequently produces toxic symptoms and/or reduced animal performance. Syndromes associated with tall fescue include fescue foot, summer syndrome, or the poor performance associated with high temperature and fat necrosis. Unfortunately, it is not known whether these syndromes are related to the same toxic factor(s). Other disorders that can develop when cattle are consuming tall fescue are ergot toxicity and hypomagnesemia (grass tetany).

Signs of fescue foot begin with reduced weight gain or even a loss of weight, rough coat, arched back, and soreness in one or both rear limbs (Garner and Cornell, 1978). Swelling occurs between the dewclaw and hooves, and if the animal is not removed from the toxin, the swelling becomes more severe, and the hooves may slough off. Other signs are discoloration of the tail (purple-black) and loss of the switch. This disorder occurs most often during the late fall and winter. A fungus identified as either *Epichloe typhina* or *Acremonium coenophialum* is clearly associated with summer syndrome and has been implicated in fescue foot. The fungus causes a low serum prolactin level in cattle.

Summer syndrome (summer toxicosis) has some of the same clinical signs associated with fescue foot, but has its own distinct characteristics. Signs include reduced growth rate, decreased milk production, decreased dry matter intake, rough coat, reduced prolactin levels, and reduced reproductive performance. Alkaloids have been implicated with the appearance of summer toxicosis syndrome. Other evidence indicates that compounds other than alkaloids may be involved.

Tall fescue varieties with low levels of the fungus are now being developed for consumption by livestock. Isolation and identification of the toxic compound(s) and factors that contribute to the growth and spread of the fungus are needed to assure long-term toxic-free tall fescue pastures for livestock.

REFERENCES

Beaty, E.R. G.V. Calvert, and J.E. Engel. 1982. Forage good enough for cattle production: When! J. Range Management, *35:*133.

Carlson, J.R., and R.G. Breeze. 1983. Cause and prevention of acute pulmonary edema and emphysema in cattle. Proc., Range Beef Cow Symposium VIII, Sterling, CO. December 5–7.

Council for Agricultural Science and Technology. 1980. Food from animals: Quantity, quality and safety. Ames, IA. Report No. 82.

Garner, G.B., and C.N. Cornell. 1978. Fescue foot in cattle. *In* Mycotoxic Fungi, Mycotoxins, Mycotoxicosis: An Encyclopedic Handbook. Vol. 2. Edited by T.D. Wyllie, and L.G. Morehouse. Marcel Dekker, New York, p. 45.

Hemken, R.W., J.A. Jackson, Jr., and J.A. Boling. 1984. Toxic factors in tall fescue. J. Anim. Sci. *58:*1011.

Quinn, C. 1982. Nutritive quality of native forage. Proc., Ranching Seminar, Optimum Production Through Management. Rapid City, SD. September 13–14.

Selected Readings

Ensign, R.D., and H.L. Harris. 1975. Idaho Forage Crop Handbook. Bulletin No. 547. Idaho Agricultural Experiment Station, University of Idaho, College of Agriculture. Moscow, ID.

Gerken, J. 1979. Various grazing systems offer numerous advantages. Beef Digest, August 1979. p. 28.

Johnson, K.D., C.L. Rhykerd, J.O. Trott, and B.J. Hankins. 1983. Forage selection guidelines. Proc., Indiana Beef Cattle Day, Purdue University. W. Lafayette, IN. December 2.

Owensby, C. 1977. Which grazing system should you use? Beef Digest, July 1977. p. 22.

Reese, P.E. 1981. Poisonous plants—Problem and management. Proc., Range Beef Cow Symposium VII, Rapid City, SD. December 7–9.

Stoddart, L.A., A.D. Smith, and T.W. Box. 1975. Range Management. 3rd Ed. McGraw-Hill, New York.

QUESTIONS FOR STUDY AND DISCUSSION

1. Why may a grass-legume mixture be advantageous in comparison with a pure stand of legume or grass?
2. What are the prerequisites for successful forage establishment?
3. Outline the cyclic growth pattern of a typical forage plant.
4. At which stage(s) of development would overgrazing interfere with the future vigor and productivity of a forage stand? Why?
5. Identify the following forage species:
   a. Legume that requires a well-drained soil that can be used for hay, pasture, or silage.

b. Clover that withstands wet, acid, poorly drained soils.

c. Legume that may be used for late summer pasture or hay and must reseed itself annually to persist.

d. Grass that requires well-drained fertile soils but tolerates wet soils, and that is slow to recover after grazing.

e. Forage for wet, fertile, medium to deep soils, providing good pasture at immature stages.

f. Forage adapted to cool, humid climates that is used in pasture mixtures with legumes and provides pure stands for pasture.

6. Describe the following grazing systems:
   a. Continuous
   b. Rotational
   c. Deferred rotation
   d. Rest rotation
   e. Short-duration, high-intensity
   f. Strip

7. Describe a situation in which it would be feasible to make use of each of the grazing systems described in question 6.

8. What factors should be considered when determining which grazing system to select?

9. How does stage of growth influence forage quality?

10. Calculate the approximate digestibility of a forage plant that contains 10% green leaves, 20% dead leaves, and 70% stem on an as-fed basis.

11. What limits dry matter intake of forages by beef cattle?

12. Discuss the cause, clinical signs, and prevention of each of the following health problems:
    a. Prussic acid poisoning
    b. Nitrate toxicity
    c. Blind staggers
    d. Bloat
    e. Grass tetany
    f. Pulmonary emphysema

Chapter Twelve

Management of Feeder Cattle (Stocker Production)

The growth and development of cattle from weaning to placement in the feedlot for finishing are referred to as *stocker production*. This is a period of growth, not fattening, whereby cattle are fed forage and a minimum amount of grain to make moderate gains. The stocker program is sometimes used as the only program on a ranch, but often, calves are grown out on the ranch at which they were born. *Backgrounding* is a specialized type of stocker program in which weaned calves are fed more grain and less forage for a faster rate of gain. This program prepares them for feedlot finishing at an earlier age than cattle that are wintered on a high forage diet for a slower rate of gain. Figure 12–1 shows the marketing options for calves after weaning.

The choice between backgrounding or stocker production should be based on economic advantage. Available feed supplies, cost of purchased feed, and cattle prices must be considered by the cattleman before he decides to sell. Cattlemen that choose backgrounding for cattle should have a potential market in mind. Feedlots prefer cattle of the same quality, weight, sex, and usually, breed.

The Grazing Advantage (Stocker Cattle Resource Manual), copyright 1981 by Elanco Products Company, Indianapolis, IN, was used extensively in the preparation of this chapter, courtesy of Elanco.

RECEIVING SHIPPED-IN CATTLE

The most critical time period for newly purchased cattle is the first 3 to 4 weeks in the new environment. Stocker cattle come from a wide range of management systems; thus, their nutritional histories vary. Stress associated with marketing and transit creates health problems. These stress factors include dehydration; unfamiliarity with feed, feed bunks, or water troughs; elevated body temperatures; runny noses; watery eyes; scours; and internal and external parasites. Variations in size, age, management, and disease exposure plus diversity of marketing systems and movement of cattle to different geographic regions add to the complexity of recommending specific handling procedures for cattle.

Unloading Cattle

Cattle should be handled as little as possible from the ranch to the feedlot. Excessive handling increases stress and the incidence of shipping fever. The trucker should be instructed to load the cattle as quietly as possible into a clean truck and not to overcrowd or load too loosely. A minimum time en route minimizes stress and shipping weight loss.

On arrival, the cattle should be quietly unloaded from the trucks. It is desirable to have one man on each side of the unloading chute to look for signs of injured cattle, e.g., those that have been down in

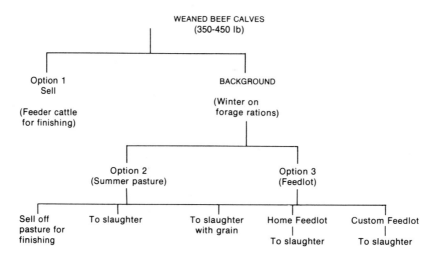

FIG. 12-1. Marketing options for weaned beef calves. (Adapted from Corbett, R.R. 1981. Feeding and management guide for growing beef cattle. British Columbia, Ministry of Agriculture and Food. Victoria, B.C. 81-9.)

the truck, cripples, or those with pinkeye. Treatment of the problem, refusal of the animal, or partial compensation is best done at this time. After being unloaded, the cattle are fed. The cattle can be observed for sickness while they are feeding. The time to begin working the calves depends on the age and size of the calves, stress during shipping, distance of haul, and whether the cattle were preconditioned or not.

When possible, departure should be scheduled so that cattle arrive during the daylight hours to allow adequate time for observing and treating sick or injured cattle.

Clean, fresh water should be available at all times. Water helps to restore dehydrated calves and helps to stimulate feed intake. Water troughs and tanks should be drained, scrubbed, and refilled with clean, fresh water. Calves used to drinking from running streams may not find an automatic waterer. Faucets and floats should be adjusted to allow a small trickle to draw the new cattle.

Feed bunks and troughs should be cleaned prior to new cattle arrival. Old, moldy feed should be removed, and disposal should take place away from the feeding area.

Methods of Feeding

Recommended feeding programs vary from area to area, depending on the source of the cattle, type of feedstuffs available, weight gain, previous feeding programs, and other factors. Shipped-in cattle eat grass hay or a grass-legume hay more readily than any other feed. In general, one should not feed cattle a good-quality legume or alfalfa hay at this time because it tends to cause scouring and results in more sickness. Cattle that have been shipped in from pasture usually do not eat corn silage as readily as grass hay, but corn silage can be fed successfully as the only roughage.

Grass or mixed grass-legume hay fed *ad libitum* and a milled concentrate provide good results. Corn silage fed *ad libitum* (4 to 6% of body weight, as-fed) plus a milled concentrate also works. A urea protein supplement should not be fed during the conditioning period. Cattle are too susceptible at this time to a urea toxicity. Also, the rumen environment is upset, owing to starvation, and cattle cannot utilize urea effectively during this adjustment period.

Research has shown that antibiotics increase daily gain and reduce the incidence of shipping fever and death loss in feeder cattle during the adjustment period. An approved antibiotic should be used, and label instructions should be followed. Shipped-in cattle often have depleted or low vitamin A reserves. Cattle should be fed 50,000 I.U. of vitamin A daily for the first 3 to 4 weeks. This dosage should then be reduced to 20,000 to 30,000 I.U. of vitamin A for the remainder of the feeding period. If cattle have been grazing on dry range grass or are from drought areas, an intramuscular injection of 1 million I.U. of vitamin A should be administered on arrival.

Producers should be cautioned that heavy stocker calves can overeat on a high grain ration, causing serious digestive upsets.

Processing Cattle

In general, cattle that are in good condition, that show no early signs of disease, and that have undergone a normal amount of shrinkage can be processed in one operation 24 hours after arrival. Calves that have undergone excessive shrinkage or that show signs of sickness should be later vaccinated for infectious bovine rhinotracheitis; processing can be completed 2 to 3 weeks later, when they have recovered from all illness.

Processing procedures vary from area to area. Table 12–1 is a checklist to be completed with the help of a veterinarian to help to tailor the program to a local situation.

Spotting Sick Animals

Newly arrived animals should be observed at least twice daily (in the morning and late in the afternoon). Cattle should be observed from a distance, and the observer should move slowly to avoid exciting them. Signs to look for are summarized in Table 12–2.

TABLE 12–1. Processing and Vaccination Checklist[a]

Item	Recommended Type/Dosage or Practice
Vaccinations	
Clostridial diseases	_____
Leptospirosis	_____
IBR	_____
BVD	_____
Pasteurellosis	_____
Corynebacterial infections	_____
Other vaccines or combinations	_____
Internal Parasite Control	
Bolus	_____
Paste	_____
Injection	_____
Drench	_____
External Parasite Control	
Grubs	_____
Lice	_____
Ear ticks	_____
Flies	_____
Growth Promoting Agents	_____
Other Treatment	
Foot rot	_____
Dehorning or tipping	_____
Castration	_____
Branding	_____
Pinkeye	_____
Vitamin A	_____

[a] Consult with local veterinarian to determine recommended program.

MARKETING

Buying and selling decisions are the most important factors affecting economic returns in a backgrounding or stocker program. Stocker programs should be planned to take advantage of seasonal fluctuations in supply and price. The cow-calf producer can use a yearling program to provide more

TABLE 12–2. Visible Signs of Sickness

Dejection
Drooping head
Drooping ears
Nasal discharge
Rapid respiration
Shallow breathing
Scours
Swelling of scrotum of castrated animals

flexibility in both marketing and grazing management programs. Areas of importance in buying or selling cattle are discussed in the following sections.

Transportation Costs

Most cattle truckers haul cattle on the basis of cost per loaded mile. A loaded truck usually hauls 44,000 lb of net weight. Shipping cattle on a less-than-full-load basis is extremely expensive. Table 12–3 shows the transportation costs for hauls greater than 100 miles. Shorter shipping distances are usually negotiated on a cost per hundredweight (cwt) basis. To calculate transportation costs, the cost per loaded mile is multiplied by the number of miles and then divided by net hundredweights on truck.

Interest Rates

Interest costs have received increased attention in the past few years because of escalating interest rates. Different interest rates have a dramatic impact on cost per pound of gain or on the total cost of grazing.

Tables 12–4 and 12–5 reflect the interest cost on a $100 investment and can be easily adapted to any situation. Table 12–6 shows the effect of interest rate on cost of gain. Interest cost on feed is usually figured by estimating the total feed requirement and then calculating the cost of interest on one half of the feed.

Death Loss

Health programs should be designed to minimize death loss. Stocker operations and backgrounding programs deal with lightweight calves, and therefore, the cost of death loss can have a significant effect on the economic returns of growing out calves and preparing them for the feedlot.

Two methods of calculating death loss cost are shown in the following.

Method 1. Multiply total cost of the animal by the percentage of the expected death loss.

Assumptions:
  Total delivered cost of animal = $400.00
  Expected death loss   = 3%

Calculations:
  Cost of death loss = $400.00 × .03
                     = $12.00/head

Method 2. Divide total delivered cost of the animal by the percentage of animals

TABLE 12-3. Transportation Costs for Cattle in Dollars per Hundredweight[a]

Miles	Cost/Loaded Mile ($)					
	1.80	2.00	2.20	2.40	2.60	2.80
100	0.41	0.45	0.50	0.55	0.59	0.64
200	0.82	0.91	1.00	1.09	1.18	1.27
300	1.23	1.36	1.50	1.64	1.77	1.91
400	1.64	1.82	2.00	2.18	2.36	2.55
500	2.05	2.27	2.580	2.73	2.95	3.18
600	2.45	2.72	3.00	3.27	3.55	3.82
700	2.86	3.18	3.5	3.82	4.14	4.45
800	3.27	3.63	4.00	4.36	4.73	5.09
900	3.68	4.09	4.5	4.90	5.32	5.73
1,000	4.09	4.54	5.00	5.45	5.91	6.36
1,100	4.50	4.99	5.50	6.00	6.50	7.00
1,200	4.91	5.45	6.00	6.55	7.09	7.64
1,300	5.32	5.90	6.50	7.09	7.68	8.27
1,400	5.73	6.36	7.00	7.64	8.27	8.91
1,500	6.14	6.81	7.50	8.18	8.86	9.55

[a] Based on 44,000-lb hauls.

TABLE 12-4. Interest Costs per Pound of Gain per $100 Investment[a]

Gain/day (lb)	Rate of Interest (%)						
	10	12	14	16	18	20	22
0.25	$0.108	$0.132	$0.152	$0.176	$0.196	$0.220	$0.240
0.50	0.054	0.066	0.076	0.088	0.098	0.110	0.120
0.75	0.036	0.044	0.051	0.059	0.065	0.073	0.080
1.00	0.027	0.033	0.038	0.044	0.049	0.055	0.060
1.20	0.023	0.028	0.032	0.037	0.041	0.046	0.050
1.40	0.019	0.024	0.027	0.031	0.035	0.039	0.043
1.60	0.017	0.021	0.024	0.028	0.031	0.034	0.038
1.80	0.015	0.018	0.021	0.024	0.027	0.0321	0.033
2.00	0.014	0.017	0.019	0.022	0.025	0.028	0.030

[a] Based on 365-day year.

Example. 100% financed: The interest cost per pound of gain for $450 steer gaining 1.8 lb per day with a 14% interest rate is:

$$\$450.00 \div \$100.00 = 4.5 \text{ increments}$$
$$4.5 \text{ increments} \times \$0.021 = \$0.095/lb$$

remaining (100% minus the expected death loss).

Assumptions:
Total delivered cost of animal = $400.00
Expected death loss = 3%

Calculations:
Cost of death loss
= $400.00 ÷ .97
= $412.37

Cost adjusted for death loss:
= $412.37 − $400.00
= $12.37/head

Shrinkage

Shrinkage, or weight loss, is an important variable affecting the economic return from cattle sales. Weight loss in cattle may be classified as either excretory shrinkage or tissue shrinkage. Excretory shrinkage is the loss in weight from the excretion of manure and urine, and this shrinkage can be regained in a short period of time. Animals held off feed and water for a 12-hour period usually undergo only excretory shrinkage.

Tissue shrinkage is the actual loss of flesh and body water (carcass weight). Tissue shrinkage occurs on extended hauls or during long periods of fasting. Recovery from tissue shrinkage takes more time than recovery from excretory shrinkage.

FACTORS AFFECTING SHRINKAGE

1. Type of Feed. Loss of weight from overnight shrinkage or from a 12-hour standing period is greater when cattle are fed grass, wet beet pulp, silage, or other high-moisture feeds as opposed to dry roughages and concentrates. When no feed or water is available, cattle fed high-moisture feeds usually undergo 4% shrinkage, while fat cattle fed concentrates undergo 2.5 to 3% shrinkage. When feed and water are available on a free-choise basis, morning weights of fat cattle are approximately 2% less than evening weights. Range cattle that are not familiar with enclosures may shrink more than 5% when held in the drylot overnight.

2. Place of Purchase. In an Iowa study involving 4685 feeder cattle (Self and Nelson, 1972), an average shrinkage of 7.2% was found to occur with cattle purchased from a rancher as contrasted with 9.1% average shrinkage, which occurred with those purchased from a sale yard. These cattle were shipped varying distances, ranging from 150 to 1133 miles. The researchers found 0.61% shrinkage for each 400 miles in transit.

3. Environmental Conditions. Temperatures that are extremely hot or cold can affect shrinkage, but wind, rain, snow,

TABLE 12-5.   Interest Costs for Grazing Period per $100 Investment[a]

Grazing Period (days)	Rate of Interest (%)						
	10	12	14	16	18	20	22
100	$2.74	$3.29	$3.84	$4.38	$4.93	$5.48	$6.03
105	2.88	3.45	4.03	4.60	5.18	5.75	6.33
110	3.01	3.62	4.22	4.82	5.42	6.03	6.63
115	3.15	3.78	4.42	5.04	4.67	6.30	6.93
120	3.29	3.95	4.61	5.26	5.92	6.58	7.24
125	3.43	4.11	4.80	5.48	6.06	6.85	7.54
130	3.56	4.28	4.99	5.69	6.41	7.12	7.84
135	3.70	4.44	5.18	5.91	6.66	7.40	8.14
140	3.84	4.61	5.38	6.13	6.90	7.67	8.44
145	3.97	4.77	5.57	6.35	7.15	7.95	8.74
150	4.11	4.94	5.76	6.57	7.40	8.22	9.05
155	4.25	5.10	5.95	6.79	7.64	8.49	9.35
160	4.38	5.26	6.14	7.01	7.89	8.77	9.65
165	4.52	5.43	6.34	7.23	8.13	9.04	9.95
170	4.66	5.59	6.53	7.45	8.38	9.32	10.25
175	4.80	5.76	6.72	7.67	8.63	9.59	10.55
180	4.93	5.92	6.91	7.88	8.87	9.86	10.85
185	5.07	6.09	7.10	8.10	9.12	10.14	11.16
190	5.21	6.25	7.3	8.32	9.37	10.41	11.46
195	5.34	6.42	7.49	8.54	9.61	10.69	11.76
200	5.48	6.58	7.68	8.76	9.86	10.96	12.06

[a] Based on 365-day year.

Example. 100% financed: For a $457 steer on a small winter grain pasture for 135 days with a finance rate of 14%, the interest cost per head (animal only) for the grazing period is:

$$\$475.00 \div \$100.00 = 4.75 \text{ increments}$$
$$4.75 \text{ increments} \times \$5.18 = \$24.61$$

humidity, and other wet weather conditions have a greater effect than temperature alone.

4. Time in Transit. Cattle lose the most weight during the second or third hour in transit. Shrinkage continues thereafter, but at a declining rate. Table 12–7 gives estimates of shrinkage due to different conditions.

5. Age and Weight. Young animals shrink proportionately more than older animals. Fat cattle shrink less than those with a higher proportion of body protein and a lower proportion of body fat.

6. Handling. Weight losses during transit are caused by loading, unloading, and jostling about in a moving truck or rail car, and by change of environment or handlers.

TABLE 12-6.   Interest Cost per Pound of Gain

Average Daily Gain (lb)	Cost/lb[a]
0.25	$0.79
0.50	0.39
0.75	0.26
1.00	0.20
1.25	0.16
1.50	0.13
1.75	0.11
2.00	0.10

[a] Based on a $400 steer with an 18% interest rate.

TABLE 12-7.   Estimated Shrinkage Due to Different Conditions

Conditions	Shrinkage (%)
8 hours in drylot	3.3
16 hours in drylot	6.2
24 hours in drylot	6.6
8 hours in moving truck	5.5
16 hours in moving truck	7.9
24 hours in moving truck	8.9

[a] Based on an average weight of 600 lb.

MINIMIZING SHRINKAGE

The following guidelines can be used to help to minimize cattle shrinkage during transit.

1. Check weather forecasts when planning cattle shipments.
2. Prevent rough handling, poor feed, dirty water, and excessive delays en route or at the market.
3. Avoid overloading and underloading.
4. Do not overfill cattle before hauling. Feed cattle a good dry hay ration before shipping. Avoid high-moisture feeds.
5. Provide adequate protection from extreme cold, warm, or wet weather. Covered trucks, proper bedding, and shade are helpful.

SHRINKAGE AGREEMENTS

One must understand how to calculate a shrinkage agreement when buying and selling cattle. For example, cattle sold for $70 per cost per hundred weight (cwt) with a 3% "pencil shrinkage" agreement, would give the seller a $67.90 pay weight (Table 12–8).

Pencil shrinkage is commonly used in cattle transactions. It may vary from 2 to 5% depending on the geographic area. Excessive shrinkage should be prevented when buying or selling cattle. If pencil shrinkage is applied to weights at a point some distance away, double shrinkage occurs. Larger shrinkage occurs during loading, during unloading, and in the first hour of hauling. Shrinkage increases at a slower rate with additional miles. Thus, even though cattle may be hauled a short distance, significant shrinkage occurs when cattle are loaded and unloaded.

If cattle are loaded and delivered as early as possible in the morning before they are fed, an additional 2% shrinkage may result from their being kept overnight in the drylot. Therefore, one would expect 2% shrinkage because of the overnight period, and if 2 to 3% pencil shrinkage is added, shrinkage becomes excessive. The custom of specifying pencil shrinkage was developed in lieu of the overnight period without feed and water.

TABLE 12–8. Pay Weight on $70/cwt Calves With Different Shrinkage Agreements

Shrinkage (%)	Pay Weight ($/cwt)
1% (70 × 0.99)	$69.30
2% (70 × 0.98)	68.60
3% (70 × 0.97)	67.90
4% (70 × 0.96)	67.20
5% (70 × 0.95)	66.50

Price Spreads

As stocker cattle reach heavier weights, their value per pound often decreases. Comparison of cattle of equal quality within a 100-lb weight range reveals that lighter cattle usually sell for more per hundredweight. These differences are usually referred to as "price spreads." These spreads are often used by both buyers and sellers of stocker or backgrounded cattle when they estimate what price cattle will be worth at the end of the stocker feeding period. The price for the cattle is scaled up or down depending on the ending weight as it relates to the current market price per hundredweight.

A market quotation sheet, which usually quotes regionalized prices, lists prices as shown in Table 12–9. Market quotations are intended for uniform groups of cattle. A portion of the price spread, however, is always associated with such quality factors as condition, breed, color, and bulls versus steers. Price ranges for 400 to 500-lb cattle may not be the same as for cattle in a different weight range. Factors such as weighing conditions and feed prices may also influence market price quotation.

The information in Table 12–9 can be used to determine realistic values for various weights of cattle. From use of Table 12–10, the price of 500-lb steers can be established as $67.50 per hundredweight ($69.75 + 65.25 ÷ 2 = $67.50). Also, a price spread system can be developed to make adjustments if cattle do not weigh exactly 500 lb, as shown in Table 12–11.

TABLE 12–9. Price Quotations for Feeder Cattle

Sex	Feeder Cattle Grade	Weight Range (lb)	Price Range ($/cwt)
Steers	Medium frame–1	400–500	66.50–73.00
Steers	Medium frame–1	500–600	62.00–68.50
Steers	Medium frame–1	600–700	59.25–66.50
Steers	Medium frame–1	700–800	58.00–63.50
Heifers	Medium frame–1	300–400	56.50–63.50
Heifers	Medium frame–1	400–500	55.50–60.50
Heifers	Medium frame–1	500–600	54.25–57.25

FEEDER CATTLE GRADES

Livestock market reporters began using the new USDA feeder cattle grading system on September 1, 1979. The old feeder cattle grades (prime, choice, good, and standard) were developed to designate potential quality grade when finished. Recent research, however, has shown little relationship between feeder calf and carcass quality grade.

The new system of grading is based on frame size and muscle thickness. The new grades reflect the expected weight at low-choice-yield grade 3 for various frame sizes: small, medium, and large (Table 12–12; Fig. 12–2). There are also three muscling scores for each frame size (Fig. 12–3).

The use of frame codes, muscling scores, and breed provides a usable system under which any feeder animal can be graded and in which the grade designation provides a meaningful description of that animal.

FEEDING MANAGEMENT

Many different feeding programs are available for growing weaned calves. Selection depends on the amount and quality of available feedstuffs, type of cattle, length of feeding period desired, marketing objectives, and skill of the operator. The "best" program for one farm or ranch may not be the best for another. Also, the feeding program best suited for a ranch one year may not be suitable the next year. Once a workable program is obtained, however, the cattle feeder should not make dramatic changes from year to year, but rather, should modify the basic plan as economic conditions warrant.

Feeding programs most frequently used to grow calves may be categorized into three groups:

1. Daily gain of less than 1 lb. (A loss in flesh condition occurs.)
2. Daily gain of 1 to 1.5 lb.
3. Daily gain of 1.5 lb or more.

Calves grown to gain less than 1 lb daily are fed only roughage or a limited amount of feed. Feed costs should be kept as low as possible when using a feeding program that features slow rates of gain. Low feed costs and compensatory gain in the following period may be necessary to offset the high cost of gain when calves are grown at

TABLE 12–10. Adjusted Market Quotations for Medium Frame-1 Feeder Steers

Weight Range (lb)	Price Range ($/cwt)	Price Spread ($/cwt)	Avg Price ($/cwt)	Avg Weight
400–500	66.50–73.00	6.50	69.75	450
500–600	62.00–68.50	6.50	65.25	550
600–700	59.25–66.50	7.25	62.87	650

a slow rate. Compensatory gain is the phenomenon whereby cattle that gain less weight during one growth period gain relatively more weight during a subsequent growth period in which the plane of nutrition is improved. Calves that are fed during winter to gain less than 1 lb daily and fed during summer on pastures will probably not be as heavy in the fall as calves wintered at high rates of gain. Net profits may not justify wintering at higher rates of gain, however.

Table 12–13 summarizes weight gain of steers fed coastal Bermuda grass hay and four levels of corn during a 149-day wintering period. Cattle gained more weight as the amount of corn increased from 0 to 2 lb, but cattle fed 3 lb of corn gained no more than steers fed on the 2-lb level. Cattle fed 2 lb of corn required less corn per pound of gain than those fed 1 or 3 lb of corn.

Following the wintering period, cattle from the four winter treatment groups were grazed together on coastal Bermuda grass pasture for 175 days. Cattle fed 1 or 2 lb of corn daily during the wintering period gained as much weight as cattle fed no corn. Cattle fed 3 lb of corn, however, gained less in summer than the other

TABLE 12-11. Figuring Price Spreads for Feeder Cattle[a]

Steer Weight (lb)	Price (6.50 cwt Spread)
500	67.50
520	66.20
540	64.90
560	63.60
561	63.54
580	62.30
600	61.00

[a] A price spread of \$6.50 cwt is equal to \$0.065/lb (6.50 ÷ 100 lb = \$0.065).

groups. The cattle fed 2 lb of corn daily during the winter phase had the highest combined winter and summer gain (Table 12–14). In this study, increased winter gain did not suppress subsequent summer growth of cattle, except in those fed 3 lb of corn during the winter.

Feeding programs in which calves are wintered to gain 1.0 to 1.5 lb daily are the most common. Usually, these calves are fed 2 to 4 lb of grain per head daily and/or commercial supplement, plus good-quality hay on a "free-choice" basis. Wintering cattle to gain 1.0 to 1.5 lb per day provides

FIG. 12-2. Feeder cattle grades based on frame size. (Courtesy of the United States Department of Agriculture, Washington, DC.)

TABLE 12–12.   Expected Weights at Low Choice Grade

Breed Type	Weight Range (lb)		Feeder Grade Frame Type
	Steers	Heifers	
Small-frame British breeds	800–1,000	640–800	Small
Average-frame British breeds	1,000–1,100	800–880	Medium
Large-frame British breeds	1,100–1,200	880–960	Medium
Average-frame European breeds and Holsteins	1,200–1,300	960–1,040	Large
Large-frame European breeds	1,300–1,500	1,040–1,200	Large

the alternative to transferring cattle to grass or to a feedlot in the spring of the year. Heifers fed to gain 1.0 to 1.5 lb per day usually reach puberty in time to conceive at 13 to 15 months of age. Examples of rations for wintering calves are shown in Table 12–15.

Stocker production is prevalent in areas of the United States where winter cereal grain pasture is available. Traditionally, the stocker season in these areas (Oklahoma, Texas, Missouri, and Kansas) is from mid-October to mid-April. Often, the season is split into an October-to-February group and a February-to-April group. The first season is stocked with approximately one calf per acre of cereal grain pasture; the second set is stocked with up to 3 calves per acre. The split season eliminates the mid-winter slump in gains that would otherwise occur and makes it possible to match stocking rate to pasture growth rate.

Turnips are used routinely in the Pacific Northwest as a feed for wintering calves. Newly purchased calves are processed (implanted, vaccinated, etc.) and are then put

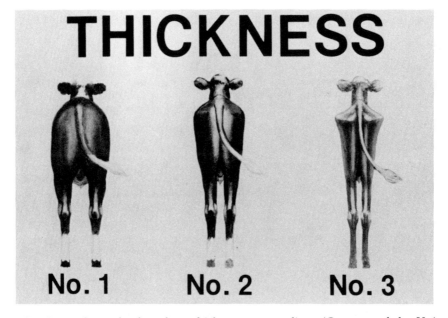

FIG. 12–3.   Feeder cattle grades based on thickness or muscling. (Courtesy of the United States Department of Agriculture, Washington, DC.)

TABLE 12-13. Weights, Gains, and Corn per Pound of Gain of Steers
Fed Coastal Bermuda Grass Hay and Four Levels of Corn
During a 149-Day Wintering Period

Corn per Head (lb/day)	Initial Weight (lb)	149-Day Gain (lb)		Pounds Corn per Pound Gain
		Head	Daily	
0	386	37.5	.25	—
1	386	76.4	.51	3.83
2	367	120.6	.80	3.58
3	395	130.2	.87	4.82

From West Louisiana Experiment Station. 1979. Winter gains affect summer stocker gains. Beef Digest. October, pp. 28–30.

onto turnip fields. Animals that have not grazed turnips previously may need several days to become accustomed to them. Turnips contain a high moisture content, and dry roughage may need to be provided. The dry roughage slows the rate at which the turnips pass through the cattle's digestive tracts, permitting better utilization. Poor-quality grass, alfalfa hay, and by-product feeds such as bluegrass straw or corn stalks have been used successfully. A daily intake of 2 to 3 lb of roughage is required. Gains during the adjustment period may be low, but if managed properly, steers can be expected to gain an average of 1.5 to 2.0 lb.

Turnips are usually seeded by the first day of August, and the grazing season begins during the month of October. The grazing period lasts approximately 90 to 100 days, depending on stocking rate. Pasture finishing programs for post-weaned calves are discussed in Chapter 13.

SUPPLEMENTATION CONSIDERATIONS

Response to Supplementation

WINTER PASTURE

Cattle being wintered on low-quality forage respond dramatically to protein supplementation. Data indicate that crude protein is the first limiting nutrient in native dormant forage. If cattle weighing 550 lb consume 2% of their body weight daily

TABLE 12-14. Gains of Steers Grazed on Coastal Bermuda Grass Pasture for
175 Days Following Winter Feeding on Four Levels of Corn

Corn Fed Daily in Winter (lb/head)	175-Day Summer Gain		324-Day Total Weight Gain (lb/head)	Percentage of Total Gain Made During Summer (%)
	(lb/head)	(lb/day)		
0	187	1.07	225	83
1	176	1.00	252	70
2	178	1.02	299	60
3	149	0.85	279	53

From West Louisiana Experiment Station. 1979. Winter gains affect summer stocker gains. Beef Digest. October, pp. 28–30.

TABLE 12–15. Wintering Rations for 500-lb Steer Calves to Gain 1.25 to 1.50 lb per Day

Item	Ration Number (lb/day as fed)			
	1	2	3	4
Feed ingredient				
Alfalfa hay	12.0	7.0	—	—
Haylage, alfalfa	—	, —	25.0	—
Corn silage	—	28.0	—	25.0
Grain	3.0	—	5.0	5.0
Protein supplement, 32%	—	—	—	2.0
Mineral mixture[a]	FC-2	FC-2	FC-2	FC-1
Total daily feed	15.0	35.0	30.0	32.0
Nutrient intake				
Dry matter	13.4	14.1	13.2	13.3
Crude protein	2.2	1.7	2.0	1.7
NE gain (Mcal/lb)	0.31	0.35	0.38	0.40

[a] Mineral Mixture No. 1: 1 part limestone, 1 part dicalcium phosphate, 1 part trace mineralized salt. Mineral Mixture No. 2: 1 part dicalcium phosphate, 1 part trace mineralized salt. FC—Free-choice.

in forage that contains 5% crude protein (Table 12–16), then they are receiving 0.55 lb of crude protein from forage. The crude protein requirement for a 500-lb steer gaining 1.5 lb is approximately 1.33 lb daily (Table 12–17a,b). Therefore, the amount of supplemental protein required is equal to 0.78 lb daily.

EARLY SPRING PASTURE

Early spring or cool-season grasses (e.g., bromegrass or orchard grass) are high in protein but provide only moderate amounts of energy. Therefore, supplemen-

TABLE 12–16. Estimate of Dry Matter Intake as a Percentage of Body Weight

Season	Type of Forage	Dry Matter Consumption (% body wt)
Spring	Lush-growing	1.75–2.50
Summer	Growing	2.00–2.30
Fall/winter	Mature, dormant	1.25–2.00
Winter	High-quality	2.00–2.60
	Low-quality	1.50–2.00

From Quinn, C.R. Nutritive quality of native forage. Proc., Ranching Seminar, Optimum Production Through Management. Elanco Products, Rapid City, SD. September, 1982.

tal energy should be given to improve average daily gains during this period. In contrast, warm-season native grasses tend to be lower in protein than cool-season grasses during the early spring period. If high gains are desired, supplementation of protein and energy should be considered.

In general, early spring forages supply an adequate amount of protein, and supplemental energy maximizes animal performance. Energy supplementation may also have a dramatic effect on the intake and digestibility of high-moisture forage.

SUMMER AND FALL PASTURES

In general, the average increase in daily gain produced by feeding 2.0 lb of an energy supplement daily is approximately 0.24 lb during the summer (Table 12–18). The protein content and digestibility of late summer and fall pastures are low. Cattle respond to supplemental protein. Yearlings require additional protein to maximize gain and digestibility.

Economics of Supplementation

Supplementation data indicate that the response from feeding 1 lb of an energy supplement is expected to be approxi-

TABLE 12–17a. Nutrient Requirements of Growing-Finishing Medium-Frame Steer Calves

Weight (lb)	Daily Gain (lb)	Minimum Dry Matter Consumption (lb)	Net Energy for Maintenance (Mcal/lb)	Net Energy for Gain (Mcal/lb)	Crude Protein		Calcium (% DM)	Phosphorus (% DM)
					(lb/day)	(% DM)		
440	0.5	9.7	0.50	0.25	0.87	8.9	0.27	0.18
	1.0	10.4	0.57	0.31	1.06	10.3	0.38	0.21
	1.5	10.8	0.64	0.38	1.24	11.5	0.47	0.25
	2.0	11.0	0.70	0.44	1.41	12.7	0.56	0.26
500	0.5	11.5	0.50	0.25	0.98	8.5	0.25	0.17
	1.0	12.3	0.57	0.31	1.16	9.5	0.32	0.20
	1.5	12.8	0.64	0.28	1.33	10.5	0.40	0.22
	2.0	13.1	0.70	0.44	1.49	11.9	0.47	0.24
600	0.5	13.2	0.50	0.25	1.08	8.2	0.23	0.18
	1.0	14.1	0.57	0.31	1.26	9.0	0.28	0.19
	1.5	14.7	0.64	0.38	1.92	9.8	0.35	0.21
	2.0	15.0	0.70	0.44	1.57	10.5	0.40	0.22
700	0.5	19.8	0.50	0.25	1.18	7.9	0.22	0.18
	1.0	15.8	0.57	0.31	1.35	8.6	0.27	0.18
	1.5	16.5	0.64	0.28	4.50	9.2	0.31	0.20
	2.0	16.8	0.70	0.44	1.65	9.8	0.34	0.21

Adapted from NRC. 1984. Nutrient Requirements of Beef Cattle. 6th Ed. National Research Council. National Academy of Sciences, Washington, DC.

mately 0.1 lb of gain. For protein-deficient cattle, a gain response of 0.3 lb per pound of 30% crude protein supplement may be expected up to the level of approximately 3 lb.

Cattle receiving a supplement usually weigh more than cattle not receiving a supplement, and thus bring less per pound;

however, additional weight gain may offset the lower price received.

A typical situation in supplemental feeding is presented by Tables 12–19 and 12–20 to show how the "price spread" discussed earlier in the chapter can be used to determine whether supplemental feeding on pasture would be economical. In these ta-

TABLE 12–17b. Nutrient Requirements of Growing-Finishing Medium-Frame Heifer Calves

Weight (lb)	Daily Gain (lb)	Minimum Dry Matter Consumption (lb)	Net Energy for Maintenance (Mcal/lb)	Net Energy for Gain (Mcal/lb)	Crude Protein		Calcium (% DM)	Phosphorus (% DM)
					(lb/day)	(% DM)		
400	0.5	9.3	0.54	0.28	0.84	8.9	0.26	0.19
	1.0	9.9	0.63	0.36	1.01	10.2	0.36	0.20
	1.5	10.2	0.72	0.44	1.17	11.4	0.45	0.24
500	0.5	11.0	0.54	0.28	0.94	8.5	0.24	0.18
	1.0	11.8	0.63	0.36	1.11	9.4	0.30	0.21
	1.5	12.1	0.72	0.44	1.25	10.3	0.38	0.22
600	0.5	12.6	0.54	0.28	1.04	8.1	0.23	0.18
	1.0	13.5	0.63	0.36	1.19	8.8	0.28	0.20
	1.5	13.8	0.72	0.44	1.32	9.5	0.32	0.21

Adapted from NRC. 1984. Nutrient Requirements of Beef Cattle. 6th Ed. National Research Council. National Academy of Sciences, Washington, DC.

TABLE 12–18. Expected Additional Gain of
Growing Pasture Cattle With
Supplementation

| Season | Supplementation | | Additional Gain (lb/day) |
	Type	Level (lb)	
Spring	Energy	2	0.60
Summer	Energy	1	0.12
		2	0.24
		3	0.34
Fall	Protein	2	0.62
Winter	Protein	2	0.74

ᵃ Adapted from Figure 8, p. 20, of The Grazing
Advantage—Stocker Cattle Resource Manual, copyright 1981, Elanco Products Company.

bles, a $6.50 price spread is used in figur-
ing the value of gain in an animal selling in
the 500-lb weight range. The price spread
system can be used in the same manner
when cattle fall into two different weight
ranges. Using these tables, the reader can
set up a base price and can scale the price
either upward or downward to arrive at
final values. In this example, the additional
gain of the supplemented cattle just offset
the cost of the supplement. Different price
spreads, feed costs, and daily gains affect
the final outcome.

FEED ADDITIVES
AND IMPLANTS

In a wintering program, each producer
should take full advantage of feed additives and implants. Among the growth-
promoting implants available are Ralgro,
Synovex S or Synovex H, Compudose, and
Steer-oid. Most implants lead to a 5 to 10%
improvement in gains and feed efficiency.
These implants have been described in
Chapter 11.

Ionophores (Bovatec and Rumensin) are
feed additives approved for pasture cattle
that improve the efficiency of rumen fermentation by reducing losses associated
with volatile fatty acid (VFA) formation.
They increase the amount of propionic acid
produced and decrease the quantities of
acetic and butyric acids, which results in
substantial energy savings since energy
losses associated with waste gas production are reduced.

Daily feed intake of cattle that are fed
low-energy rations, including most pastures, is thought to be limited by the bulk
of the diet. Ionophores enable such cattle
to obtain more energy from the feed consumed daily. This increase in available energy enables the animal to gain weight at a
faster rate.

BREAK-EVEN SELLING PRICE

Table 12–21 offers an example of how to
calculate the amount that a stocker operator
or backgrounder can afford to pay for
feeder cattle.

Table 12–22 shows that to break even,
the cattle producer must receive $69.48/cwt
for non-supplemented cattle and $66.50/
cwt for the supplemented cattle. Beef cattle
producers who contemplate background-

TABLE 12–19.   Economic Assumptions

1. Pasture type	Coastal Bermuda grass (late summer; fall)
2. Grazing period	120 days
3. Initial weight	400 lb
4. Daily gain	
Non-supplemented	0.9 lb
Supplemented with 2 lb grain	1.3 lb
5. Final weight	
Non-supplemented	508 lb
Supplemented with 2 lb grain	556 lb
6. Feed cost	
2 lb/day at $.06/lb	$14.40

TABLE 12-20. Economic Returns Using Price Spread

Base price of 500-lb steers	= $ 67.50 cwt (Table 12–10)
Price of 600-lb steers	= $ 61.00 cwt (Table 12–10)
Price spread ($/cwt)	= $ 6.50 cwt (Table 12–9)
Value of non-supplemented steers (508 lb): $67.50 − (8 lb × $0.065) = $66.98 $66.98/cwt × 5.08 cwt	= $340.26
Value of supplemented steers (556 lb): $67.50 − (56 lb × $0.065) = $63.86 $63.86/cwt × 5.56 cwt	= $355.06
Additional return of supplement $335.06 − $340.26	= $ 14.80
Less cost of supplement	= $ 14.40
Return from supplement	= $ 0.40

ing calves or growing yearlings as an individual enterprise (stocker production) should perform this calculation routinely, prior to purchasing calves or growing out cattle.

REFERENCES

Corbett, R.R. 1981. Feeding and management guide for growing beef cattle. British Columbia, Ministry of Agriculture and Food. Victoria, B.C. 81-9.

Elanco Products Company. 1981. The Grazing Advantage—Stocker Cattle Resource Manual. Elanco Products Company, Indianapolis, IN.

NCR, 1984. Nutrient Requirements of Beef Cattle. 6th Ed. National Research Council. National Academy of Sciences, Washington, DC.

Quinn, C.R. 1982. Nutritive quality of native forage. Proc., Ranching Seminar, Optimum Production Through Management. Extension Service—WY, CO, SD, NE. Rapid City, SD. September 1982.

Self, H.L., and G. Nelson. 1972. Shrink during shipment of feeder cattle. J. Anim. Sci. *35*:489.

West Louisiana Experiment Station. 1979. Winter gains affect summer stocker gains. Beef Digest. October, pp. 28–30.

Selected Readings

Brownson, R. 1982. Rations for wintering calves. 33rd Montana Livestock Nutrition Conference. Montana State University, Bozeman, MT. January 1982.

TABLE 12-21. Assumptions for a Sample Budget

Length of grazing period	= 150 days
Cost of 500-lb steer	= $67.50 cwt
Distance to delivery point	= 500 miles
Buying commission	= $0.50/cwt
Transportation cost (44,000 lb)	= $1.80/mile
Interest rate	= 14%
Death loss	= 3%
Veterinary cost	= $5.00/head
Labor cost Non-supplemented Supplemented	 = $0.03/day = $0.05/day
Pasture cost	= $10.00/head/month
Feed cost Grain supplement ($120/ton) Feeding 2 lb/head/day Interest in ½ feed at 14% Salt and mineral	 — = $18.00/head = $1.26/head = $1.50/head
Performance of cattle Non-supplemented Supplemented	 = 0.9 lb/head/day = 1.3 lb/head/day
Value of cattle at end of growing period 600 lb at $63/cwt (spread of $0.05/lb)	 = $378.00

TABLE 12-22. Calculating Break-even Selling Price

Item	Description	No Supplement	Supplement
Beginning value	5 cwt at $67.50/cwt	$337.50	$337.50
Ending value			
Non-supplemented	6.35 cwt at $61.25/cwt	338.94	
Supplemented	6.95 cwt at $58.25/cwt		404.84
Feed costs			
Pasture	5 months at $10/month	50.00	50.00
Supplement	2 lb/day × 150 days at $0.06/lb	—	18.00
Salt and mineral		1.50	1.50
Total feed costs		$ 51.50	$ 69.50
Non-feed costs			
Veterinary cost		5.00	5.00
Labor			
Non-supplemented		4.50	
Supplemented			7.50
Interest			
Cattle	14% at 5 months	19.44	19.44
Supplement[a]	14% at ½ feed		0.52
Buying Costs			
Transportation cost per head		10.25	10.25
Buying fees, brand and health inspections etc.,/hd		2.50	2.50
Death loss[b]			
Total delivered cost	$350.25 × % death loss	10.51	10.51
Total non-feed costs		$ 52.20	$ 55.72
Total costs		$103.70	$125.22
Feed cost per lb gain[c]		$ 38.14/cwt	$ 35.64/cwt
Total cost per lb gain[d]		$ 76.81/cwt	$ 64.22/cwt
Break-even selling price[e]		$ 69.48/cwt	$ 66.58/cwt

[a] $18 ÷ 2 = $9 × 0.14 × 5/12 = $0.52.
[b] $337.50 + $10.25 + $2.50 × 0.03 = $10.51.
[c] $51.50 ÷ 1.35 cwt gain = $38.14; $69.50 ÷ 1.95 cwt gain = $35.64.
[d] $103.70 ÷ 1.35 cwt gain = $76.81; $125.22 ÷ 1.95 cwt gain = $64.22.
[e] $337.50 + 103.70 ÷ 6.35 cwt = $69.48; $337.50 + 125.22 ÷ 6.95 cwt = 66.58.

Brownson, R.M. 1975. Shrinkage in beef cattle. Great Plains Beef Cow-Calf Handbook. Stillwater, OK. GPE-4002.

Brownson, R.M. 1970. The importance of cattle shrinkage. Montana State University Bulletin 1080. Bozeman, MT.

Clanton, D.C. 1982. Nutrient requirements of growing cattle on range. Proc., Ranching Seminar, Optimum Production through Management. Rapid City, SD. September 13–14, 1982.

QUESTIONS FOR STUDY AND DISCUSSION

1. Distinguish between backgrounding and stocker production.
2. Why is it important to provide clean, fresh water for shipped-in calves?
3. Outline a simplified feeding program for shipped-in cattle.
4. How soon after arrival should cattle be processed?
5. List visible signs of a sick animal.
6. Calculate the transportation cost in dollars per hundredweight (cwt) for cattle being shipped 300 miles at a cost per loaded mile of $2.00.
7. Determine the interest cost for a grazing period of 140 days at an interest rate of 14% when the purchase price of the steer was $300.
8. What would be the expected death loss cost in the following situation?

Cost of delivered animal = $400.00
Expected death loss = 2%

9. Define excretory and tissue shrinkage.
10. Discuss the factors affecting shrinkage.
11. Outline a management program to minimize shrinkage.
12. Define "price spread."
13. Calculate the dollar value of a 553-lb calf with the average price received for 500- and 600-lb calves being $70/cwt and $63.50/cwt, respectively.
14. Define compensatory gain.
15. Describe two feeding management systems used when wintering stocker cattle.
16. How do ionophores improve average daily gain of cattle on pasture?
17. Given the following information, calculate the return from administering a supplement to pasture cattle:

Base price of 500-lb steers = $73.00/cwt
Price of 600-lb steers = $67.00/cwt
Price of 700-lb steers = $62.50/cwt
Daily gain
   Non-supplemented = 0.6 lb
   Supplemented = 1.2 lb
Grazing period = 120 days
Supplement fed
1 lb/day at $0.08/lb

18. Given the following information, calculate the break-even selling price:

Length grazing period = 150 days
Cost of 500-lb steer = $73/cwt
Distance to delivery point = 200 miles
Buying commission = $0.50/cwt
Transportation cost (44,000 lb) = $1.80/mile
Interest = 12%
Death loss = 1.5%
Veterinary cost = $3.00/head
Labor cost = $0.03/head
Pasture cost = $8/head/month
Performance cattle = 1.3 lb/head/day

19. What would be the estimated dry matter intake of steers grazing the following forages?
  a. Lush, growing spring pasture
  b. Growing summer pasture
  c. Fall/winter, mature dormant pasture
20. Why should stocker cattle grazing fall or winter pastures be fed a protein supplement while cattle grazing summer pasture should be supplemented with energy?

Chapter Thirteen

Feedlot Management

Cattle feeding is a dynamic industry. Since World War II, the industry has experienced significant developments characterized by differential growth rates among geographic areas. Also, a shift has occurred in the size of operation—from farmer-feeders feeding less than 500 head to large-scale feedlots with capacities of more than 100,000 head. Only 1 to 2% of the feedlots in the 23 major cattle feeding states have capacities of 1000 or more, yet these feedlots marketed 58% of total U.S. fed-beef products in 1977 (Gee, et al., 1979).

The growth and structural changes in the cattle feeding industry resulted from increased consumer demand for fed beef after World War II. Differential rates of growth among regions in population and an increase in per capita income altered the regional distribution of meat consumption and changed the level and character of demand (Williams and Dietrich, 1966). Also, greater feed supplies brought about by irrigation enhanced cattle feeding in the western and southwestern United States. The development of specialized equipment and machinery for feed processing and distribution has made it more economical to feed large numbers of cattle at one location.

Increased per capita consumption of beef from 1950 to 1976 was due in part to the growing taste for grain-fed beef. Grain feeding of cattle began in the midwestern region of the country in the mid-1800s and gradually spread to the southwestern states (Thompson, 1983). The military food service during World War II is credited with spreading the popularity of grain-fed beef into other regions of the country. Consumers began to demand grain-fed beef in the mid-1950s, which ended the practice of extensive backgrounding of 3- to 4-year periods of forage feeding for cattle in the southwestern region of the United States.

Grain will continue to be fed to cattle so long as grain-fed beef continues to be marketed profitably. Faster gains associated with grain feeding result in reduced non-feed costs by allowing shorter feeding periods to reach acceptable market weights. Today, however, consumers prefer leaner beef. Because of this change in consumer preference, changes in cattle feeding programs are needed to produce the taste of grain-fed beef with less fat (Thompson, 1983).

New technology is the key to ensuring a strong future for the cattle feeding industry. New production developments in the areas of meat technology and marketing will have an impact on all segments of the beef cattle industry. The ability to adapt feeding operations to meet the challenges and changes of the industry will determine its future success or failure.

FEEDS AND FEEDING PROGRAMS

Feeds

CEREAL GRAINS

Cereal grains are high in starch and low in fiber. They are widely used as energy sources in beef cattle rations. The principal cereal grains are corn, wheat, oats, barley, and sorghum. In general, grains are not good sources of protein, but when fed in large amounts, they may contribute a major portion of the protein in the ration. Crude protein content ranges from approximately 7 to 15%. Corn is the most popular cereal grain grown in the United States, because it produces more digestible energy per unit of land area than any other grain crop. Also, corn is a digestible and palatable feed and rarely causes nutritional problems when fed to cattle. The energy value of corn is used as a standard for comparison with other cereal grains (Table 13–1).

Grain sorghums are important sources of grain for cattle feeding in semi-arid regions of the country that are too dry for corn production, and in other areas where corn cannot be grown. The grain sorghums have a nutrient makeup similar to corn, but require more rigorous processing. The seeds of grain sorghums are hard and must be crushed mechanically prior to cattle feeding to maximize digestibility and utilization. Chemically, grain sorghums contain less energy and more protein in comparison with corn.

Barley is a cool weather, drought-resistant crop. It contains more protein but less energy than corn, owing to its lower starch and higher fiber content. It is palatable and can be used to constitute 100% of the grain component of rations if properly supplemented with vitamins and minerals. Feeding value of barley depends on its test weight. To maximize animal performance, barley should be processed before being fed in high grain rations to feedlot cattle.

The amount of oats fed in feedlot rations is limited, owing to their high fiber content and low energy value. Because oats are fibrous and palatable, they are used frequently in creep and backgrounding rations.

Wheat is grown primarily for flour, but it can also be fed to cattle when such use is economically justified. Feeding wheat to cattle requires better management than feeding other grains. The starch in wheat is digested at a rapid rate in the rumen, which produces large quantities of acid in a short period of time. Cattle that develop "acidosis" are usually manifesting an accumulation of lactic acid in the rumen. When managed properly, however, wheat is an excellent feed for feedlot cattle. Cattle should be adapted gradually to a wheat-based ration. Recommended management practices for feeding wheat include the following:*

1. Limit wheat to no more than 40% of the dry matter ration or 50% of the corn, whichever is higher. Forty percent of the dry matter equals approximately 1% of body weight.
2. Add wheat to rations gradually, especially if cattle are on a low-roughage diet.
3. Mix wheat with other ingredients, such as silage, haylage, or corn grain. This practice

---

* Adapted from a presentation by K. S. Hendrix, Purdue University, W. Lafayette, IN, before the 1978 Purdue Feed Industry Conference, February 22.

TABLE 13–1. Chemical Composition of Common Cereal Grains

Grain	International Feed Number	Percentage (%) of Dry Matter				
		Crude Protein	Crude Fiber	TDN	Calcium	Phosphorus
Corn	402–931	10.1	2.2	90.0	0.02	0.35
Barley	400–549	13.5	5.7	84.0	0.05	0.38
Oats	403–309	13.3	12.1	77.0	0.07	0.38
Sorghum	420–894	12.5	2.6	83.0	0.04	0.36
Wheat	405–337	11.3	2.6	89.0	0.07	0.36

Adapted from NRC. 1984. Nutrient requirements of beef cattle. 6th Ed. National Research Council, National Academy of Sciences, Washington, DC.

reduces the chance of individual animals consuming too much wheat at one time.

4. Process the wheat before feeding. Coarse rolling or grinding is preferred. Avoid fine grinding.

5. To take full advantage of cost savings, reduce supplemental protein when wheat is fed. Minerals, vitamin A, and additives should not be changed, however.

6. Observe whether a slight reduction in grain consumption occurs when wheat is fed in place of corn. If this reduction exceeds 10%, cut back on the wheat, increase the roughage slightly, and feed 0.10 to 0.15 lb of sodium bicarbonate daily per head. Also, administration of Rumensin or Bovatec at recommended levels may be helpful when the diet is shifted from corn to wheat.

MOLASSES

Molasses is a by-product of sugar. Beet and cane molasses contain, respectively, 78% and 75% dry matter, 79% and 72% TDN, and 8.5% and 5.8% crude protein on a 100% dry matter basis (NRC, 1984). Molasses is used as a carrier of micro-ingredients in liquid supplements, as a flavoring agent to mask the unpalatable characteristics of other feed ingredients, as a means of reducing the dustiness of rations, and as a pellet binder. Molasses has been fed as a supplement on a free-choice basis in countries of the world where it is relatively inexpensive in comparison with more traditional energy sources such as cereal grains.

FAT

Fat or feed-grade tallow has been used extensively by the feed industry. Fat is added to feedlot rations to provide an energy source, to reduce dustiness of rations, and to improve ration palatability. Fat is commonly added to feedlot rations in amounts constituting 2 to 5% of the total ration. Feed manufacturers incorporate some fat into rations to act as a lubricant in feed processing.

PROTEIN SUPPLEMENTS

Protein supplements are defined as those feeds that contain greater than 20% crude protein. Selection of a protein supplement for feedlot cattle that are being fed high-energy grain diets is determined by (1) availability and cost; (2) presence of undesirable or toxic compounds; (3) content of other nutrients; and (4) utilization of the supplement with selected feed ingredients (Church, 1977). Protein supplements are derived from animal, marine, and plant sources as well as from non-protein nitrogen sources such as urea from chemical manufacturing.

The most important plant protein supplements in North America are derived from cottonseed and soybeans. Less important crops are peanuts, flax linseed, sunflower, safflower, rapeseed, and various legume seeds. Meals derived from the above crops, with the exception of legume seeds, are called *oilseed meals*, because the seeds have a high oil content.

Soybean meal contains 44 to 50% crude protein and is an excellent source of protein for feedlot cattle. It is highly palatable and digestible and is equivalent to most cereal grains in energy content.

Cottonseed meal is a by-product of cotton that contains approximately 41% crude protein. Cottonseed meal contains gossypol, which can be toxic to monogastric animals but is not toxic to cattle with a fully developed rumen. Cost and availability are the two factors that limit the amount of cottonseed meal in feedlot rations.

Animal protein supplements are derived from animal tissues. The quality of these products is highly variable and is affected by processing method. Tankage and meat meal are made from trimmings that originate on the floor of the meat packing plant—inedible parts and organs, cleaned entrails, fetuses, some condemned carcasses, and parts of carcasses. *Tankage* is the unground product produced by rendering (fat removal), and *meat meal* is the dry, ground residue. When tankage or meat meal contain greater than 4.4% phosphorus, the word "bone" must be inserted into the name of the product. These products usually contain more than 50% protein and are highly digestible and rich in minerals.

Blood meal is ground-dried blood that contains more than 80% protein. The protein in blood meal tends to be less digestible and of poorer quality than that in tankage or meat meal. Other protein supplements of animal origin that are used in limited quantities are fish meals, feather and hair meals, and condensed and dried fish solubles.

Urea is the most common non-protein nitrogen source fed to feedlot cattle. Urea contributes a source of nitrogen to the ration that microorganisms present in the rumen can convert into usable protein. Urea must be fed with a readily available source of energy such as grain or molasses. One pound of urea is equivalent to 2.82 lb of protein. Because urea is difficult to mix thoroughly and can be toxic in small quantities, these recommendations should be followed:

1. Mix thoroughly in the ration.
2. Do not feed urea or urea protein supplements to newly arrived or shipped-in cattle until 21 to 28 days after arrival.
3. Do not feed urea to cattle that have been off their feed until they have had a chance to resume a full diet.
4. Feed no more than 3½ oz (100 g) of urea per head daily to cattle on fattening rations.
5. Formulate rations so that no more than one third of the protein in the total ration is derived from urea.
6. Change to urea-based feeds gradually.

HAY

Hays are primarily a source of energy for cattle; however, high-quality grass and legume hays can be excellent sources of protein, calcium, and vitamins. Most hays contain 50 to 60% TDN as compared with 70 to 90% for most cereal grains. Alfalfa is the predominate hay fed to feedlot cattle because of its availability and cost per unit of energy and protein. In most finishing rations, hay makes up less than 25% of the total ration on a dry matter basis. This is due to its lower energy content in comparison with cereal grains, and to the bulky nature of hay, which limits dry matter in-take. Most feedlots prefer to use some hay in the ration, however, as a "scratch factor" to provide bulk and thereby minimize indigestion, founder, and animals going off feed.

SILAGE

Silage is the product of a controlled anaerobic fermentation (i.e., fermentation occurring without oxygen). High-quality silage is an excellent source of feed for cattle. Many different types of crops are used to make satisfactory silage. Corn and sorghum are the two most important silage crops grown in the United States. Other crops raised for silage are cereal grains, sunflowers, grasses, and legumes. Good silage is nutritious and palatable. Most silage contains approximately 25 to 40% dry matter. The bulk and moisture content of silage limit dry matter intake as well as the distance that it can be transported.

Corn silage usually contains 65 to 70% TDN and 7 to 9% crude protein (Table 13–2). Silage made from well-eared crops may contain as much as 50% grain. The sorghums are grown in dry, unirrigated areas of the western and southwestern United States. Legume and grass silages contain 50 to 60% TDN and 12 to 20% crude protein. All silages are excellent sources of energy for feedlot cattle. In addition, grass and legume silages require less protein supplementation, owing to their higher protein content.

Feed Processing

GRAIN PROCESSING

Grain is processed by either dry or wet methods. Dry methods include grinding, rolling, popping, micronizing, extruding, roasting, and pelletization. Wet processing methods include steamrolling or flaking, reconstitution, and ensilage of high-moisture grain. All grain processing methods affect feeding value by improving either dry matter intake or ration palatability and digestibility.

TABLE 13-2. Chemical Composition of Silages

Grain	International Feed Number	Percentage of Dry Matter				
		Crude Protein	Crude Fiber	TDN	Calcium	Phosphorus
Corn	3-28-250	8.1	23.7	70.0	0.23	0.22
Alfalfa	3-00-216	17.0	28.0	60.0	—	—
Barley	3-00-152	10.3	30.0	51.0	0.34	0.13
Sorghum	3-04-323	7.5	27.9	60.0	0.35	0.21
Wheat	3-05-185	8.1	30.9	59.0	0.18	0.05

Adapted from NRC. 1984. Nutrient requirements of beef cattle. 6th Ed. National Research Council. National Academy of Sciences, Washington, DC.

Grinding. Grinding is one of the cheapest and simplest grain processing methods. Most ground grains fed to feedlot cattle should be coarsely ground. It has been demonstrated that when cattle are fed diets that are high in corn, grinding has no advantage over feeding the whole grain.

Dry Rolling. Dry rolling involves passing the grains between a closely fitted set of steel rollers without the use of steam. This smashes the grain, breaking the hull. The end product usually lacks uniformity and is similar to coarse grinding with many fines and large pieces. Primary advantages of dry rolling over steamrolling are that the former method is cheaper per ton of grain processed and that it has a higher processing capacity per hour.

Popping. Popping is caused by rapid application of dry heat to the grain kernel. Grain containing 15 to 20% moisture is heated to 300 to 310°F for 15 to 20 seconds. The moisture in the kernel turns to steam and causes the kernel to pop. Popped grain is a dry, low-density feed. Popping is used extensively with grain sorghums.

Micronizing. The term micronizing is used to describe a dry heat treatment of grain by microwaves emitted from infrared burners. As in popping, grain is heated to 300°F; it is then dropped into rollers with spiral grooves that place pressure on the kernel. The grain does not pop, and the final product is low in moisture. Micronizing is primarily used on sorghum for feedlot cattle.

Extruding. Extruding is a process in which ground grain is passed through an orifice in a tapered spiral screw. The resulting ribbonlike strip from the orifice breaks up into flakes of different shapes and lengths. Steam may be added to soften the feed. This process is used on pet foods and on breakfast cereals for humans. The effects of this method on cattle on high-grain rations are similar to those produced by other processing methods.

Pelletization. Pelletization is accomplished by grinding the feed and forcing it through die openings. Feeds are usually steamed before being pelletized. Pellets can be made in different sizes. This process results in a dry, extremely dense feed. Pellets are dust-free and tend to decrease feed wastage, but they are more expensive to produce than other forms of processed grain. Usually, feedlot cattle are fed only a pelleted protein supplement, with the rest of their ration fed in loose form.

Roasting. Roasting is a simple process whereby the grain is heated to 280 to 320°F for 5 to 15 minutes. Corn is probably the most popular roasted grain. Roasting decreases moisture content by 5 to 8% and improves feed efficiency by 5 to 10%. All processing techniques that require heat are probably more expensive to use than ensilage of high-moisture grains (Perry, 1980).

Steamrolling. Steamrolling exposes the grain to steam for a short period of time (1 to 5 minutes) and is followed by rolling. Steamrolling produces less fines than coarse grinding or dry rolling. Steam-processed flaked grains are prepared in a manner similar to steamrolling. The grain is exposed to steam for 15 to 30 minutes under atmospheric conditions, or under pressure of about 50 psi for 1 to 2 minutes, and is then rolled to produce a thin flake. The moisture content is raised to between 18 and 20%. Steam flaking is used with sorghum and improves performance by 5 to 10%. It is expensive and has a high-energy requirement.

Reconstitution. Reconstitution is the addition of water to dry grain to increase the moisture content to between 25 and 30%, followed by storage in the whole form for 14 to 21 days prior to feeding. Research indicates that this method is most advantageous when used with corn or sorghum. To prevent mold, this product must be placed in an oxygen-limiting silo after reconstitution, or organic acids (propionic acid) may be added.

Ensilage of High-Moisture Grain. Grain is harvested when it contains between 20 and 35% moisture, and is then stored in a silo. Usually, high-moisture grains are rolled or ground prior to feeding. Their use is advantageous because the grain can be harvested when weather conditions do not allow normal drying, and the fuel saved from avoiding the drying process probably offsets the more expensive storage facilities.

Feed Additives and Implants

Feed additives and growth-promoting implants are common management tools used in beef cattle production. Feed additives include antibiotics, hormones, and hormone-like drugs. They improve the health of the animal and/or improve cattle performance.

ANTIBIOTICS

Continuous low-level feeding of antibiotics (35 to 100 mg/head/day) helps control foot rot and diarrhea, improves rate of gain and feed efficiency, and aids in reducing the incidence of liver abscesses in cattle that are fed high-concentrate diets. The improvement in performance has been greater for newly arrived cattle that are stressed and are fed lower-energy rations.

MELENGESTROL ACETATE (MGA)

Melengestrol acetate (MGA) is an orally administered, progesterone-like compound. MGA stimulates the ovaries of heifers to produce their natural hormones, which stimulate growth and suppress estrus. For MGA to exert its growth-promoting effect, heifers must not be pregnant and must have intact, functional ovaries. A 48-hour period of withdrawal from MGA prior to slaughter is required. Marketing heifers shortly after this withdrawal period is preferable; otherwise, heifers may come into heat and go off their feed. The following guidelines should be followed when feeding MGA:

1. Feed a non-medicated withdrawal supplement for 24 to 36 hours (depending on expected time of slaughter) prior to removing the cattle from the pen. The required 48-hour withdrawal time must be observed, however. After the cattle are removed from the pen, the MGA ration can once again be fed to those cattle remaining. Virtually no estrus is seen in the remaining cattle for 24 to 36 hours after withdrawal from the MGA.
2. Remove the entire supplement completely for 24 to 36 hours prior to shipping finished slaughter heifers. Once again, the MGA supplement can be reinstituted with few or none of the heifers showing signs of estrus.
3. Feed a withdrawal supplement continuously until all animals are marketed. This practice results in some animals coming into heat during the prolonged withdrawal period, but research trials have shown no detrimental effect on carcass quality, even when heifers have been withdrawn from MGA for 22 days. In other words, with a prolonged withdrawal period, the MGA-fed animals were equal to the controls in grade, yield, and cut, and they had no greater incidence of dark cutters or bruises.

IONOPHORES

Bovatec (lasalocid) and Rumensin (monensin sodium) are two feed additives initially marketed as anticoccidial drugs for poultry under the names Avatec and

Coban. Both Rumensin and Bovatec alter rumen microflora (bacteria and protozoa) to favor the production of propionic acid. Propionic acid synthesis in the rumen is 70% more efficient and is utilized more effectively after absorption than are acetic or butyric acids. Bovatec and Rumensin have been shown to improve feedlot performance and are routinely fed in feedlot rations.

GROWTH-PROMOTING IMPLANTS

Growth-promoting implants currently approved for beef cattle are Compudose, Ralgro, Steer-oid and Synovex S and H. These products have consistently improved gains in feedlot cattle by 5 to 10% and improved feed efficiency by 7 to 10%. The implanting procedure for these products is as follows:

1. Read the instructions for proper implanting that are included with the product.
2. Use the type of implanting gun specified in the instructions. Different implanting guns are required for different products. The cost of each gun ranges from $5 to $18 depending on the product.
3. Follow these general steps when implanting growth stimulants:
   a. Restrain each animal in a squeeze chute so that the ear can be held in the hand without danger to the implanter.
   b. Place the implants underneath the skin on back of either the right or left ear near the base. Make sure that the pellets are deposited in the ear and not in the neck.
   c. Disinfect the site of implantation with some type of antiseptic such as alcohol. One convenient way of dispensing the antiseptic is to use a plastic bottle with a spray nozzle attached.
   d. To make sure the proper number of pellets have been deposited in the ear, use the index finger and feel the pellets beneath the skin.
   e. Follow recommendations on the product label for subsequent reimplanting. Use the other ear for the second implant.

Management of Shipped-In Cattle

The most critical and important period for feeder cattle is the first two to three weeks after arrival. During this time, cattle adjust to their new environment and recover from the stress of shipping. Proper feeding and management practices reduce losses due to shipping fever and ensure adequate gains during the finishing period. Use of the following recommendations, which are open to modification, can help to reduce death loss and improve the condition of cattle from the outset:

1. Handle cattle as little as possible from the ranch to the feedlot. Excessive handling increases stress and the incidence of shipping fever. Load cattle quietly into a dry, clean truck for their ride to the feedlot; do not overcrowd or load too loosely.
2. Unload cattle quietly, and move them into a well-bedded, sheltered pen that is equipped with separate feed and water supplies. The time at which to begin to work calves depends on age and size, stress during shipping, distance of haul, and whether cattle have been preconditioned.
3. If necessary, allow water to splash into the tank so that calves can locate the water.
4. Put calves into a large pen (a minimum of 40 square feet per animal), and allow them to walk and bawl themselves out.
5. Because shipped-in cattle often have depleted or low vitamin A reserves, feed 50,000 IU of vitamin A during the adjustment period, and reduce to 20,000 to 30,000 IU of vitamin A daily for the remainder of the feeding period.
6. Give cattle routine immunizations upon arrival if they appear to be strong and healthy. Feedlot diseases vary from one locale to another; therefore, it is important to establish a vaccination program with the help of a veterinarian familiar with the area.
7. Remove sick animals to a "hospital area," and treat them according to a veterinarian's advice.
8. Treat for lice and grubs with a pour-on material when the cattle are unloaded from the truck. Spraying and dipping are also effective in controlling external parasites.
9. Worm all animals with an effective product recommended by a veterinarian.
10. Delay castration and horn tipping if possible.
11. Implant cattle with a recommended growth-promoting agent.

FEEDING MANAGEMENT

Nutrients most likely to be deficient in receiving rations are energy, protein, and potassium. An adequate level of nutrient intake should be obtained as soon as possible. Therefore, feed must be palatable and fresh to encourage consumption.

Commercial starter feeds are available. In most cases, they are pelleted, are readily consumed by cattle, and are an excellent means of starting cattle on feed.

Grass or mixed grass-legume hay is an ideal feed for new cattle. High-quality legume hay should be avoided because of its tendency to cause scours. Putting hay on the ground and in bunks encourages consumption. Heavily stressed cattle should be allowed to fill up on hay 2 to 4 hours before having access to water. This prevents cattle from filling up on water and reducing hay consumption. Once calves begin to feed in the bunk, adding 1 to 2 lb of grain plus 1 to 2 lb of a protein supplement provides needed energy and protein. Grain should gradually be increased to 1.0 to 1.5% of body weight.

If corn silage is fed, it should be sprinkled on top of the hay. As cattle become adjusted in the feedlot and begin to consume silage, hay can be gradually decreased.

For calves weighing over 400 lb, 50% concentrate appears to be ideal. Lighter calves appear to perform better when given 70% concentrate. Sick calves do not eat at normal levels, and it is sometimes best to feed protein supplement and grain on a per head basis, with hay or silage fed on a free-choice basis.

One should not rely on non-protein nitrogen as a sole source of supplemental protein during this period. Natural protein such as soybean meal should be used until the calves are in good condition or weigh 600 lb.

If the receiving ration contains limited grass or grass-legume forage, supplemental potassium may be beneficial.

No single program is best for all operations; however, attention to management details is essential to obtain optimum performance.

Cattle Feeding Programs*

There are many different feed programs or systems for producing slaughter cattle. Selection depends on amount and quality of feedstuffs, kind of cattle, length of feeding period desired, marketing objectives, facilities, and skill of the operator. Management involves fitting these variables together with labor into an efficient cattle feeding operation to obtain maximum returns.

Because of these many influencing factors, one can understand why the best program for one farm may not be the best for another. In addition, because of economic changes in price relationships between feedstuffs, feeder cattle, and slaughter cattle, the program best suited to a particular farm one year may not be the optimum one for that same farm in another year. Once a workable program is obtained, however, the cattle feeder should not make major changes from year to year, but rather modify the basic plan as economic conditions warrant.

MATCHING FEEDING PROGRAM TO TYPE OF CATTLE

Most cattle feeding programs require a certain kind of cattle if they are to be successful. Type or kind of feeder cattle is usually defined by:

1. Age—calves, yearlings, and two-year-olds.
2. Condition—thin, medium, and fleshy.
3. Weight—light (300 to 500 lb), medium (500 to 700 lb), and heavy (700+ lb).
4. Sex—steers, heifers, and bulls.
5. Quality grade.
6. Other factors—e.g., breed and origin of cattle.

AGE, WEIGHT, AND CONDITION

Since weight increases with age, it is logical that these be considered together. The influence of age, weight, and condi-

* Adapted with permission from Hendrix, K.S. 1975. Cattle feeding programs. Purdue University Cattle Feeders Day, W. Lafayette, IN. April 4.

tion upon rate and efficiency of gain depends on whether the cattle have been fed sufficient energy early on to provide for both growing and finishing or whether they had previously been on a nutritional plane that allowed growth but no fattening.

If cattle are fed a ration high in energy from weaning to slaughter, they tend to gain less rapidly and less efficiently as they get older and heavier. The composition of gain contains more fat and proportionally less bone and muscle as the animal increases in age and weight. More energy is needed for the deposition of fat than for bone or muscle. In addition, because of their larger weight, older cattle have a higher maintenance requirement; thus, more energy is needed for maintenance, leaving less for gain.

If cattle have been fed on a lower plane of nutrition (forages with little or no grain) while young and lightweight, they will have attained much of their skeletal and muscular growth but will be relatively thin. When this type of feeding regime is followed by a higher energy feeding level, daily gains and efficiency of feed utilization increase, mainly because of the larger capacity of this type of animal in relation to that of calves. Therefore, considerable quantities of the energy consumed are used for gain ("compensatory gain"). As a rule, cattle that were previously fed at lower planes of nutrition for less than maximum gains, and that are later fed a higher energy ration, gain 10 to 15% more rapidly

than cattle fed a similar ration for maximum gain from weaning.

SEX

More steers than heifers are available for feeding since a portion of the heifers are saved back for replacements. If fed for similar lengths of time, heifers gain about 10% less than steers and with 10 to 15% lower efficiency of gains. If fed to the same degree of finish, however, heifers gain nearly as rapidly and efficiently as steers. Heifers usually attain a similar degree of finish 40 to 60 days earlier, and at weights that are 100 to 150 lb lighter, than steers of comparable breeding. Therefore, heifers are well suited to shorter feeding periods and to areas where light carcasses are desirable.

Recently, bull feeding has received more attention. In general, intact males produce beef more efficiently than either steers or heifers (Table 13–3). They are weaned at heavier weights, convert feed to gain more efficiently, and produce carcasses of higher yield than steers or heifers. Disadvantages include the following: they need to be kept separate from other cattle; stress must be minimized prior to slaughter, to prevent dark cutters; many graders discount bullock beef; the fat cover is often too thin; and special feeding facilities are required in some situations.

SPAYED VS. OPEN HEIFERS

The spaying of heifers is a relatively easy operation involving removal of the ovaries

TABLE 13–3. Animal Performance of Bulls vs. Steers

Item	Bulls	Steers
Age in days at slaughter	559	555
Initial weight (lb)	790	743
Final weight (lb)	1085	1005
Daily gain (lb)	2.21	1.83
Feed per lb gain (lb)	7.15	8.02

Adapted from Jacobs, J.A., et al. 1975. Profitability and consumer acceptability of beef from feedlot bulls versus steers. University of Idaho. Moscow, ID. Bulletin 556.

from the abdominal cavity. Spaying is performed to avoid the disturbances caused by heifers in heat. At present, the principal reasons for spaying heifers are:

1. To allow neutered heifers to run with steers or with the breeding herd without disturbance or risk of being bred.
2. To assure the feedlot operator that the heifers he purchases are open and carrying a calf.
3. To give heifers a more tranquil disposition in the feedlot, for which feedlot operators may pay a premium.

Based on research results from the early 1950's and more recent data, however, spayed heifers seldom have made gains in the feedlot as large and as economical as those made by open heifers. Apparently, removal of the ovaries retards the growth and development of young heifers (Neumann, 1977). Also, in the past, spaying became impracticable because of its high cost, and the market would not pay a premium for spayed heifers.

Spaying cannot be recommended, on the basis that it lessens activity and improves feedlot performance, since the performance of spayed heifers is poorer than that of open heifers. Therefore, the only advantage of spaying must be the certainty that heifers are open and not in danger of becoming pregnant by the time they are ready for market.

The feed additive MGA suppresses estrus in feedlot heifers and also acts as a growth-promoting agent. For MGA to be effective, however, the heifers must be spayed and must have functional ovaries. Other anabolic growth-promoting products such as Ralgro (resorcyclic acid lactone) and Synovex-H (200 mg testosterone and 20 mg estradiol benzoate) are effective in spayed heifers and can counteract their depressed performance.

Feeding Systems

In general, there are three types of feeding programs for producing slaughter cattle: immediate finishing, deferred finishing, and pasture feeding. The following sections discuss each program and give an example of how it works.

IMMEDIATE FINISHING PROGRAM

Under this plan, cattle are brought up to high-energy rations and are finished for slaughter in the shortest possible time (Table 13–4). Length of feeding period depends on the initial age, weight, and condition of cattle on the degree of finish desired when slaughtered. Calves usually require 240 to 260 days, yearlings require 120 to 150 days, and two-year-olds require 90 to 100 days.

Immediate finishing programs are usually best suited for:

1. Cattle in fleshy condition, such as creep-fed calves or yearlings that may have been fed grain while on pasture.
2. High-quality choice feeder steers to be fed to achieve a similar slaughter grade.
3. Cattle with rapid gain potential and heavy, mature weights.
4. Older, heavier steers (800 lb or more).
5. Feeding operations that have large amounts of grain but limited roughage or that do not have facilities for storing and feeding large quantities of roughage.
6. Commercial feeding operations or operations with expensive facilities where a rapid turnover rate is important for maximum returns.

DEFERRED FINISHING PROGRAM

This program involves a growing phase and a finishing phase. The growing phase may extend from 2 to 5 months, during which time the cattle are fed to gain from 0.5 to 2.0 lb per day. Rations during this period are based mainly on roughages (Table 13–5). If grains are fed, they are usually limited to no more than 1% of body weight. The growing period is often referred to as a stocker or backgrounding program, and is well suited for the marketing of harvested roughages, pasture, and crop residues through cattle.

Often, the growing program is the only feeding program followed on a particular ranch. It can be used to complement a cow

TABLE 13-4. Feed Requirements for Feedlot Cattle

Type of Cattle and Total Gain	Feeding Program						
	Corn silage; no corn	Corn silage; shelled corn (1.0% body wt)	Corn silage; shelled corn (1.5% body wt)	Corn silage; shelled corn (2.0% body wt)	Hay; shelled corn (1.0% body wt)	Hay; shelled corn (1.5% body wt)	Hay; shelled corn (2.0% body wt)
450-lb Steer calf to gain 700 lb							
Market wt: 1100 lb[a]							
Silage of hay (tons)	8.3	4.0	2.2	0.75	2.1	1.2	0.5
Shelled corn (bu)	none	43.5	61.0	77.0	50.0	64.0	79.0
Supplement (lb)	350.0	300.0	285.0	270.0	175.0	225.0	275.0
Projected daily gain (lb)	2.0	2.3	2.45	2.6	2.0	2.3	2.55
400-lb Heifer calf to gain 550 lt							
Market wt: 910 lb[a]							
Silage or hay (tons)	7.0	4.2	3.0	2.0	1.7	1.0	0.5
Shelled corn (bu)	none	29.0	41.0	52.0	43.0	56.0	65.0
Supplement (lb)	290.0	244.0	230.0	215.0	1070.0	220.0	262.0
Projected daily gain (lb)	1.9	2.25	2.4	2.55	1.6	1.85	2.1
650-lb Yearling steer to gain 500 lt							
Market wt: 1100 lb[a]							
Silage or hay (tons)	6.4	3.3	2.4	0.95	1.61	1.0	0.5
Shelled corn (bu)	none	32.0	42.0	56.0	38.0	50.0	60.0
Supplement (lb)	220.0	200.0	185.0	175.0	120.0	156.0	185.0
Projected daily gain (lb)	2.3	2.5	2.65	2.8	2.1	2.4	2.7
550-lb Yearling heifer to gain 400 l							
Market wt: 910 lb[a]							
Silage or hay (tons)	5.6	3.4	2.4	1.6	1.5	1.1	0.6
Shelled corn (bu)	none	23.0	33.0	41.0	30.5	45.0	50.0
Supplement (lb)	190.0	174.0	163.0	114.0	171.0	186.0	none
Projected daily gain (lb)	2.1	2.3	2.45	2.6	1.75	1.95	2.15

[a] This is net market weight. Allowance is made for 4.5% of the slaughter animal to cover shrinkage in transport to feedlot and to market.

TABLE 13-5. Suggested Deferred Finishing Programs

Program I
Description: Steer calf weighing 450 lb wintered on good quality hay to 750 lb (170-day growing period for 300-lb gain) and finished to 1050 lb with alfalfa grass hay and barley (120-day finishing period for 300 lb gain).

| | Time On Feed | | Total Feed Required |
Ration Fed	Days 1 to 170	Days 171 to 290	
Hay (alfalfa grass)	16 lb	5–6 lb	1.66 tons
Barley (grain mixture)	—	18 lb	1.10 tons
Protein supplement	1 lb	1 lb	290 lb

Program II
Description: Light yearling steer weighing 600 lb grazed on stalk field for 60 days to 650 lb and wintered on corn silage to 880 lb (135-day growing period for 200-lb gain), then finished to 1050 lb with corn and corn silage (100-day finishing period for 250-lb gain).

| | Time on Feed | | | Total Feed Required |
Ration Fed	Days 1 to 60	Days 61 to 135	Days 136 to 235	
Grazing stalks	Free choice	—	—	1–2 tons
Corn silage	—	45 lb	10–15 lb	2.2 tons
Corn	—	—	18 lb	0.9 ton
Protein supplement	2 lb	2 lb	2 lb	470 lb

Adapted with permission from Hendrix, K.S. 1975. Cattle feeding programs. Purdue University Feeders Day, W. Lafayette, IN. April 4.

herd when additional roughage is available. Instead of selling the calf crop at weaning, the producer keeps the calves over the winter to sell them in the spring, or to graze them during the summer and sell them as yearlings the following fall. This program permits selection of replacement heifers as yearlings rather than as weanling calves. If a cow herd is not involved, the operator may purchase cattle in the fall for winter feeding in drylot, or may purchase light yearlings in the spring and graze them all summer or until pastures mature or become dormant.

COMBINATION GROWING-FINISHING OPERATION

Calves or light yearlings are purchased in late summer or fall and are grazed on pasture aftermath, small grain stubble, or stalk fields. After most of these feeds have been salvaged, the cattle are usually put in drylot and fed corn silage or high-quality hay plus limited grain. They eventually attain a condition whereby gains are reduced below a level that permits economical conversion of roughages to gains. This point is usually reached after 60 to 90 days for yearlings and after 120 to 150 days for calves. Then, the cattle are finished on a full ration of grain (2% of body weight) plus sufficient roughage to keep them on feed and to avoid digestive upsets.

PASTURE FEEDING PROGRAMS

During the grazing season, pasture can provide essentially all of the nutrients for the growing phase or can serve as a supplement to grain for the finishing phase. Pasture is especially adapted to the small farmer-feeder or to the feeder who maintains a cow herd.

Yearling cattle should gain 150 to 200 lb during the grazing season without supplemental grain. Performance depends on amount and quality of pasture available plus condition of the cattle at the beginning of the grazing season. Pastures mature and become dormant during July and August, resulting in relatively low-quality grazing during that time. Grass pastures containing 25 to 50% legumes are of higher quality and produce faster gains than all-grass pastures.

Condition of the cattle depends mainly on their rate of gain during the previous winter. The more rapid the winter gains, the slower the gains are on pasture, provided that no supplemental grain is fed.

The effect of winter gain upon subsequent pasture gains varies considerably; however, a general rule states that for every additional pound gained during the winter, gains on pasture are reduced by ½ lb (Table 13–6).

If cattle are to be given full rations of

TABLE 13-6. Three Suggested Pasture Feeding Programs

Program I

Description: Steer calf weighing 450 lb wintered on hay for 150 days to 660 lb and grazed without grain for 180 days to 800 lb (330-day growing period for 350-lb gain), then finished in drylot to 1100 lb on hay and corn (110-day finishing period for 300-lb gain).

	Time on Feed			Total Feed Required
Ration Fed	Days 1 to 150	Days 151 to 330	Days 331 to 440	
Pasture	—	Free choice	—	1 acre
Hay	FF (12 lb)[a]	—	4 lb	1.1 tons
Corn or barley	2 lb	—	18 lb	1.1 tons
Protein supplement	0.1 lb	—	1 lb	110–250 lb

Program II

Description: Steer calf weighing 450 lb wintered on hay and corn for 150 days to 600 lb and grazed without grain for 60 days to 700 lb (210-day growing period for 250-lb gain), then finished on pasture with full ration of grain to 1050 lb (140-day finishing period for 350-lb gain).

	Time on Feed			Total Feed Required
Ration Fed	Days 1 to 150	Days 151 to 210	Days 211 to 350	
Pasture	—	Free choice	Free choice	0.51 acre
Hay	FF (12 lb)[a]	—	—	0.9 ton
Corn or barley	2 lb	—	SF (16 lb)[a]	1.2 tons
Protein supplement	0.1 lb	—	0–1 lb	140–290 lb

Program III

Description: Steer calf weighing 450 lb wintered on corn silage to 750 lb (170-day growing period for 300-lb gain), then finished on pasture with full ration of corn to 1050 lb (120-day finishing period for 300-lb gain).

	Time on Feed		Total Feed Required
Ration Fed	Days 1 to 170	Days 171 to 290	
Pasture	—	Free choice	0.5 acre
Corn silage	35 lb	—	3.0 tons
Corn or barley	—	SF (16 lb)[a]	0.9 ton
Protein supplement	2 lb	0 to 1 lb	340–460 lb

[a] Full-fed (FF); self-fed (SF). Amounts in parentheses are estimates of the average feed consumed daily during the period indicated.

Adapted with permission from Hendrix, K.S. 1975. Cattle feeding programs. Purdue University Cattle Feeders Day, W. Lafayette, IN. April 4.

grain on pasture, wintering at a higher rate of gain is usually best, so that cattle can be finished for slaughter at the end of the grazing season. Cattle fed considerable quantities of grain prior to grazing should be put on pasture early, before it becomes lush. A good practice is to feed cattle a full ration of shelled grain, put them on pasture in March or by mid-April, and then allow them to feed on grain on a free-choice basis while on pasture.

If grain feeding is begun after the cattle are on pasture, they can be started on a self-feeding program by mixing in 40 to 50% ground hay with the grain, then gradually reducing the roughage so that they are on grain in approximately 3 weeks. Studies have shown that no supplemental protein is needed if pasture is still growing and contains some legumes. During midsummer, when pasture quality deteriorates, supplemental protein should be provided.

Because of differences in soil fertility and pasture growth, stocking rate recommendations are difficult to make. A general guideline is that soils capable of producing 2.0 to 2.5 tons of hay per acre can carry one yearling steer on 1.0 to 1.5 acres during the grazing season if some grain is fed. Stocking rate could be increased to 2 to 3 head per acre if the cattle are fed grain while on pasture. In fact, some studies show that with good management, nearly four head of cattle can be grazed successfully on one acre when they are fed grain.

Formulation of Rations

Ration formulation may seem complex when one considers the number of possible ingredients available (roughages, grains, and supplements). Formulation can be simplified, however, with use of a systematic approach. To formulate most rations, one must know the weight of the animal, its nutrient requirements, and the available feeds and their nutrient composition.

The steps to follow when formulating a ration for beef cattle are:

1. Select proper nutrient requirements (NRC, 1984).
2. Evaluate available feeds.
   a. Determine nutrient composition.
   b. Determine availability of important nutrients.
   c. Determine nutrients needed to balance the ration.
   d. Determine cost per unit of nutrient.
   e. Determine value of other nutrients in feedstuff not specifically considered in balancing the diet (meat-and-bone meal is used as a protein supplement but they also provide significant amounts of phosphorus).
3. Select a suitable set of ingredients so that the ration is nutritionally balanced, safe, palatable, and economical.
4. Calculate the amount of each feed needed to make up the desired ration.

The following items are usually considered in formulating rations for beef cattle:

1. Dry matter intake
2. Protein
3. Energy
4. Calcium
5. Phosphorus
6. Vitamin A or carotene
7. Potassium

METHODS OF FORMULATION

At present, most rations are formulated using the Pearson Square method, algebraic or simultaneous equations, or computer least cost formulation.

The Pearson Square method can only be used with two feed ingredients or mixtures. Its use is beneficial when only one or two nutrients are being considered in the formulation.

Example. Formulate a ration containing 13% crude protein (CP) using corn containing 9.6% CP and soybean meal containing 49% CP.

a. Place the percentage of CP desired in the combination of the two feeds in the center of a square and the percentage of CP of each feed at the left corners as shown:

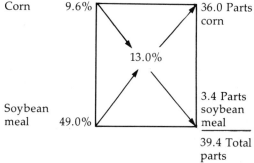

Corn   9.6%   36.0 Parts corn

13.0%

Soybean meal   49.0%   3.4 Parts soybean meal

                             39.4 Total parts

b. Subtract diagonally across the square, the smaller number from the larger, without regard to sign, and record the difference at the right corners.

c. The parts of each feed can be expressed as a percentage of the total, and these percentages can be applied to any quantity.

$$\frac{36.0 \text{ Parts corn}}{39.4 \text{ Total parts}} \times 100 = 91.37\% \text{ Corn}$$

$$\frac{3.4 \text{ Parts soybean meal}}{39.4 \text{ Total parts}} \times 100 = 8.64\% \text{ Soybean meal}$$

d. Check:

91.37% Corn × 9.6% CP = 8.77% CP

8.64% Soybean meal × 49.0% CP = 4.23% CP

                                  13.00% CP

The problem can also be solved using algebraic equations:

a. x = lb Corn

y = lb Soybean meal

    x + y = 100 lb ration

    0.096x + 0.49y = 13.0 lb CP
    (13% of 100 lb)

b. Develop a third equation to be subtracted from the second equation to cancel out either x or y. Multiply the first equation by a factor of 0.096 to form:

     0.096x + 0.49y = 13.0

    −0.096x − 0.096y = −9.6

            +0.394y = +3.4

c. $y = \dfrac{3.4}{0.394} = 8.36$ lb (SBM)

    x = 100 − 8.63 = 91.37 lb (Corn)

Simultaneous equations can be used for a mixture of two feeds to meet two requirements.

a. Nutrient requirements: 1.4 lb CP
                             8.5 lb TDN

b. Feed composition
Hay: 16% CP   Grain: 11% CP
       60% TDN   80% TDN

c. Mathematical procedure
x = lb Hay
y = lb Grain

d. 0.16x + 0.11y = 1.4 lb CP
   0.60x + 0.80y = 8.5 lb TDN

e. Develop a factor to cancel out x or y:
$$\frac{0.6}{0.16} = 3.75$$

f. (0.16x + 0.11y = 1.4)(3.75)

g.   0.60x + 0.80y   = 8.50

   −0.60x − 0.4125y = 5.25

     0   + 0.3875y = 3.25

h. $\dfrac{3.25}{0.3875} = 8.39$ lb Grain

i.   0.16 + 0.11 (8.39) = 1.4
   0.16 + 0.9229 = 1.4
   0.16x = 1.4 − 0.9229
       $x = \dfrac{0.4771}{0.16}$
      x = 2.98 lb Hay

j. Check:
   0.16 (2.98) + 0.11 (8.39) = 1.4 lb CP
   0.60 (2.98) + 0.80 (8.39) = 8.5 lb TDN

At present, more than 70% of the total feed in the United States is formulated by least cost linear programming. Technologic improvements in the computer industry and the development of the microcomputer have placed the capabilities of least cost formulation within the range of feeders, who previously could not afford such a service.

Least cost formulation accepts or rejects feed ingredients based on costs and nutritive value. This method enables the nutritionist to compare a wide range of feedstuffs to determine which will blend together to provide desired nutrient levels at the least cost. Linear programming is a mathematical technique for determining the optimum allocation of resources (feedstuffs) to obtain a particular objective when there are alternate uses for the resources. In addition to formulating the ration with the least cost, this method can be used to analyze the economics of alternate availability of resources or to explore the effect of changes in nutrient specifications of the diet on the selection of feed ingredients.

To formulate a ration using linear programming, the following information must be known:

1. Nutrient composition of feed ingredients.
2. Cost of feed ingredients.
3. Minimum and maximum allowable amounts of all feed ingredients in the ration (restrictions).
4. Nutrient requirements of the animal.
5. Minimum and maximum allowable amounts of each nutrient in the ration (e.g., 10.5% CP minimum; 13.0% CP maximum).

The minimum and maximum amounts of a feed ingredient allowed in a ration are determined by the nutritionist based on the ingredient's usefulness in a formulation. Nutrient composition of a feed ingredient reveals little about its palatability and its ability to blend with other feed ingredients.

HEALTH PROGRAMS

Many disease and pest problems associated with feedlot cattle have been discussed in earlier chapters. A discussion of all diseases related to feedlot cattle is beyond the scope of this text. Jensen and Mackey (1979) provide an excellent text for those interested in such a detailed discussion (see "Selected Readings").

Health problems are associated with the management and nutritional facets of the feedlot. Most feedlots employ the services of resident veterinarians, who give instructions regarding the proper use of drugs, their administration, and their dosage levels. One of their responsibilities is to keep the management personnel up-

dated on the latest research, new products, and disease conditions.

The veterinarian uses a record-keeping system to measure the performance of animal health in the feedlot. Average death loss provides an index current with the feedlot cattle inventory. Large feedlots usually keep average death loss on a per thousand head per month basis. This figure is calculated as follows:

1. Establish the average daily cattle inventory for the month.
2. Divide this figure into the death loss for the month, move the decimal three places to the right to express the final value on a per thousand head basis. For example:

June average daily cattle inventory
= 5000 head

June death loss
= 15 head

$$\text{Loss per 1000 head} = \frac{15}{5000} \times 1000$$

$$= 3 \text{ head}$$

This figure is current and gives the veterinarian a good indication of the health status of the feedlot.

Many diseases in the feedlot are of nutritional origin. The following paragraphs offer brief descriptions of these diseases.

Acute indigestion is caused by the ingestion of larger than normal amounts of grain that contain high levels of readily fermented carbohydrate (grain). Cereal grains such as wheat are often associated with this condition.

Clinical signs appear from 2 to 3 hours to 2 to 3 days after the ingestion of a larger than normal amount of grain. Death losses may occur as early as 6 to 7 hours, or up to an interval of several days, after ingestion. Animals lose their appetite (anorexia), rumen motility stops, and diarrhea follows. The pH of the ruminal contents may fall to a range of 4.8 to 5.0. This condition is referred to as "acidosis."

The epithelia of the rumen exhibit localized necrosis (death of tissue), edema, and hemorrhage caused by penetration by such bacteria as *Sphaerophorus necrophorus*. In addition, these bacteria may move farther into the lining of the ruminal epithelia and eventually enter the portal veins, which draw blood from the rumen to the liver. The organisms become lodged in the liver, and they set up secondary sites of infection (liver abscesses). Data collected from Nebraska from 1965 to 1969 indicated that out of 25,222 steers fed to slaughter weight, approximately 24.8% of the livers were condemned owing to abscesses. Cattle with abscesses had a significantly decreased quality grade, dressing percentage, and average daily gain (2.57 versus 2.42 lb) as compared with non-abscessed cattle.

Some animals with acute indigestion also develop acute laminitis or founder. The development of founder is poorly understood. Presumably, blood vessels inside the hooves dilate and become engorged with blood. As a result of a local increase in blood pressure, transudate (fluid passing through a membrane) accumulates in the connective tissue. Pressure of the nerves inside the hooves from engorgement and edema causes pain and rapid growth of the hoof wall. Feedlot animals that develop founder have severely depressed performance, and therefore are slaughtered since treatment is seldom satisfactory.

Acute indigestion is avoided by preventing cattle from having accidental access to excessive amounts of concentrate or grain. In the feedlot, the ration should be changed gradually from a low proportion to a high proportion of concentrate.

Polioencephalomalacia (PEM), also called polio or circling disease, is associated with feedlot cattle that consume rations high in concentrate. It is characterized by a central nervous system disorder, circling, and in some but not all cases, blindness. The disease is noninfectious and reponds to thiamine injection. Animals with PEM should be switched to a ration containing more roughage. Studies indicate that gradual adaptation to rations

high in concentrate may help to prevent the disease.

"Water belly," or urinary calculi, is a noninfectious disease of cattle, characterized by the formation within the urinary tract of stones composed of salt crystals (phosphates, silicates). The stones are formed in the bladder and cause no discomfort unless they enter the urethra and prevent the normal discharge of urine. Animals with calculi are extremely nervous, refuse to eat, and get up and down at frequent intervals. If the bladder ruptures, temporary relief occurs, but peritonitis caused by urine in the abdomen develops.

BEEF CARCASS EVALUATION

Beef is a prestige meat item. It is and has long been a quality product. At present, consumers look for high-quality lean meat with a minimum of waste (either fat or bone), and producing that product should be the goal of the beef industry.

Factors of Economic Importance

LIVE WEIGHT

High-priced carcasses weigh from 600 to 700 lb cold weight. Therefore, steers weighing 1000 to 1150 pounds are worth the most money.

DRESSING PERCENTAGE

$$\frac{\text{Hot carcass weight}}{\text{Live weight}} \times 100 = \text{Dressing percentage}$$

Most choice 1000- to 1500-lb animals dress between 62 and 64%.

QUALITY GRADE

Quality evaluation represents an attempt to predict palatability characteristics; therefore, factors related to tenderness, juiciness, flavor, and appearance are given major consideration.

Maturity and marbling are the two main factors used in the USDA quality grading of beef.

Maturity

Maturity is an estimation of the physiologic age of the animal that produced the beef carcass. The maturity of the carcass is determined by evaluating the size, shape, and ossification of the bones and cartilage as well as the color and texture of lean flesh. Maturity is one of the best indicators of meat palatability because it is closely related to meat tenderness. Five maturity groups are identified in Figure 13–1 as A, B, C, D, and E in order of increasing maturity. The term "hard bone" is sometimes used to describe carcasses that have been downgraded because of advanced maturity.

Marbling

Marbling is the visible intramuscular fat that is observed as flecks of fat in the cross-sectional surface of the rib eye. Marbling tends to improve the juiciness and flavor of meat, and it may also have a positive effect on meat tenderness. Descriptive terms are used to describe the degrees of marbling as illustrated in Figure 13–1.

Color of Lean

Color of lean affects the appearance of meat cuts, which in turn affects consumer acceptability. Dark color in meat is usually associated with the more advanced stages of maturity in animals. Dark cutters, however, may occur in young cattle. When dark cutters do occur, the USDA grade of the beef may be reduced as much as one full grade, depending on the severity of the dark color. Highly excitable animals, or normal animals subjected to severe stress, may produce dark-cutting carcasses.

Texture

Coarse texture of meat is usually associated with the more advanced stages of maturity. Coarse texture in the rib eye results in a slightly older maturity designation.

Firmness of Lean

Occasionally, lean meat tends to be soft and watery. In this condition, the ability of lean tissues to bind the natural moisture within the fibers is low. Meat from this type of muscle may be drier than normal.

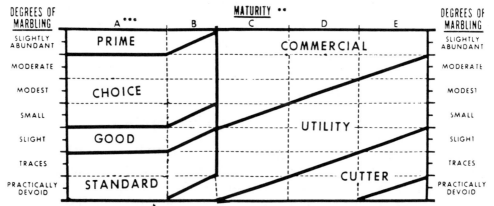

RELATIONSHIP BETWEEN MARBLING, MATURITY, AND CARCASS QUALITY GRADE •

*Assumes that firmness of lean is comparably developed with the degree of marbling and that the carcass is not a "dark cutter."
**Maturity increases from left to right (A through E).
***The A maturity portion of the Figure is the only portion applicable to bullock carcasses.

FIG. 13–1. USDA quality beef grade standards. (Courtesy of the United States Department of Agriculture.)

YIELD GRADE

The USDA yield grades for beef can be used to estimate the yield (amount) of salable retail meat from a beef carcass. There are five USDA yield grades numbered 1 through 5. Yield Grade 1 carcasses have the highest expected yield of retail cuts, while Yield Grade 5 carcasses have the lowest retail yield and the highest percentage of fat.

The following four factors are used to establish the USDA yield grade:

1. *Hot carcass weight.* The carcass is weighed immediately after slaughter but prior to shrouding and chilling. If a chilled weight must be used, it should be adjusted for a 2 to 3% cooler shrinkage.

2. *Rib eye area at the 12th rib.* The cross-sectional area of the longissimus muscle is measured in square inches. The evaluator should be careful not to include the small muscles surrounding the longissimus muscle in the measurement of rib eye area.

3. *Adjust fat thickness.* Fat thickness is measured over the rib eye at the 12th rib location at a point 3/4 of the length of the longissimus muscle from the chine bone end, and perpendicular to the outside surface of the fat. This measurement should be adjusted upward or downward to reflect unusual amounts of fat on other parts of the carcass, or to compensate for excessive trimming of fat over the rib eye.

4. *Estimated percentage of kidney, pelvic, and heart (KPH) fat.* The amount of fat in the kidney knob (kidney and surrounding fat), the amount of lumbar and pelvic fat in the loin and round, and the amount of heart fat in the chuck and brisket area, which are normally removed in making loosely trimmed retail cuts, are evaluated subjectively and expressed as a percentage of the carcass weight. A carcass with an average amount of fat in these locations equals approximately 3.0 to 3.5% of the hot carcass weight. Carcasses that are extremely trim with regard to internal fat may have as little as 1% KPH fat, while ex-tremely fatty carcasses may have as much as 4.5 to 5.5% KPH fat.

Yield Grade Equation

The yield grade of a beef carcass is determined by the following equation:

USDA Yield Grade
= 2.5 + [0.0038 × hot carcass weight (lb)]
− [0.32 × rib eye area (sq. inches)]
+ [2.5 × adjusted fat thickness (inches)]
+ [0.2 × percentage of KPH fat]

The official USDA yield grade is always expressed as a whole number; any fractional part of the designation is dropped. For example, if a computation results in a designation of 3.86, the official USDA Yield Grade is 3.0—it is not rounded up to 4.

Table 13–7 shows that the expected yield of salable cuts from a dressed beef carcass changes by 4.6% with each change of one full yield grade. For example, a 700-lb, yield grade 2.5 carcass would be expected to yield 32.2 lb more salable retail cuts than a 700-lb, yield grade 3.5 carcass. If the average retail value of beef is about $2.40 lb for choice beef, the difference in retail value between the two carcasses would be $77.20.

FEEDLOT ECONOMICS AND MARKETING

Costs of Feeder Cattle*

Feedlot operators must carefully estimate the prospects for making a profit before buying animals. Although market price at selling time is difficult to predict, one can determine the price necessary to recover the cost of the feeder animal plus feed and other costs of production. Comparison of this break-even selling price with future cattle price predictions can help in deciding whether one should buy a particular group of feeders or take alternative actions. Table 13–8 describes seven

* Adapted with permission from Petritz, D., and K. Hendrix. 1979. How much can I afford to pay for feeder cattle? Cooperative Extension Service, Purdue University. W. Lafayette, IN. ID-108.

TABLE 13-7. Relation of Yield Grade to Expected Retail Yield

Yield Grade	Predicted Retail Yield (%)
1.0	84.30
1.5	82.00
2.0	79.70
2.5	77.40
3.0	75.10
3.5	72.80
4.0	70.50
4.5	68.20
5.0	65.90
5.5	63.60

feeding systems. Tables 13–9 and 13–10 present the necessary returns and costs for each system. Table 13–11 is a budget form that can be used to estimate the production costs and the necessary selling prices for a particular feeding program.

Marketing*

The three major methods of marketing fed cattle in the United States are direct country, terminal, and auction. In 1979, 60% of the slaughter steers and heifers were marketed directly to the packer (Wirak, 1983). Terminals accounted for 8% and auctions accounted for 32%.

Twenty-five years ago, terminal markets were the major outlet for fed cattle. Since 1969, however, the percentage of fed cattle marketed through terminals has declined from 21% to 8%.

Public auctions have never been as important for fed cattle as for other species and classes of cattle. Auction markets are the major method of marketing slaughter cows and bulls.

Meat packers prefer buying fed cattle by direct methods, which allow them more control over weighing conditions. Also,

* Adapted with permission from Wirak, O. S. 1983. Buying feeders and selling fed cattle. *In* The Feedlot. Edited by C.B. Thompson and C. O'Mary. Lea & Febiger, Philadelphia, 1983.

direct buying allows packers to coordinate the flow of cattle to their plants and to improve slaughter efficiency through better utilization of labor and facilities. At terminal and auction market, buyers can bid only on cattle available on that particular day.

Cattle feeders also can benefit from direct marketing. The cattle feeder's bargaining position is improved when the price is negotiated and settled prior to the time at which cattle are shipped.

Other methods of marketing cattle include country feedlot sales, producer-owned cooperatives, and joint-venture agreements. The advantage of country feedlot sales is that it increases the number of bids on a given group of cattle. The advantage of producer-owned cooperatives and joint-venture agreements is that it allows sharing of profits and losses, and greater control over marketing.

Negotiating Cattle Prices

Cattle feeders are usually paid according to these criteria: live weight, dressed weight, and grade. Most cows and bulls are purchased on a live-weight basis, as are most cattle bought at auction and terminal markets.

The price that a meat packing firm bids on a particular lot of cattle is influenced by (1) the price for a specific sex-weight-grade-yield grade (SWGYG) carcass as reported by the National Provisioner Daily Market News Service (the yellow sheet), (2) the reported price differences from the base type carcasses, (3) the by-product value, (4) the packer's own slaughtering cost, and (5) a profit target.

The field buyer purchases cattle based on the purchase order after adjusting the price for the percentage of cattle in the SWGYG categories, dressing percentage, cost of transporting to the plant, weighing conditions and pencil shrinkage, bargaining skills of the seller, cattle health, buying competition, and other factors.

TABLE 13-8. Approximate Feed Requirements and Rates of Gain for Seven Alternative Cattle Feeding Programs (Average Farm Conditions)[a]

Item	Choice Steer Calves		Choice Heifer Calves		Choice Yearling Steers		Dairy Steers Fed to Good
	High Grain	High Silage	High Grain	High Silage	High Grain	High Silage	High Silage
System number	1	2	3	4	5	6	7
Purchase weight[b]	450 lb	450 lb	400 lb	400 lb	700 lb	700 lb	450 lb
Market weight[c]	1,050 lb	1,050 lb	900 lb	900 lb	1,100 lb	1,100 lb	1,100 lb
Gain (payweight to payweight)	600 lb	600 lb	500 lb	500 lb	400 lb	400 lb	650 lb
Days on farm	250 days	270 days	230 days	150 days	150 days	170 days	270 days
Daily gain[d]	2.66 lb	2.46 lb	2.39 lb	2.20 lb	3.13 lb	2.76 lb	2.65 lb
Feed requirements[e]							
Corn	60 bu	34 bu	52 bu	29 bu	50 bu	27 bu	41 bu
Corn silage[f]	1.25 tons	3.5 tons	1.15 tons	3.1 tons	0.75 tons	2.7 tons	4.00 tons
Supplement (40%)	375 lb	405 lb	345 lb	375 lb	225 lb	225 lb	405 lb
Hay[f]	200 lb	200 lb	200 lb	200 lb	150 lb	150 lb	200 lb
Salt and mineral	25 lb	27 lb	23 lb	25 lb	20 lb	23 lb	35 lb

[a] With high-silage programs, the first half of gain is accomplished by feeding a full ration of corn silage and protein supplement, and the last half of grain is accomplished by feeding a full ration of corn grain, limited silage (10 to 15 lb per day), and protein supplement. Once cattle are on feed in high-grain programs, the total gain is made by feeding a full ration of corn grain, limited silage, and protein supplement. All cattle are assumed to be of average frame size, with calves in moderate condition and yearlings below moderate condition when started on feed. Cattle of larger or smaller frame size are slaughtered at weights 100 to 150 lb heavier or lighter, respectively, and days on feed and feed requirements are adjusted accordingly. All cattle are assumed to be implanted at optimum intervals with a growth stimulant. If a growth stimulant is not used, daily gains are reduced by 10%, and days on feed are increased accordingly. If Rumensin is fed at recommended levels, grain and silage requirements may be reduced by approximately 10%. If lot conditions are muddy, daily gains are reduced by 10 to 15%, and feed requirements are increased by 10 to 15%.

[b] Equal to payweight. Beginning feedlot weight is equal to payweight minus 5.5% shrinkage for steer calves, heifer calves, and dairy steers, and minus 4.5% shrinkage for yearling steers.

[c] Equal to payweight, which is final feedlot weight minus 3.5% shrinkage.

[d] Daily gain necessary to achieve payweight to payweight gain and account for shrinkage of cattle.

[e] Feed requirements given on an as-fed basis: corn grain, 87% dry matter; corn silage, 35% dry matter; hay and supplement, 90% dry matter.

[f] Legume hay or haylage may be substituted into these rations: 500 lb of legume hay or 1000 lb of haylage (17% protein) plus 7 bu of corn grain can replace 1 ton of corn silage and 125 lb of protein supplement.

Adapted with permission from Petritz, D., and K. Hendrix. 1979.

TABLE 13-9. Approximate Production Costs and Necessary Returns for Seven Alternative Cattle Feeding Programs

Item	Choice Steer Calves		Choice Heifer Calves		Choice Yearling Steers		Dairy Steers Fed to Good
	High Grain	High Silage	High Grain	High Silage	High Grain	High Silage	High Silage
System number	1	2	3	4	5	6	7
Market weight	1,050 lb	1,050 lb	900 lb	900 lb	1,100 lb	1,100 lb	1,100 lb
A. *Cost per head*							
Systems 1 and 2: 450 lb at $0.90	$405.00	$405.00					
Systems 3 and 4: 400 lb at $0.80			$320.00	$320.00			
Systems 5 and 6: 700 lb at $0.82					$577.50	$577.50	
System 7: 450 lb at $0.80							$360.00
B. *Cost of gain per head*							
Feed (corn, $2.60/bu; corn silage, $20.50/ton; supplement, $12.50/cwt; salt and mineral, $12.50/cwt; hay, $50/ton)	$236.60	$219.15	$209.80	$193.95	$179.75	$164.05	$248.60
Other variable costs (death, veterinarian, interest, taxes, gas, oil, repairs, market costs)	$ 92.60	$ 94.70	$ 75.50	$ 77.20	$ 74.50	$ 77.90	$ 94.50
Fixed costs (labor, buildings, equipment, machinery depreciation)[a]	$ 70.00	$ 70.00	$ 60.00	$ 60.00	$ 36.00	$ 36.00	$ 70.00
Total costs (A & B)	$804.20	$788.85	$665.30	$651.15	$867.75	$855.45	$773.10
C. *Selling price per cwt to return*							
Feed and other variable costs	$ 69.95	$ 68.45	$ 67.25	$ 65.70	$ 75.60	$ 74.50	$ 63.90
All costs (including fixed)	$ 76.60	$ 75.10	$ 73.90	$ 72.35	$ 78.90	$ 77.75	$ 70.30

[a] These values are adjusted for average age of buildings and equipment, and they are equal to 75% of annual fixed costs for new facilities. Labor wage rate used was $4.00 per hour.

Adapted from Petritz and Hendrix, 1979.

TABLE 13-10. Breakdown of Variable Costs (Dollars per Head) for Seven Alternative Cattle Feeding Programs

Item	Choice Steer Calves		Choice Heifer Calves		Choice Yearling Steers		Dairy Steers Fed to Good
	High Grain	High Silage	High Grain	High Silage	High Grain	High Silage	High Silage
Veterinary and medicinal costs	$ 8.00	$ 8.00	$ 8.00	$ 8.00	$ 6.00	$ 6.00	$ 8.00
Death loss[a]	8.10	8.10	6.40	6.00	5.80	5.80	7.20
Marketing costs[b]	18.00	18.00	15.00	15.00	20.00	20.00	20.00
Power, fuel, repairs	10.00	10.00	9.00	9.00	7.50	7.50	10.00
Interest, insurance, taxes on feed and cattle[c]	46.50	48.60	35.10	36.80	33.70	37.10	47.30
Miscellaneous	2.00	2.00	2.00	2.00	2.00	2.00	2.00
Total variable costs	$92.60	$94.70	$75.50	$76.80	$74.50	$77.90	$94.50

[a] Two percent of purchase value for calves and 1% for yearlings.

[b] Includes commission cost of purchasing feeder animal and hauling to farm, and commission cost of marketing fed animal and hauling to market.

[c] Equal to 11.0% (.11) × the following:

Purchase price of animal + value of corn fed + value of corn silage fed + value of hay fed + one half of the value of supplement, salt, and mineral fed. This value is then multiplied by the fraction of the year that is in the feeding period.

Adapted with permission from Petritz and Hendrix, 1979.

TABLE 13–11. Budget Form for Estimating Break-Even Selling Price for Cattle

Variable costs

1. Purchase price   _____ lb at   _____ /lb =   _____

2. Feed cost

   a. Corn   _____ bu at   _____ /bu = _____

   b. Corn silage   _____ tons at _____ /ton = _____

   c. Supplement   _____ lb at   _____ /bu = _____

   d. Other   _____ at   _____ / = _____

   e. Other   _____ at   _____ / = _____

   Total feed cost   _____

3. Other variable costs

   a. Veterinarian and medicine   _____

   b. Death loss (2% of line 1 for calves,
      1% of line 1 for yearlings)   _____

   c. Marketing, commission charges,
      trucking expense[a]   _____

   d. Power, fuel, equipment repair[b]   _____

   e. Interest, insurance, taxes on feed
      cattle[c]   _____

   f. Miscellaneous   _____

   Total variable cost   _____

4. Fixed costs

   a. Buildings and equipment[d]   _____

   b. Labor   _____ hrs at   _____ /hr[e] = _____

   Total fixed cost

5. Total of all costs (1 + 2 + 3 + 4)

6. Necessary selling price to return

	Total Cost per Head	Necessary Selling Price per Cwt[f]
Cost of feeder and feed costs (1 + 2)	_____	_____
All variable costs (1 + 2 + 3)	_____	_____
All costs (1 + 2 + 3 + 4)	_____	_____

[a] Include commission cost of purchasing feeder animal and hauling to farm, and commission cost of marketing fed animal and hauling to market.

[b] Power, fuel, and equipment repair costs are estimated to be $7 to $10 per head.

[c] Equal to 11% (0.11) multiplied by the following: purchase price of steer, plus value of corn fed, plus value of corn silage fed, plus one half of the value of supplement, salt, and mineral fed. This value is then multiplied by the fraction of the year that represents the feeding period.

[d] Equal to 14% of the current investment in shelter, silage storage, and feed handling, and other equipment associated with the cattle feeding enterprise. Storage for corn is assumed to be included in the price of corn.

[e] Estimated to be 3 to 6 hours per head typically, with cattle having long feeding periods requiring the most amount of time.

[f] Divide costs per head by market weight (pay weight) per head.

Adapted with permission from Petritz and Hendrix, 1979.

Live weight pricing is most common and is based on the live weight of the cattle. The buyer estimates the number of cattle that are in quality and yield grade classes, the percentage of cattle producing heavy or light carcasses, the live weight, and the dressing percentage.

Cattle may be sold on either a dress weight basis or an "in the beef" basis. The latter method is more common in farm feeding areas than in commercial feedlots. "In the beef" pricing may be of two types. The first type is a flat price with no discounts or premiums. All cattle in the sale return the same amount. The second type is a multiple price system. In effect, this system is an "in the beef" bid with discounts or premiums for cattle other than choice grade. Yield grade 3 steers produce 600- to 700-lb carcasses.

Cattle sold on the basis of dress weight and grade are priced on their actual weight, quality grades, and yield grades. The base price, premiums and discounts are usually those reported on the yellow sheet. Because they are not estimates by buyers, the buyers assume less risk, and feeders assume more risk.

Forward-contracting cattle, which means selling some cattle prior to the time of delivery, is used by some cattle feeders. Price is usually specified in the contract and may be tied to a future market price.

REFERENCES

Church, D.C. 1977. Livestock Feeds and Feeding. O & B Books, Corvallis, OR.

Gee, K.C., R.N. Vanarsdall, and R.A. Gustafson. 1979. U.S. fed-beef production costs, 1976–77, and industry structure. Agricultural Economics Report 424. ESCS-USDA, Washington, DC.

Hendrix, K.S. 1975. Cattle feeding programs. Purdue University Cattle Feeders Day, W. Lafayette, IN. April 4.

Hendrix, K.S. 1978. Presentation at the 1978 Purdue Feed Industry Conference. Purdue University, W. Lafayette, IN. February 22.

Jacobs, J.A., A.D. Howes, J.C. Miller, et al. 1975. Profitability and consumer acceptance of beef from feedlot bulls versus steers. University of Idaho. Moscow, ID. Bulletin 556.

Neumann, A.L. 1977. Beef cattle. 7th Ed. John Wiley and Sons, New York.

NRC. 1984. Nutrient requirements of beef cattle. 7th Ed. National Research Council, National Academy of Sciences, Washington, DC.

Perry, T.W. 1980. Beef cattle feeding and nutrition. Academic Press, New York.

Petritz, D., and K. Hendrix. 1979. How much can I afford to pay for feeder cattle? Cooperative Extension Service, Purdue University. W. Lafayette, IN. ID-108.

Thompson, G.B. 1983. The future of cattle feeding. In The Feedlot. Edited by G.B. Thompson and C.C. O'Mary. Lea & Febiger, Philadelphia.

Williams, W.F., and R.A. Dietrich. 1966. An international analysis of the fed-beef economy. Agricultural Economics Report 88. Economics Research Service, Oklahoma and Texas Agricultural Experiment Station, USDA, Washington, DC.

Wirak, O.S. 1983. Buying feeders and selling fed cattle. In The Feedlot. Edited by G.B. Thompson and C.C. O'Mary. Lea & Febiger, Philadelphia.

Selected Readings

Jensen, R., and D.R. Mackey. 1979. Diseases of Feedlot Cattle. 3rd Ed. Lea & Febiger, Philadelphia.

QUESTIONS FOR STUDY AND DISCUSSION

1. What is the most popular cereal grain fed to cattle in the United States?

2. Why is it important to process sorghum grain before feeding it to feedlot cattle?

3. What precautions should be observed when feeding wheat to feedlot cattle?

4. a. Define protein supplement.
   b. What factors influence the selection of a protein supplement?

5. Outline a management program for feeding urea in beef cattle rations.

6. What limits the amount of hay and silage fed in feedlot rations?

7. Briefly describe the following grain processing methods:
   a. Grinding
   b. Dry rolling
   c. Popping
   d. Roasting
   e. Micronizing
   f. Extruding
   g. Pelleting
   h. Reconstitution
   i. Ensilage of high-moisture grain

8. Define the following and discuss their use in feedlot programs:
   a. MGA
   b. Antibiotics
   c. Bovatec
   d. Rumensin
   e. Compudose
   f. Ralgro
   g. Synovex S and H
9. Outline a management program for handling shipped-in feeder cattle.
10. What is meant by the term "compensatory gain"?
11. How do the following affect feedlot performance?
    a. Sex
    b. Spaying
    c. Age, weight, and body condition
12. Give examples of when each of the following programs should be used:
    a. Immediate finishing
    b. Deferred finishing
    c. Pasture finishing
13. Outline the steps to follow when formulating rations for feedlot cattle.
14. a. Formulate a 12.5% CP ration using barley grain that contains 10.1% CP and a 32% CP protein supplement.
    b. Why are simultaneous equations used in some formulations instead of the Pearson Square method?
    c. What information must be known to formulate a ration using linear programming?
15. List the cause(s), clinical signs, and method(s) of prevention for the following diseases:
    a. Acidosis
    b. Founder
    c. Polioencephalomalacia
16. What criteria are used in determining the quality and yield grade of a carcass?
17. Why do meat packers prefer buying fed cattle by direct methods?
18. What factors influence the prices that packers pay for feedlot cattle?
19. What is meant by "forward-contracting" cattle?

Chapter Fourteen

Business Aspects of Beef Production

Today's beef cattle producer must have a strong background in business management to survive. Most beef cattle producers take pride in their "free-enterprise" environment and in being "independent"; however, some do not realize that producing beef cattle is a business and not merely a way of life. Many do not want to recognize that in a free enterprise society they must compete with all other businesses for resources (i.e., money, land, and labor), and for the returns that can be realized from these resources. The "cost" of money (interest rate) is determined by the supply and demand for money—not only for agriculture, but for agriculture in competition with manufacturers, builders, retailers, government, and all others who need money as a resource. The price fluctuations in beef are an obvious example of free-enterprise supply and demand, and of the changing return from resources.

Many other businesses have recognized this need to compete for resources and have developed methods and tools to help them compete more effectively. Agriculture has been slow in adopting these tools. To compete effectively, however, the cattleman must understand the language, methods, and tools used by others who are competing for the same resources.

The business of raising cattle is becoming more complex, and the business challenges facing cattlemen are greater than ever before. Most cattlemen are happier working with cattle than with the managerial aspects of their business; however, survival depends on their ability to compete and to manage successfully.

Good managers take pride in what they are doing, make the right decisions most of the time, enjoy interacting with people, have a good applied business background, keep abreast of new technology, and are hungry for new ideas. The basic and/or applied aspects of beef cattle production, such as feeding and breeding management, castration, and dehorning, are also important, but the difference between success and failure depends more on business management than on any other factor. Today's beef cattle production students must understand and appreciate the importance of a good business background if they wish to become involved in the beef cattle industry. This chapter provides the student with background information related to the business of raising beef cattle.

MANAGEMENT*

Ranch management involves the making of ranch business decisions that increase the probability of reaching the producer's (or family's) objectives or goals (Hewitt, 1980a).

Examples of objectives:

1. To protect health.
2. To provide time for recreation.
3. To preserve ranch lifestyle.

---

* Mr. Gilbert Ball, Jr., MBA at the University of Idaho and currently a livestock producer in Weiser, ID, contributed significantly to this chapter.

4. To provide time for community activities.
5. To provide net income equal to a given amount of purchasing power.
6. To build net worth.

A 16-hour day may tend to maximize net income but may not be consistent with good health. Maximizing net income in a given year is sometimes inconsistent with conservation and with income maximization over a longer period.

Management is applicable to business; it is equally applicable to ranching. Management involves the abilities to control, to direct, and to make things happen. "Good management" refers to the proper application of these abilities (Evans, 1980a). Hard work in itself does not necessarily bring success to a business operation; accomplishing the wrong things may require the same amount of effort as accomplishing the right things. Managing is productive and can actually reduce work. Successful management, however, is measured less by the amount of work than by the type of work. Work that leads toward predetermined goals is productive work.

Income of ranchers differs greatly, even when the location, size, and type of ranch are the same. Studies have shown that the main factor responsible for most of the differences in earnings is management.

Who makes a good manager? Studies have shown that a successful rancher:

1. Takes pride in his ranch and work.
2. Has ambition.
3. Thinks problems through before taking action.
4. Plans his work and works from his plan (and budget).
5. Keeps adequate records to measure progress.
6. Makes good investments—spends money wisely, bases decisions on functional and practical lines, and prevents emotional and traditional feelings from entering the decision-making process.
7. Displays expertise in modern ranching and keeps abreast of changes in the industry.

Decision Making

As with any business, ranching requires making decisions based on sound judgment. Making decisions that accomplish the *right* things is at the heart of the management process.

There are two types of decisions:

1. *Organizational*—Broad planning for a period of years (policy).
2. *Operational*—Specific implementation of plans (strategy).

Examples of organizational decisions include:

1. Livestock operation (purebred or commercial)
2. Machinery
3. Labor (long-term commitments for input to use in production)
4. Buildings

Examples of operational decisions include:

1. Daily work schedule
2. Purchase of specific tractor
3. Choice of feed ration

Another way of describing the broad types of management decisions to be made is shown in the following list:

A manager must decide:

1. What to produce
2. How much to produce
3. The kinds and amounts of resources to use
4. The technology to employ
5. When and where to sell and buy
6. How to finance the operation

How do I know I have a problem? How do I make good decisions? These are tough questions that can be answered through a systematic series of steps in making and implementing decisions. These steps are described in the following paragraphs (Rue and Byars, 1980).

1. Recognition of a problem. This is probably the most difficult task facing the manager. A problem is said to exist if there is a perceived difficulty, but the real problems may lie in an area other than where the difficulty is felt. For example, the lack of income may be the felt difficulty, but the problems may lie deep within the business organization or even with the spending patterns of the rancher or his family. The detection of difficult management problems is aided by good records.

2. Observation of relevant facts. This is the data-gathering phase as it relates to the problem under study, that is, the learning process. The nature of the data is dictated by the problem. Pertinent information may lie within the records and accounts of the business.

3. Analysis and specification of alternatives. The tools of analysis are primarily those of economic principles and budgeting procedures. Because comparison is fundamental to any decision-making activity, accurate recordkeeping is essential.

4. Choice of alternatives. Most decisions are of an either/or type and require measurements involving some criteria based on welfare (perhaps other than dollar measurements). The benefits and consequences must be weighed for each problem decision.

5. Action and supervision. To act on a given problem, the manager must be properly motivated. Many intelligent and clever people fail because they are unable or unwilling to make decisions and take action.

6. Evaluation and responsibility. A problem is never really solved until the results of decisions have been evaluated. Such evaluation may not be possible until a year or so after the decisions have been put into practice. Usually, the results appear as changes in production, income, net worth, or in some other section of the rancher's recordkeeping set. Evaluation is made by comparing the records before and after the change as well as by comparing the records after the change with the budgeted projections that were used to analyze the alternatives. If such analyses are not made, the farmer may never know whether his management is effective or what further actions are needed. Taking on responsibility without evaluating performance endangers the life of any business.

BUSINESS CHARACTERISTICS OF THE BEEF CATTLE INDUSTRY

Decision-Making Environment

A producer must have a basic knowledge of the broad boundaries of the environment in which his decisions and actions must take place. Some of the more significant characteristics of the cattle business are as follows:

1. The biologic cycle of the cow is for the most part fixed and certain. A fixed amount of time is usually required for production, and this time is difficult to shorten.
2. The rate of turnover in production is relatively slow.
3. The nature of the work is seasonal. Also, ranch labor is less specialized. Ranch work demands a familiarity with many different skills. This situation makes it difficult to obtain high labor efficiency and also makes the job of training new workers for the ranch more complex.
4. In general, the different areas of production are not separated. A producer usually provides labor, capital, and management to the operation.
5. Fixed costs are usually high and constitute a large proportion of the total costs.
6. Ranching involves risk. Exposure to weather and disease adds to the basic risk. The biologic cycle of cattle tends to predetermine many decisions.

Although the cattle industry has several unique qualities, a cattle business is similar to any other business in the following ways:

1. All businesses need to be operated efficiently and effectively if they are to be successful.
2. An "adequate" volume is required for any business to succeed.
3. The business must be studied continuously to bring about improvements in efficiency or size.
4. Sound financial management is important.
5. All businesses are affected by the remainder of the economy.

RECORDS

The foundation of any successful business is a well-organized set of records and accounts to aid in the decision-making process. Successful ranchers claim that one secret of good management is that a recordkeeping system eliminates the need to remember *details*. Agriculture has grown into a businessman's industry. Hard physical labor in itself no longer ensures success. Exercise of the mind is necessary to bring business success. The test of a record system's efficiency is whether it provides the information needed to make management decisions in a form that can be quickly and easily understood and acted upon. The small "business computer" or "home computer" is a tool that can be of aid in keeping and organizing records, but one should be sure that the records meet the necessary standards, and one should not expect the computer to make managerial decisions. Records are only one of the tools that should be used in making managerial decisions, and they must be used in connection with such other tools as budgeting, forecasting, and general economic principles.

Records and accounts are historical by definition; they predict the future only in terms of the past. If a rancher is trying to predict next fall's cattle prices, he heeds information not included in his bookkeeping records, such as outlook reports. Accounts can be useful, however, in identifying the ailing as well as the healthy segments of the business, and in some instances, they can be used in prescribing the appropriate solution. Even though they do not predict the future, they may serve as useful guides.

Recordkeeping Objectives

Major uses of records and accounts on the ranch are described in the following paragraphs (Herbst, 1976).

1. *Management tool.* Records allow the rancher to measure his efficiency in using the factors of production—land, buildings, labor, and other items—and in producing livestock products for sale at a profit. He must gather and evaluate the information from the records if he hopes to improve management and increase profits.

2. *Preparation of income tax reports.* All taxpayers must keep records for accurate preparation of tax returns. While most producers see the need of records for tax reporting, a considerably smaller number

recognize their need for tax management. Experience suggests that farmers who keep poor records pay more taxes than farmers, whose records are accurate and efficient. The objective of tax management is to maximize the after-tax income, which does not necessarily mean minimizing the amount of taxes paid.

3. *Basis for credit.* Records furnish information for planning the credit needs and repayment schedule of the producer as well as for determining his ability to repay, and they provide evidence of security for production loans and security interest agreements. A banker who has records of the borrower's business is able to compare the borrower's past performance against standards for the area. These records also become a basis for projecting and evaluating the future profitability and loan repayment capacity of the business. Records that are properly and accurately kept provide the banker with the financial information needed for prompt handling of credit requests.

4. *Additional uses.* Records also provide the basis for farm lease arrangements and other contracts, farm insurance programs, and participation in government programs.

Components of a Ranch Record and Accounting System

The following text describes some of the important records needed for good management.

Asset and Liability Account. This is a physical and financial account of all ranch resources (assets) and the claims against those resources (liabilities). Proper ordering of these accounts provides the net worth statement, or balance sheet, which is an account of the farmer's financial position at a given point in time.

Capital Account. This includes a purchase record of capital assets and improvements that cannot be fully expensed in the year purchased, and a sales record of similar items.

Credit Account. This account of farm liabilities records new loans, keeps track of principal and interest payments, and tabulates unpaid principal balance on existing loans.

Receipt and Expense Account. This account organizes financial cash flow into the business (receipts) and out of the business (expenses) over a period of time, usually one year, and it may include both cash and non-cash transactions. Subtracting expenses from receipts gives the net farm income, which measures the profitability of operating the business, i.e., the return to the operator for his labor, management, and capital.

Production and Statistical Records. These records relate to the production of crop and livestock enterprises on the farm and the resources used (for example, labor, feed, and production and performance records of livestock).

The Farm Business Analysis. The inventory and the receipt and expense accounts are combined with production records to reveal the strong and weak areas within the business. These analyses are often called "efficiency measures." The information they provide is often useful in identifying problems and directing future managerial decisions.

Enterprise Records and Accounts. All information that can be recorded for the total operation can be kept for individual enterprises. Often, only minor additional records are required beyond those necessary for the total business to make rather detailed enterprise analyses. These are often referred to as "cost accounts."

MANAGEMENT PRINCIPLES AND PRACTICES

Management to accomplish the *right* things requires directing one's efforts toward predetermined goals. It also requires planning, controlling, and organizing of one's efforts. Successful management requires the following (Evans, 1980b):

1. Setting objectives or goals
2. Analyzing resources
3. Setting a critical path
4. Measuring progress

In other words:

1. Determine *what* needs to be accomplished.
2. Determine what tools are available.
3. Plan *how* to accomplish the job, and set up a timetable (and other needed resources).
4. Check occasionally to see whether the timetable is being followed, or whether changes are needed.

Setting Goals

A goal is a statement of expected results designed to give an organization (or person) direction and purpose. Setting goals is probably the most important and most difficult part of management. Goals should be (Evans, 1980b):

1. Clearly defined and fixed.
2. Stated in terms of the results to be accomplished, not in terms of methods.
3. Measurable.
4. Challenging but achievable within a realistic time frame.
5. Restricted to desired results.
6. Ranked in order of priority.

Three areas for which goals are set within a ranch operation are family matters, business matters, and personal matters.

Setting goals is the first step of a future management plan. Examples of the three types of goals are shown in the following:

1. Family goals
   a. Take time off for vacations.
   b. Maintain a nice family car.
   c. Provide savings for children's college education.
2. Business goals
   a. Increase carrying capacity of the ranch.
   b. Increase the calf-crop percentage.
   c. Develop an optimal marketing system for calves.
3. Personal goals
   a. Become more familiar with futures trading.
   b. Read *War and Peace.*
   c. Lose 10 pounds before high school reunion.

Analyzing Resources

What are resources? In general, a resource represents wealth—something that is scarce, useful, material, and transferable. Four types of resources are:

1. Land
2. Capital
3. Labor
4. Management

Land.   This is a natural wealth used in production. It includes soil, flora, fauna, and minerals.

Capital.   This is wealth used in production, but it differs from land in that it is man-made rather than natural. Improvements to the land become a form of capital.

Labor.   This refers to primarily physical energy used in production. Usually, the term is broadly classified into the categories of operator labor, family labor, and hired labor.

Management.   This term refers to mental energy used in production as distinguished from labor. Management is primarily concerned with decision-making and activities bearing risk.

Optimum ranch operation is achieved by employing the proper mix of these four resources to produce results that are consistent with one's objectives.

The purpose of analyzing resources is to define the tools available for accomplishing goals. A manager needs to assess the following resources (Herbst, 1976):

1. Financial sources (e.g., net worth and credit access)
2. Family (e.g., labor and expertise)
3. Farm
4. Livestock
5. Land
6. Labor
7. Machinery
8. Management ability

Classifying resources by groups such as the following is recommended:

1. Family
2. Business

3. Finances
4. Farm
5. Livestock
6. Equipment
7. Personal matters

One should also try to quantify the capabilities of resources. Examples include:

1. The carrying capacity of pastures.
2. The productivity of farm ground.
3. The extent of credit and value of resources as collateral.

Honesty is needed in this evaluation. The more honest one is in listing strengths and weaknesses, the better chance one has of designing a management plan that will make it possible to reach specific goals.

Planning the Critical Path

Once the business operator knows what his goals are, what should be done, and what resources he has, the question becomes, "How do I use these resources to accomplish my goals, and in what length of time can I plan to accomplish them?" He now needs to establish a critical path. The easiest way to do that is by:

1. Writing down goals in order of priority.
2. Writing down the steps that must be taken to accomplish a goal and the amount of time in which each step must be completed.

This process involves some writing and much thinking, but it provides an outline of what has to be done, when and how it is to be done, and the additional resources needed to meet one's goals.

The critical path process can be compared to planning a trip. If someone starts out by knowing where he is, where he is trying to go (goal), and the route he expects to use to get there (critical path), the process is simplified. If he does not know the route, or worse yet, does not know his destination, he may spend much time going around in circles. In planning the critical path, one builds a "road map." Detours may be taken, and the destination might even be changed, but at least one has established the direction, and can work toward that direction, instead of wasting time, effort, and valuable resources by going around in circles. The critical path becomes the master plan and is the easiest way for the business operator to ensure that he is working to make the *right* things happen, not merely working.

Measuring Progress

After the critical path has been chosen and the resources have been committed, the actual results must be compared with the critical path and the expected results of the business operator's goals. Is he "on the path" or are there areas where he needs to catch up? To answer this question, the business operator uses controls, which are important functions of management. A *control* is a standard that indicates whether everything occurs in accordance with the original plan. The following are criteria for good control devices (Schermerhorn, 1980):

1. Adequate information is provided to the user.
   a. Historical data—financial records
   b. Future data—budgets, forecasts
2. Timely information is provided. Information must be received promptly to correct unfavorable deviations at once.
3. Information is provided in a usable, easily understandable form.
4. Data (costs) are reported on a *responsibility* or commodity basis.
5. Control measures are economical. They do not cost more than their anticipated benefits.
6. Control measures indicate not only where the problem lies but also what should be done to remedy the problem.

Use of a control system requires (1) established standards, (2) appraised performance, and (3) correction of deviations.

STANDARDS

The purpose of using standards is to be able to detect deviations from expected results in advance, or at least as soon as they occur. Standards should be established only in key performance areas:

1. Production
2. Finances
3. Marketing
4. Personnel

APPRAISAL OF PERFORMANCE

Performance is compared with a standard, and the difference is ascertained. Here, one should be concerned with management by exception, concentrating only on areas that deviate materially from the standard. Judgment must enter this process, particularly regarding the significance of the deviation from the standards.

CORRECTION OF DEVIATIONS

Whenever significant deviations are realized, *action* is required to correct them. Types of standards to use in key performance areas are given in the following outline:

1. Procurement of input (the availability of a continuous, dependable supply of input at a reasonable price).
   a. Cost per unit of input
   b. Lowest price for acceptable product
2. Production (the combination of human, physical, and capital resources to produce efficiently marketable products).
   a. Average cost per unit
   b. Output per unit of input
3. Finances (the abilities to carry out business operations and to meet current and future obligations).
   a. Liquidity
   b. Solvency
   c. Profitability
   d. Return on investment
4. Marketing (the sale of products in an efficient manner and at a reasonable price).
   a. Selling costs per unit
   b. Transportation costs per unit
   c. Quality measurements
5. Personnel (the employment of an adequate labor force to carry out operations with a minimum of problems).
   a. Employee turnover
   b. Salary scales
   c. Performance reviews

Measuring progress through control standards can accomplish the following (Rue and Byars, 1980):

1. Determine where the operation stands as to a management plan.
2. Expose strengths and weaknesses in operational and managerial abilities.
3. Force the review of goals and critical path choices and consider modifications and alternatives.

FINANCIAL MANAGEMENT AND METHODS*

Once the control standards have been set up to measure progress, the question becomes, "Do I need more money as a resource to reach my goals? If so, how much money do I need?"

Most ranchers operate on a cash basis. Cash outlays for production usually occur prior to cash inflows from sale of livestock. Thus, a cash flow budget can answer how much money is needed to accomplish one's goals.

Constructing a Cash Flow Budget

To develop a budget, one must (1) identify the anticipated cash receipts by month, (2) identify the anticipated cash expenses by month, and (3) identify the deficit or surplus cash balance by month.

A cash budget provides the following information (Herbst, 1976):

1. How much cash is required
2. When the money is required
3. Whether the money is needed on a short- or long-term basis
4. How the money should be acquired (internally or externally)
5. How and when external funds can be repaid

A cash budget can be developed from using a single form; last year's records or last year's tax returns can be used if they are the best record available. The totals for

each expense and income account should be considered in the following ways:

1. What will inflation do to them?
2. Are different procedures planned for next year?
3. What changes should be made in these totals to show what is planned for next year?

Once one has an idea of the expected totals, they can be broken down by months (Evans, 1980e). For example, the labor bill will probably be higher in April than in January. This can be shown in the monthly breakdown. Then, totals can be added to show how much cash is needed each month, how much is expected to come in, and how much needs to be borrowed (Table 14–1).

The cash budget is just another planning tool or road map. One knows the destination, how to get there, when more fuel (money) is needed, and how one plans to pay for the fuel.

Partial Budgeting and Enterprise Analysis

Partial budgeting is the estimation of what it will cost to do something and what is expected in return. Partial budgeting can be used to compare a change in operations with the current procedure and measure the impact of that change in terms of dollars (profit or loss). Table 14–2 shows an example of such a change—increasing herd size by 10 cows. When a partial budgeting analysis is constructed, only the differential costs that occur from the proposed undertaking should be considered. Costs that have already been committed under the current enterprise are irrelevant in this type of analysis.

Partial budgeting is a tool to help determine how best to employ available resources. Thus, the term partial budgeting refers to the evaluation of a potential change in one part of the business to see whether it is worth carrying out.

* Adapted from Evans, E.T. 1980c, d. Cattleman's Library. Cooperative Extension Service, University of Idaho, Moscow, ID. CL-925 and CL-930.

TABLE 14-1. Cash Budget for Year Ending December 31, 1984

	Jan	Feb	Mar
Cash balance, beginning	$ 46,300	$ 48,620	$ 29,020
+ Budgeted receipts	527,470	895,900	1,145,920
Total	$873,770	$994,520	$1,174,940
− Budgeted disbursements	834,950	895,500	1,153,500
Budgeted balance, ending (from operations)	$ 38,620	$ 49,020	$ 21,440
+ Financing obtained	10,000	0	10,300
− Debts repaid	0	(20,000)	0
Budgeted cash balance, ending (from all sources)	$ 48,620	$ 29,020	$ 31,740

MAKING A SUCCESSFUL LOAN APPLICATION*

Operating credit is probably the most important area of financing requirements for ranchers. Operating loans are usually ap-

* Adapted from Hewitt, R.R. 1980. How to make a successful loan application. *In* Cattleman's Library. Cooperative Extension Service, University of Idaho, Moscow, ID. CL-945.

proved for a one-year term to cover all phases of producing a calf until it is sold. For most western ranchers, calves are born in the spring and weaned in the fall. Their sale is the primary source of income that is used to retire an operating loan.

After the calves are sold, one should analyze current financial conditions and plan and arrange financing for the next year.

TABLE 14-2. Partial Budgeting Analysis of Cow-Calf Enterprise

	With Cows	Without Cows
Expenses		
Cows at $650 × 10	$6,500	$ _____
Trucking	50	_____
Medicine and branding	30	_____
Feed (15 tons hay at $50)	750	_____
Pasture not hayed (20 tons at $50)	1,000	_____
Cost of haying saved	(−350)	350
Fence of hay to pasture	30	_____
Interest (6,500 × 10 months × 15%)	812	_____
Total cost with cows	$8,822	(−350)
Sale		
Cows (10 × 1,000 lb × $0.45)	$4,500	$ _____
Calves (9 × 450 lb × $0.75)	3,050	
Hay (15 tons now)	_____	750
Hay (20 tons next year)	_____	1,000
Total sales	$7,550	$1,750
Total income	$(1,272)	$1,400

From Evans, E.T. 1980d. Partial budgeting and enterprise analysis. *In* Cattleman's Library. Cooperative Extension Service, University of Idaho, Moscow, ID. CL-930.

Shopping for Credit

Credit varies not only in interest rates, but in service and flexibility. A rancher should become acquainted with available sources of credit in his area. When shopping for credit, he should:

1. Ask borrowers of particular lending institutions for their opinions as to how well the institution serves them.
2. Visit lending institutions, and discuss their lending programs with the manager.
3. Determine whether the manager is knowledgeable about the type of business in question, what the eligibility requirements are for borrowing, and who would handle the account.
4. Give a general description of his operation, credit requirements, and management objectives.
5. Ask about interest rates.

The final decision should be based on a lender who the rancher feels is dependable and who possesses a basic knowledge of the rancher's business. The lender should be interested in the rancher and in his programs and should understand his management objectives.

After the rancher selects a lender who he feels is most qualified to handle his credit requirements, he should attempt to consider the lender's position. As the lender considers the needs of the borrower, he looks at what bankers call the "3 C's"—character, capacity, and collateral.

Character

The lender seeks to establish the character of the borrower by use of the following questions:

1. Do you pay your bills?
2. Do you try to be a good citizen?

Questions of honesty and paying habit are crucial, but they concern the borrower's history; therefore, these aspects of character are already established.

Capacity

The rancher's cash flow budget shows his repayment or earnings capacity provided that it is honest and not out of line with capacity demonstrated in the past. Income and balance sheet can establish its accuracy (Evans, 1980c).

Collateral

The lender wants to know what other resources are available to the borrower in case the budget plan proves to be unworkable. (The net worth statement can uncover these resources.)

The rancher needs to assemble facts and figures and to make plans that he can submit to the lender to show that he has a *viable program*. He should be able to:

1. Describe the current conditions of his business operation.
2. Define management goals (what he wants to accomplish).
3. Demonstrate that his operation has the repayment capacity to service the debt that he intends to incur.

Information for Lender

The rancher should be able to give the lender the following items of information:

1. Complete real estate schedules, debts owed, terms, and repayment schedules
2. Any installment contracts for such items as equipment and payments
3. Information regarding previous loans
4. Gross income from all sources, including itemized expenses (copies of income tax returns)
5. Net worth statement
6. Cash budget

The rancher should present a realistic program based on factual operating costs that can be met by proven income. The emphasis in a loan application is on the borrower's knowledge of his business and on his managerial ability. He conveys confidence if he possesses these attributes and has faith in his program.

When the rancher has completed his requests, he should ask the lender to visit his operation. He should give the lender a tour, describing how the operation works throughout the year. By showing that he is applying successful management techniques and knows how to reach his goals on a sound basis, the rancher instills confidence in the lender and encourages a strong working relationship. He should remember that the lender is a tool working for him, not against him. His service is of value not only in lending money, but also in helping to evaluate the rancher's business so that profits are increased.

Sources of Credit*

Will it pay to borrow money? Does the anticipated return from using the borrowed money exceed the cost of borrowing, which includes interest and finance charges? If so, one should shop for a lender who provides the greatest service for the least cost.

INSTITUTIONAL LENDERS

Commercial Banks

These institutions are by far the largest in terms of number and volume of loans for agriculture. They provide short- and intermediate-term loans and offer a full line of financial services. Usually, they employ agricultural loan specialists.

Cooperative Farm Credit System

This system was created in 1916 by the federal government. Production Credit Association (PCA), Federal Land Bank (FLB), and intermediate institutions were established and have since repaid the original government investment. PCA and FLB are now completely owned by their borrowers.

Production Credit Associations

These associations provide short- and intermediate-term credit for up to 7 years. Rural housing, farm-related business, and operating and family needs of the ranch are eligible for this source of credit.

Production credit associations are credit cooperatives owned and controlled by a

board of directors who are elected by voting members. Each borrower invests in the association to help capitalize it. He invests an amount equal to 10% of the principal loan in stock (voting). Simple interest is charged on the amount of money used for the actual number of days used. Loan limits are based on the purpose and repayment capacity for which the loan is initiated.

Federal Land Bank Association (FLBA)

These institutions provide long-term loans, for 5 to 40 years, secured by first mortgages on farm real estate. The borrower buys stock in the federal land bank association (FLBA) equal to 5% of the loan though he may be required to purchase up to 10% of the loan in stock. The FLBA in turn buys a like amount of stock in a district FLBA through which the loan is made. Loans are offered in amounts of up to 85% of appraised value. (This figure is usually conservative compared with market value). Interest rates vary and are tied to current national money market rates. Borrowers may repay any part or all of the loan at any time without penalty.

Farmers Home Administration

The Farmers Home Administration is a federal government lending agency under the direction of the United States Department of Agriculture (USDA). Its purpose is to provide loans at low interest rates to ranchers who cannot obtain credit elsewhere. Most of these loans are for operating and farm ownership, but the agency also provides emergency loans for disaster areas. Operating loans are made under "supervised credit," with technical management assistance provided. Interest rates are set periodically by the agency, and repayment is scheduled according to the borrower's ability to repay. A local committee of three area residents screen applications for eligibility.

---

* Adapted from Luft, L.D. 1980. Sources of credit. *In* Cattleman's Library. Cooperative Extension Service, University of Idaho, Moscow, ID. CL-940.

Farm Ownership Loans

These loans are made to qualified buyers for up to 40 years, with 100% financing. Interest rates vary according to ability to repay. Each borrower is advised to refinance through other sources as such an option becomes financially feasible.

Insurance Companies

Insurance companies constitute a major source of farm-ranch mortgage loans. Most are long-term mortgages, and terms of the loans range from 5 to 25 years. Insurance companies typically loan large sums at one time for cases in which production is stable and risk of loss is low. There is no limit to the size of loans, although state laws may limit a loan to 75% of appraised value. Interest rates for long-term loans are responsive to the market.

NON-INSTITUTIONAL LENDERS

Dealers—Machinery

Dealers usually have their own finance corporations that provide trade credit. Dealer credit often commands a higher interest rate than credit from institutional sources.

Individuals—Land

In contracts between buyers and sellers of real estate, the seller often takes a down payment for the real estate and contracts with the buyer for delivery of the deed upon payment of the balance. This arrangement is attractive to the buyer because the down payment and interest rate are often low. The buyer can afford to offer a higher price because of the lower down payment and interest, and can therefore convert ordinary income to capital gains. Conversely, such contracts are attractive to the seller because they often attract more buyers, which in turn stimulates a higher contract price. In addition, the seller can earn interest on outstanding principal and accrue some tax benefits.

Choosing a Lender

The type of loan needed often dictates the appropriate source. If alternatives are available, they should be evaluated on the basis of the following criteria:

1. Services received
2. Cost of credit
3. Relative ease of securing the loan

The lender should be familiar with ranching and with the potential losses that can result from uncertainty. In addition, the lender should be willing to service the borrower in times of financial difficulty as well as in times of prosperity.

The behavior of credit institutions and the services offered by them depend on their prime responsibilities to such sources of funds as depositors, insurance policyholders, and bondholders in the money markets. A better credit risk brings a lower cost of borrowing and better terms for the borrower.

FINANCIAL STATEMENTS*

A financial statement is a quantified document that reduces resources to dollars (Table 14–3). A balance sheet, or net worth statement, summarizes the extent of business and investment at a specific point in time and provides a tool to determine:

1. Liquidity (the ease with which assets can be turned into cash)
2. A borrowing base
3. Operational progress

---

* Adapted from Hewitt, R.R. 1980b. Financial statements—Their preparation and use. *In* Cattleman's Library. Cooperative Extension Service, University of Idaho, Moscow, ID. CL-950.

TABLE 14–3. Financial Statement

FINANCIAL STATEMENT

Name _____ Date _____

Assets	Amount
Cash on hand	_____
Notes/accounts receivable	_____
Liquid investments	_____
Livestock (see Schedule A)	_____
Feed/crops on hand	_____
Cash invested in growing crops	_____
Supplies	_____
Other current assets	_____
Total current assets	_____
Real estate	_____
Contracts receivable	_____
Machinery/equipment	_____
Other fixed assets	_____
Total fixed assets	_____
Total assets	_____

Liabilities	Amount
Notes payable	_____
Installments due within one year	_____
State/federal income taxes due	_____
Real estate payments due/current	_____
Other current liabilities	_____
Total current liabilities	_____
Real estate mortgage/contracts	_____
Long-term portion of notes payable	_____
Other long-term liabilities	_____
Total fixed liabilities	_____
Total liabilities	_____
Net worth	_____
Total liabilities and net worth	_____

Adapted from Hewitt, R.R. 1980b. Financial statements—Their preparation and use. *In* Cattleman's Library. Cooperative Extension Service, University of Idaho, Moscow, ID. CL-950.

Assets

An *asset* is something of value that one has or owns. It is normally listed on the balance sheet along with its length of life and type. *Current assets* include cash, accounts receivable within one year, livestock, and feed. *Fixed assets* are those not normally liquidated during the course of standard operations: real estate, building, permanent improvements, machinery, and equipment.

Most statements have attached schedules for particular assets, which give additional information (e.g., livestock by class, number, and value per head) (Table 14–4).

Liabilities

A *liability* is a debt or obligation. It is usually expressed in terms of money and is handled much the same as an asset. *Current liabilities* are debts payable within one year (e.g., operating loans, equipment contracts, current portion of real estate payments, and taxes payable). *Fixed liabilities* are notes or debts payable over more than one year (e.g., real estate mortgage or contracts and equipment notes).

Large liabilities are broken down on schedules that describe the tracts of land, value of debt owed, to whom debt is owed, yearly principal, and interest payments.

Valuing Assets

In general, two approaches are used: (1) current market value and (2) book value (cost less depreciation). The current market value approach is susceptible to market fluctuations, and these values do not give a true picture of actual operating gains or losses experienced through assets. The book value approach is more often used and accepted by lenders to evaluate a financial position for loan decisions.

Valuation Procedures

Raised breeding stock has a conservative market value. Crops or livestock to be sold during the year have the current market value. Machinery and improvements are worth their cost less depreciation.

Changes in net worth that are based on these guides for valuing assets reflect the actual earning capacity of an operation more closely than changes based on paper or on changes in the potential market.

Using all of the information available in financial statements, a manager can analyze his past transactions and achievements and his current financial state, and can determine the decisions that need to be made to accomplish his goals. Comparison of the current financial statement with previous statements should indicate trends.

A manager should consider working capital, which is the amount of capital available for use in operating the ranch. It

TABLE 14-4. Financial Schedule A

Schedule A—Cattle			
Number	Kind	Value per Head	Total Value
200	Cows: 3–7 yr	$ 45	$ 90,000
10	Cull cows	400	4,000
30	Heifers: 2 yr 600 lb at $0.50/lb	300	9,000
20	Steers: 2 yr 650 lb at $0.55/lb	375.50	7,150
	Calves		
	Weaners		
170	Suckers	125	21,250
8	Bulls	800	6,400
438	Total cattle		$137,800

Adapted from Hewitt, R.R. 1980b. Financial statements—Their preparation and use. *In* Cattleman's Library. Cooperative Extension Service, University of Idaho, Moscow, ID. CL-950.

is calculated by subtracting the current liabilities from the current assets. Working capital is important to a primary lender of operating funds, and a manager should be able to monitor progress in building equity in current assets. A year-to-year decline in working capital may indicate too great an investment in fixed assets or a debt incurred at a rate greater than the rate at which it can be repaid.

Leverage

Leverage refers to the ratio of total debt to net worth:

$$\frac{\text{Total liabilities}}{\text{Net worth}}$$

A ratio of 1.5:1.0 means that for every dollar a manager has invested in the ranch, creditors have $1.50 invested. The greater that the ratio of debt to net worth becomes, the more debt is owed, and the less equity is possessed. A point can be reached at which normal production can no longer service the debt load of the operation.

A manager should learn to use a financial statement for the following reasons:

1. It supplies valuable management information.
2. It helps to reverse negative trends by allowing early recognition of them.
3. It is valuable tool for financing.

REFERENCES

Evans, E.T. 1980a. Management principles and practices. *In* Cattleman's Library. Cooperative Extension Service, University of Idaho, Moscow, ID. CL-900.

Evans, E.T. 1980b. Setting goals. *In* Cattleman's Library. Cooperative Extension Service, University of Idaho, Moscow, ID. CL-905.

Evans, E.T. 1980c. Financial management and methods. *In* Cattleman's Library. Cooperative Extension Service, University of Idaho, Moscow, ID. CL-925.

Evans, E.T. 1980d. Partial budgeting and enterprise analysis. *In* Cattleman's Library. Cooperative Extension Service, University of Idaho, Moscow, ID. CL-930.

Evans, E.T. 1980e. Cash flow forecasting and budgeting. *In* Cattleman's Library. Cooperative Extension Service, University of Idaho, Moscow, ID. CL-935.

Herbst, J.H. 1976. Farm Management: Principles, Budgets, Plans. 4th Ed. Stripes Publishing Co., Champaign, IL.

Hewitt, R.R. 1980a. How to make a successful loan application. *In* Cattleman's Library. Cooperative Extension Service, University of Idaho, Moscow, ID. CL-945.

Hewitt, R.R. 1980b. Financial statements—Their preparation and use. *In* Cattleman's Library. Cooperative Extension Service, University of Idaho, Moscow, ID. CL-950.

Luft, L.D. 1980. Sources of credit. *In* Cattleman's Library. Cooperative Extension Service, University of Idaho, Moscow, ID. CL-940.

Rue, L.W., and L.L. Byars. 1980. Management Theory and Application. 2nd Ed. Richard D. Irwin, Homewood, IL.

Schermerhorn, R.W. 1980. Management by objectives. *In* Cattleman's Library. Cooperative Extension Service, University of Idaho, Moscow, ID. CL-920.

QUESTIONS FOR STUDY AND DISCUSSION

1. What characteristics do good ranch managers have in common?
2. What is the difference between an organizational and an operational decision?
3. Outline the steps to follow in making and implementing decisions.
4. List several significant characteristics of the cattle business.
5. How are records used in ranch management?
6. Why is the establishment of goals so important to the success of a beef cattle operation?
7. What criteria should be used when setting goals?
8. Why should livestock producers analyze their available resources?
9. What is meant by the phrase, "establish your critical path?"
10. Why are controls necessary?
11. What are the components of a cash flow budget?
12. How may partial budgeting or enterprise analysis be used as a management tool?
13. What factors do agricultural lenders take into consideration when evaluating loans?
14. Define the following terms:
    a. Net worth
    b. Liquidity
    c. Asset
    d. Liability
    e. Leverage

Chapter Fifteen

Beef Cattle Facilities*

Every beef cattle operation, regardless of its size, needs adequate beef cattle facilities. Well-planned barns, corrals, and workable and practical equipment make handling cattle easier and safer and promote more efficient use of time and labor. They also help to keep animal shrinkage to a minimum. A proper handling facility is an essential investment that contributes to the success of any cattle operation.

To be effective, a working facility should consist of at least a corral, a chute, headgate, and a facility for loading and unloading cattle. These items facilitate the timely application of such management practices as vaccination, identification, castration, dehorning, worming, spaying, and pregnancy testing.

LOCATION

The handling facility should be located in a well-drained area that is convenient to the roadway and to cattle in relation to the total operation. It should be designed to give the desired direction and control of cattle movement, and to allow a logical sequence of procedures. Drainage should occur away from feed areas and driveways. The best location for a handling facility is a hillside, with feed alleys and mangers extending along the high side.

*Much of the information in this chapter was adapted with permission from Singleton, W.L., L.A. Nelson, and D.D. Jones. 1976. Handling facilities for beef cattle. Cooperative Extension Service, Purdue University, W. Lafayette, IN. ID-109.

WATER

A year-round supply of clean, fresh water is basic to any successful cattle enterprise. The water supply should be free of ice in winter, and automatic nonfreezing water fountains are a good investment in areas where frost often occurs. Water should be accessible in each lot or corral.

ELECTRICITY

Electricity in working areas and calving barns makes performance of many operations easier and faster. An open-front barn with electrical connections for heat lamps makes an excellent unit for winter calving. Many operations such as branding, dehorning, and clipping hair are made easier if electricity is available. Electricity can also extend the working day, pump water, keep water from freezing in the winter, and add to the convenience of recordkeeping.

COMPONENTS OF A HANDLING FACILITY

The primary reasons for building cattle handling facilities are:

1. To direct and control animal movement.
2. To reduce cost and labor requirements for handling animals.
3. To promote safety of workers and animals.
4. To promote humane treatment of animals on the farm.

A complete cattle handling facility provides an organized system for gathering, directing, holding, sorting, positioning, restraining, and elevating or lowering animals. It includes headgate, holding chute, working chute, crowding pen, holding pens, scale, loading chute, and several handling gates. One must remember, however, that not all ranches require all of these components. Figures 15–1 and 15–2 are examples of two complete handling facilities. Table 15–1 lists the dimensions of the various components within a beef cattle handling facility.

Headgate

The headgate should be sturdy, safe, and easy to operate and should have a quiet action. Complicated construction of a headgate should be avoided, as should a headgate with holes for the insertion or removal of pins when each animal is caught or released. There are many acceptable types of headgates. These range from simple homemade wooden constructions (Fig. 15–3) to more elaborate and expensive metal selfcatching devices, which are available commercially.

Holding Chute

The holding chute (squeeze chute) is immediately behind the headgate and is fastened to it. A headgate is adequate for routine health functions, but a holding chute allows complete control of the animal and reduces the risk of injury to both the worker and the animal. Useful features of a holding chute include the following:

1. Width adjustment for different sizes of animals
2. Squeeze action for complete restraint
3. Removable side panels for easy access to the animal
4. Floor with a nonslip surface
5. A roof over both headgate and holding chute to allow working of cattle regardless of the weather (Fig. 15–4)

An additional helpful feature is a 2-foot service gate at the rear of the chute that allows easy access to the animal's posterior for castration, dehorning, artificial insemination, or pregnancy testing.

Working Chute

The working chute is the heart of a cattle handling system. It leads from the holding

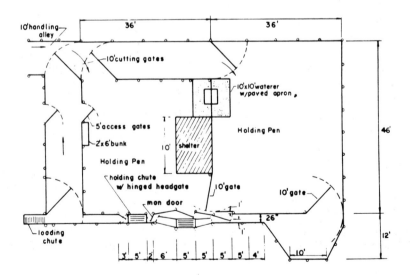

FIG. 15–1. Example of a complete handling facility for a large cattle operation. (From Singleton, W.L., L.A. Nelson, and D.D. Jones 1976. Handling facilities for beef cattle. Cooperative Extension Service, Purdue University, W. Lafayette, IN. ID-109.)

pen to the headgate and should be a maximum of 26 inches wide, at least 4 feet high, and long enough to accommodate four or five animals at a time. A straight run of working chute longer than 20 feet may cause cattle to balk when they look ahead and see the headgate or holding chute; to avoid this problem, a curved chute should be used (Fig. 14–2).

A covered horseshoe-shaped working area with proper facilities and equipment can provide a practical, efficient, and versatile system. A well-designed curved chute has a catwalk for the handler along its inner radius. As the animals walk up the chute, they move in the natural pattern of circling around the handler (Grandin, 1978). The cattle see only the handler because the solid sides of the chute prevent outside distractions.

Proper lighting of the working chute helps cattle to move smoothly. Poor lighting and shadows cause cattle to be unsure of their footing and to balk. The working chute should be dull in color and have smooth transitions for all points at which its shape changes.

The floor of the working chute should be made of concrete. To allow proper drainage, it should slope slightly away from the side where the handler walks. To facilitate cleaning, a 23-inch gap should be left between the totally enclosed sides and the chute floor.

Alley backstops that prevent cattle from backing up in the chute are recommended; however, one should select backstops that do not appear to block the alley, to avoid blocking the view of cattle any more than necessary.

Overhead restrainers are also recommended for working chutes. These restrainers prevent cattle from rearing up and turning around in the chute. Overhead restrainers need to be sturdy, and their height should be adjustable to match the height of the cattle.

A well-designed working chute and crowding area confines cattle and restricts

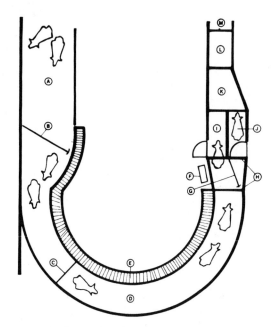

FIG. 15–2. Example of a semi-circular handling facility.

A. Holding pens
B. Squeeze gate
C. "No back-up" gate
D. Circular chute
E. Catwalk
F. Scales
G. Dividing gate
H. Rolling gates
I. Right-handed artificial insemination (AI) stall
J. Left-handed AI stall
K. Sorting area
L. Headgate
M. Passage to pens

(From How to Handle Your Cattle. 1975. Simmental Shield Publishing Co., Lindsborg, KS.)

their vision except in the directions that they are wished to move. Excessive prodding and noise does not improve working efficiency, but only excites cattle. Electric prods should not be used since they cause cattle to balk the next time they are handled.

Pressure-treated posts should be used. They should be set no more than 6 feet apart and at least 3½ feet into the ground. Sides of a working chute should be boarded up solidly from ground level to 2 to 2½ feet above the ground. Some cattle-

TABLE 15–1. Recommended Dimensions of Components for a Beef Cattle Handling Facility

Facility Component	Cattle Weight		
	Less Than 600 lb	600 to 1200 lb	Greater Than 1200 lb
Holding chute			
Height	45 in	50 in	50 in
Width	18 in	22 in	26 in
Length			
With service gate	5 ft	5 ft	5 ft
Without service gate	7 ft	7 ft	7 ft
Working chute			
For vertical sides:			
Width	18 in	22 in	26 in
Length (minimum)	20 ft	20 ft	20 ft
For sloping sides:			
Bottom inside width	15 in	15 in	16 in
Top inside width	20 in	24 in	26 in
Length (minimum)	18 ft	18 ft	18 ft
Chute fence			
Height			
Solid wall (if used)	45 in	50 in	50 in
Overall (top rail)	55 in	60 in	60 in
Post spacing	6 ft	6 ft	6 ft
Post depth in ground	36 in	36 in	36 in
Crowding pen			
Space per head	6 sq ft	10 sq ft	12 sq ft
Holding pen			
Space per head	17 sq ft	17 sq ft	20 sq ft
Pen fence			
Height	60 in	60 in	60 in
Post spacing	8 ft	8 ft	8 ft
Post depth in ground	30 in	30 in	30 in
Loading chute			
Width	26 in	26 in	26–30 in
Length (minimum)	12 ft	12 ft	12 ft
Rise (maximum)	3½ in/ft	3½ in/ft	3½ in/ft
Ramp height for			
Gooseneck trailer		15 in	
Pickup truck		28 in	
Van-type truck		40 in	
Tractor-trailer		48 in	
Double deck		100 in	
Handling alley			
Width	8 ft	10 ft	12 ft
Alley fence			
Height	45 in	50 in	50 in
Post spacing	6 ft	6 ft	6 ft
Post depth in ground	36 in	36 in	36 in

From Singleton, W.L., L.A. Nelson, and D.D. Jones. 1976. Handling facilities for beef cattle. Cooperative Extension Service, Purdue University, W. Lafayette, IN. ID-109.

men prefer a completely solid chute. A cat-walk built alongside makes it easier to drive cattle toward the holding chute; it should be approximately 18 inches wide and 24 inches off the ground. Provisions for stop bars or blocking gates should be made at approximately every 9 feet throughout the length of the chute. Sharp corners that bruise cattle can be avoided by the use of padding (Fig. 15–5).

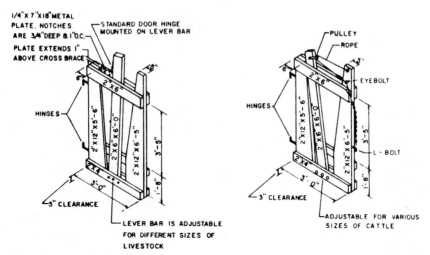

FIG. 15-3. Design suggestion for two wooden homemade headgates. (From Singleton, W.L., L.A. Nelson, and D.D. Jones. 1976. Handling facilities for beef cattle. Cooperative Extension Service, Purdue University, W. Lafayette, IN. ID-109.)

Crowding Pen

The crowding pen at the back of the working chute should be approximately 150 square feet which is large enough to accommodate 15 to 20 head at a time. Many crowding areas are shaped to form a gradual "V" along the approach to the working chute. This shape can cause problems, however, because it does not allow cattle to be forced into the working chute.

To restrict animal vision in the crowding pen, totally enclosed sides are recommended. Frequently, the crowding area walls are totally enclosed, but the crowding gate is not, which encourages cattle to look

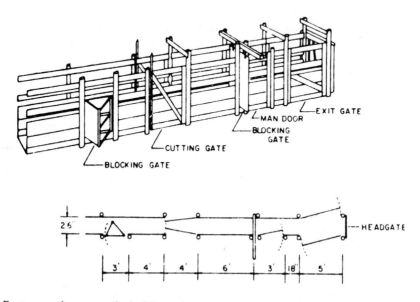

FIG. 15-4. Features of a versatile holding chute. (From Singleton, W.L., L.A. Nelson, and D.D. Jones. 1976. Handling facilities for beef cattle. Cooperative Extension Service, Purdue University, W. Lafayette, IN. ID-109.)

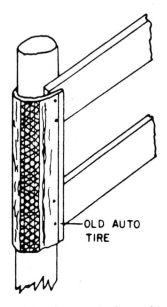

FIG. 15-5. Padding for posts in the working chute area. (From Singleton, W.L., L.A. Nelson, and D.D. Jones. 1976. Handling facilities for beef cattle. Cooperative Extension Service, Purdue University, W. Lafayette, IN. ID-109.)

toward the gate instead of toward the working chute. A good crowding area is circular with totally enclosed sides and crowding gate.

The crowding area should be designed so that when the cattle enter the holding pens or alleyways, they cannot see the handler working in the squeeze chute. Corners whose angles exceed 60 degrees should be eliminated. Many crowding areas do not work properly because of a rough transition from the crowding area into the working chute.

Figure 15–6 gives two examples of crowding area and working chute.

Holding Pens

Holding pens should be located conviently in relation to the rest of the handling facility. Each pen should provide 12 to 20 square feet of space per animal and should be large enough to handle 50 to 80 head, or one truckload, of cattle.

Fence posts should be set no more than 8 feet apart and at least 2½ feet into the ground. A sick pen should also be provided, allowing 20 square feet of space per head for about 3% of the herd.

Scale

A scale is an essential tool for performance testing, evaluating feedlot gains, and determining sale weights. Most cattlemen prefer a portable scale that can be installed ahead of the headgate or placed in line with the working chute by removing one of the gate sections.

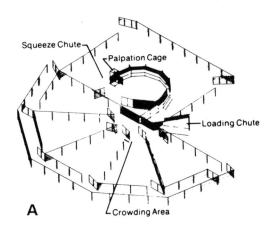
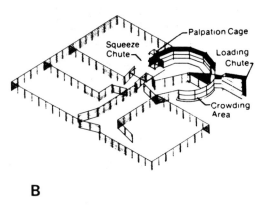

FIG. 15-6. A. Circular crowding pen and curved working chute for 10 to 400 or more head. B. Circular crowding pen and working chute provides good sorting and loading arrangement. (From Jedele, D.G. et al. 1979. Beef Housing and Equipment Handbook. 3rd Ed. Midwest Plan Service, Iowa State University, Ames, IA.)

Scales vary in size, depending on their intended use. Large scales should be of the pit type, and the pit should be made of concrete. Scales should be carefully located and properly treated. A scale house is a good investment to protect scales from the weather.

Loading Chute

The loading chute should be located outside the corrals, and should be accessible to an all-weather road (Fig. 15–7). It should be built to accommodate both rear- and side-loading equipment as well as both large and small trucks or trailers. Trucks should not have to enter corrals nor pass through gates to load or unload cattle.

When possible, the chute should be run from south to north so that cattle do not face the sun. Such positioning also permits faster thawing and drying in the winter. For quickest lot loading, the chute should be located directly off the crowding pen.

The loading chute ramp can be either sloping or stepped. The maximum incline should be 30% (or about a 3½-inch rise per foot of length). Total chute length, therefore, depends on maximum rise needed. It should be at least 12 feet, however, to ensure that each animal starting up the ramp is following at least two cattle and not walking directly into the truck or trailer.

To determine minimum chute length, the height of the truck bed is divided by the desired rise per foot of length (e.g., a truck bed height of 47.95 ft divided by a 3.5-inch rise yields a 13.7-foot chute). To improve footing 1 × 3 hardwood cleats can be nailed to the chute floor 6 inches on center.

If a stepped ramp is used, treads should not be less than 18 inches wide, and risers should not be more than 4 inches high. Steps formed with concrete should be roughened.

The sides of the chute should be at least 4 feet high and solid, to prevent cattle from seeing out. If Brahmans are being handled, the sides should be raised to 6 feet.

Gates

Strong gates that swing both ways are a must for a beef cattle handling facility. It is best to locate them in corners of pens, or other convenient places where cattle naturally congregate. Gates should be opened in the direction in which the cattle are being driven.

Handling Alleys

A handling alley that leads from the feedlot, pasture, or other feeding area to the handling facility is an important and useful item, but it is often overlooked, especially where confinement barns are used in place of open feedlots. Although cattle in confinement buildings are rather docile

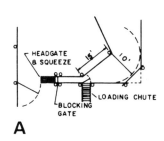

A

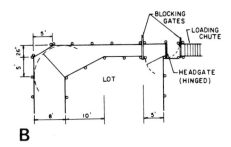

B

FIG. 15–7. Two basic handling facility layouts (*A, B*) easily constructed in the corner of an existing lot. Note that in both cases, loading chute is outside the lot. (From Singleton, W.L., L.A. Nelson, and D.D. Jones. 1976. Handling facilities for beef cattle. Cooperative Extension Service, Purdue University, W. Lafayette, IN. ID-109.)

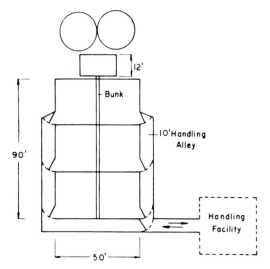

and can often be treated in a pen, handling facilities are still needed for receiving, for loading for departure, or for treating the entire herd. An example of a handling alley arrangement is shown in Figure 15–8.

Handling alleys should be at least 10 feet wide and laid out to provide a desired flow of traffic. Before any construction, the facility should be laid out on paper, and the traffic pattern mentally followed through the handling area and back to the feedlot, confinement barn, pasture, or the truck loadout. This procedure helps to locate the gates properly and to ensure an unobstructed flow of cattle traffic (Fig. 15–9).

FIG. 15–8. Example of a handling alley arrangement for a beef confinement building. (From Singleton, W.L., L.A. Nelson, and D.D. Jones. 1976. Handling facilities for beef cattle. Cooperative Extension Service, Purdue University, W. Lafayette, IN. ID-109.)

DETERMINING YOUR HANDLING FACILITY NEEDS

No one handling facility layout fits all cattle operations. Therefore, a manager should determine the components needed and their design according to his particular

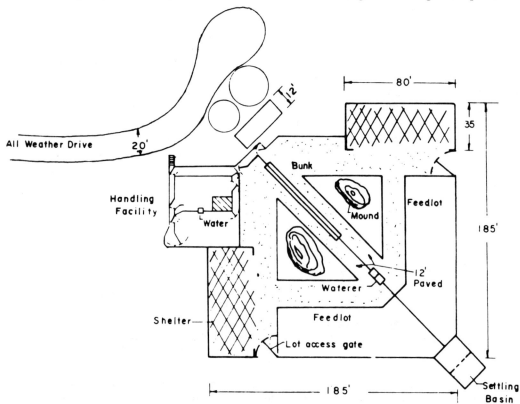

FIG. 15–9. Plain view of feedlot with adjacent handling facility permitting easy movement from lot to handling area. (From Singleton, W.L., L.A. Nelson, and D.D. Jones. 1976. Handling facilities for beef cattle. Cooperative Extension Service, Purdue University, W. Lafayette, IN. ID-109.)

type of operation, the herd size, the existing facilities, and the materials available. Obviously, the requirements for 20 head of cattle would be much different than those for 600 head. The objective should be to have a facility that allows handlers to sort, weigh, restrain, receive, and/or ship cattle as efficiently and economically as possible.

As a guide, the components most often desired in a cattle facility may be listed according to herd size:

Up to 100 head:
1. Headgate
2. Holding chute (simply constructed)
3. Portable scale
4. Crowding pen
5. Loading chute

More than 100 head:
1. Headgate
2. Holding chute (squeeze-type)
3. Working chute
4. Crowding pen
5. Holding pens
6. Scale
7. Loading chute

Adequate handling facilities need not be elaborate or expensive. Often, existing fences or buildings can be used to provide one side of a chute or pen. With a little ingenuity, most or all of the components can be home-constructed from readily available material. Figure 15–10 gives an example of a calving barn and corral system that has been used successfully.

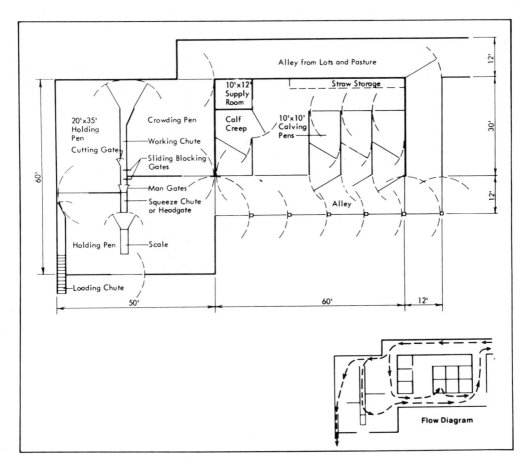

FIG. 15–10. Example of a calving barn and handling facility. (From Jedele, D.G., et al. 1979. Beef Housing and Equipment Handbook. 3rd Ed. Midwest Plan Service, Iowa State University, Ames, IA.)

CONSIDERATIONS IN DESIGNING A HANDLING FACILITY

The following points may be used as guidelines when designing cattle facilities:

1. Design the facility with an eye toward safety of both animals and operator.
2. Plan for economy. Expensive facilities are not needed for smaller herds.
3. Design for convenience in sorting and handling animals.
4. Locate the facility in a well-drained area.
5. Build all components strongly and solidly enough to hold the animals.
6. Use decay-resistant materials. Paint with a wood preservative solution every 3 to 5 years.
7. Avoid constructing any corners or projections that could bruise or cripple animals, or at least cover such areas with padding.
8. Locate headgate at the highest elevation (i.e., have cattle walk slightly uphill through the handling facility).
9. Ensure that headgate and other equipment work quietly. Loud noises frighten cattle and make them more difficult to control.
10. Do not suspend catwalks above the working chute. Cattle have poor depth perception and may balk.

The time and effort spent in planning and building a beef cattle handling facility pay large dividends by permitting timely treatment of health problems and the proper application of management practices, by preventing injuries, by minimizing labor, and by reducing animal stress and shrinkage.

REFERENCES

American Simmental Association. 1975. How to handle your cattle. Simmental Shield, Lindsborg, KS.

Grandin, T. 1978. Transportation from an animal's point of view. American Society of Agricultural Engineers. ASAS Technical Paper 78-6013. St. Joseph, MI.

Jedele, D.G., D.S. Bundy, R.L. Maddex, and J.H. Pedersen. 1979. Beef Housing and Equipment Handbook. Midwest Plan Service, Iowa State University, Ames, IA. 0-89373-003-3.

Singleton, W.L., L.A. Nelson, and D.D. Jones. 1976. Handling facilities for beef cattle. Cooperative Extension Service, Purdue University, W. Lafayette, IN. ID-109.

QUESTIONS FOR STUDY AND DISCUSSION

1. What are the primary reasons for building cattle handling facilities?
2. Name some useful features of a squeeze chute.
3. What is the advantage of building a curved working chute rather than a straight one?
4. How large should crowding and holding pens be?
5. Why is a scale such an important management tool in a beef cattle operation?
6. Where should the loading chute be located in relationship to corrals and roads?
7. How does one determine the minimum loading chute length?
8. What handling facility components are recommended for herds of up to 100 head? For herds of more than 100 head?
9. List the factors to consider when building or designing a handling facility.

Glossary

Anabolic—Of or promoting anabolism; the process in plant or animal by which food is changed into living tissue.

Artificial insemination (AI)—The process of mechanically introducing semen into the reproductive tract and uterus of a female for the purpose of producing fertilization.

Balance—The harmonious relationship in cattle of all body parts, blended for symmetry and pleasing appearance.

Balling gun—Metal instrument inserted into the throat of an animal to discharge pills.

Barren—Sterile (refers to the female).

Black baldy—A cow or bull having a white face and black body.

Blocky—Of deep, wide and low-set conformation; used to describe a compact animal.

Bloom—An inclusive term used to describe the general appearance of a healthy, clean, and lustrous coat.

Brand—Permanent identification marking on animals made by scar tissue on skin, usually accomplished by hot iron or chemicals.

Branding—The act of marking with a brand.

Bred—Pregnant.

Breed—A group of animals that possess certain well-defined distinguishing characteristics and that reproduce these characteristics in their offspring with reasonable regularity.

Breed character—A combination of masculine or feminine qualities with ideal breed type features. Head and color markings are given considerable attention in estimating breed character.

Breed type—A particular group of traits typical of the breed, including distinguishing characteristics in head and color markings.

Brockle-faced—White-faced with other colors splotched about the head and face.

Bull—Uncastrated male of bovine species regardless of age.

Bulling—Term describing cow in heat.

Bullock—An English term for fat steer.

Calf—Young animal of cattle species, usually less than one year of age.

Cancer eye—Cancerous growth on or around the eyelid.

Carcass—The dressed body of a meat animal, with the usual items of offal removed.

Casting of the withers, or *eversion of the womb*—Process by which the womb turns inside out and protrudes from the vagina as a result of excessive straining during calving.

Castrate—To remove the testicles from a male animal.

Cattle—More than one animal of bovine species.

Chute—A narrow set of panels constructed for ease in handling cattle, often used for loading.

Clip—To shear the hair of an animal, especially around the head and udder.

Close breeding—Inbreeding, i.e., breeding sire to daughter, son to dam, or brother to full sister.

Cod—Steer's scrotum with its content of fat.

Colostrum—The thick viscous milk produced by the dam during the first week of lactation following parturition.

Concentrate—Feedstuffs such as grains, which are low in fiber and high in total digestible nutrients.

Condition—Degree of fat in meat animals.

Conformation—The shape and design of an animal.

Cow—Mature female of cattle species.

Cow-hocked—Refers to animal whose hocks turn in to each other and often rub together during walking.

Creep feeding—The process by which animals (usually calves) are given extra feed by means of small openings in panels that permit only smaller-sized animals to enter.

Crest—Back of neck, most pronounced on mature bulls.

Crops—The part of a beef animal just behind the upper half of the shoulders, extending from the top line to about halfway down the side.

Crossbreeding—Mating of purebreds of the same species but of different breeds.

Cryptorchid—Male bovine animal with undescended testes.

Cud—Regurgitated feed or bolus from stomach.

Cull—To take an animal out of a herd because it is below herd standards.

Currycomb—Comb made of several circular blades.

Cutability—Carcass cutout value, or yield of salable meat, sometimes designated as yield grade by USDA meat graders.

Cutting—Removal of testicles by castration; also, separation of one or more animals from a herd.

Dam—Mother of a calf.

Dehorn—To remove horns or horn buds so as to prevent any horn growth.

Declaw—Rudimentary toes, of which there are two at the rear and two above each pastern joint.

Dewlap—Loose skin found on brisket and neck of some cattle.

Disposition, or *temperament*—A tendency to act in a certain way.

Dropped—Born, as in *dropping* of a calf.

Dry cow—Cow that is not lactating.

Dry weight—Weight of cattle after water is withheld for 12 hours.

Dual-purpose—Bred and used for both beef and milk production.

Dwarf—In beef cattle, a small abnormal bovine animal, usually a calf, recognized by a short broad head, a bulging forehead, stunted growth, heavy breathing, and protruding lower jaw, resulting from a hereditary factor.

Dystocia—Difficult birth.

Ear notching—Making slits or perforations in an animal's ears for identification purposes.

Ear tagging—Act of placing a tag in an animal's ear for identification purposes.

Early maturity—1. Full development at an early age. 2. Quick achievement of market size and finishing. 3. Tendency to grow and fatten at the same time.

Estrus—The recurrent period of sexual excitement in mature cows, when the cow will accept sexual advances of the bull; period of heat.

Fecundity—Ability to produce eggs or sperm regularly.

Feeder cattle—Beef cattle, often steers, carrying enough age and flesh so that they are ready to be placed in the feedlot and fattened on grain.

Femininity—Possession of well-developed secondary female sex characteristics, such as refinement of head and neck and udder development.

Fertile—Able to reproduce regularly.

Fill—The amount of feed and water in an animal.

Finish—Fatness. *Highly finished* means very fat.

Fitting—The proper feeding, grooming, and handling of an animal to make it appear at its best.

Flab—Fleshing in beef cattle that lacks firmness.

Foot—Horny box and its contents excluding the remainder of leg.

Founder—A nutritional ailment resulting from overeating. Foundered animals become lame with sore front feet and excessive hoof growth.

Freemartin—A sterile heifer twin born with a bull. Occasionally, heifers twinned with bulls are not freemartins, that is, they are able to reproduce.

Full weight—Weight with no special preparation; weight as is.

Gene—One of the biologic units of heredity contained in the chromosome, each of which controls the inheritance of one or more characteristics.

Get—Calves sired by the same bull.

Gobby—Lumpy in fleshing; extremely poor distribution of fat.

Grade—An animal not eligible for registration though one of its parents is purebred.

Grooming—Brushing or combing of animals to exercise their skin and improve their appearance.

Grubs—Larvae of heel fly found under the skin on the backs of cattle.

Heifer—A female of the cattle species usually under three years of age. The term is usually applied to those that have not yet produced a calf.

Hiplock—A condition at parturition whereby the hips of the calf cannot get past the pelvis of the cow, causing the hips to "lock."

Horn weight—Weight attached to horn to shape it properly.

Hybrid—The offspring obtained by mating two different species.

Hybrid vigor—Increased vigor that accompanies matings that result in hybrids.

Impaction—Overfilling of one or more compartments of the stomach with dry feed.

Inbreeding—The mating of related animals. Most breeders use the term to refer only to the mating of closely related animals, such as sire-to-daughter or son-to-dam.

Lactation—Period in which an animal is producing milk.

Leggy—Having legs that are too long and too high from the ground.

Linebreeding—Inbreeding that results when the blood of an animal is introduced more than once, such as grandsire-to-granddaughter.

Long fed—Having a feeding period of 120 to 180 days or longer.

Long yearling—Cattle between 18 months and 2 years of age.

Longevity—Refers to the life span of an animal, especially to a long life span.

Lousy—Infested with lice.

Lumpy jaw—Enlarged jaw caused by a foreign object or infection.

Marbling—Distribution of fat in irregular streaks in lean meat, giving meat the appearance of marble.

Masculinity—Possession of well-developed secondary male sex characteristics in head, neck, shoulders, and crest.

Maturity—Age at which there is little subsequent increase in the size of an animal and animal is capable of reproducing or has produced its first offspring.

Maverick—An unbranded animal on the range.

Muley—An animal that is naturally hornless.

Natural fleshing—Lean meat or muscle.

Nicks (niches)—Results of certain matings that produce an animal of a high order of excellence although parents may be mediocre.

Nurse cow—Lactating cow that is used to nurse calves other than her own.

Off feed—Having stopped eating or eating very little. This condition occurs most often with fattening cattle.

Offal—All organs or tissues removed from inside of the carcass in slaughtering.

On the prod—Angry. The term usually refers to a cow cross enough to pursue humans.

Open—Not pregnant.

Outcross—To mate members of a herd that is more or less interrelated with cattle that are from some outside and unrelated source but from the same breed.

Overshot—Having the upper jaw protrude over the lower jaw.

Parturition—The birth process.

Patchy—Having lumps of exterior fat, which prevent the smooth finish desired.

Paunchy—Having too much belly.

Pedigree—A written statement giving the record of an animal's ancestry.

Performance test—Measure of individual performance—specifically, rate and efficiency of growth, and carcass traits.

Plain-headed—Having an undesirable type of head.

Poll—Top of head.

Polled—Naturally hornless.

Prepotency—The ability of an animal to transmit its own qualities to its offspring.

Prolapsed vagina—Inverted womb.

Prolific—The ability to reproduce regularly and numerously.

Purebred—Of pure breeding, eligible for registration in a breed association.

Quality—Fineness of texture and freedom from coarseness, which usually indicate superior breeding or merit.

Range—Area of land that produces natural feed that is harvested by beef cattle.

Rangy—Having too much length of body, associated with too much length of leg.

Ration—The amount of feed allowed an animal in one 24-hour period.

Registered—Having a pedigree that is recorded in the breed registry.

Replacement—Animals selected to keep in the breeding herd.

Reproduction—The process, sexual or asexual, by which plants or animals produce new individuals or offspring.

Roman nose—A convex facial profile.

Roughage—Coarse feeds that are relatively high in fiber and low in total digestible nutrients (e.g., hay).

Rugged—Big, strong, and sturdy.

Scale—Size.

Scotch comb—Straight or single-blade comb.

Scours—Diarrhea or loose running feces.

Scrub—An animal that is inferior in breeding and conformation.

Scurs—Small, imperfectly formed horns not attached to the skull.

Seed stock—Foundation animals for establishing a herd.

Self-feeding equipment—Equipment designed to supply feed to animals so that they can eat at will.

Served—Bred, but not guaranteed pregnant.

Shank—In the live meat animal, it is the part of the leg between the knee and ankle and between the hock and ankle.

Short fed—Having a feeding period 90 to 120 days or shorter.

Short two—Cattle just under two years of age.

Shrinkage, or *"shrink"*—Weight lost by cattle during transit or other handling processes.

Shy breeder—A female who has difficulty becoming pregnant.

Sickle-hocked—Having a bent hock joint so that animal is forced to stand with hind feet forward and under the body.

Sire—Paternal parent of a calf.

Spayed heifer—A heifer whose ovaries have been removed surgically.

Spotter bull—A bull that has been given a vasectomy and is used to detect females in heat.

Springing—Showing considerable evidence of approaching parturition.

Squeeze—Piece of equipment similar to a chute except that one side is hinged to permit more thorough restraint of the animal during operations such as branding.

Stag—A male that has been castrated after reaching breeding age, when secondary sex characteristics are more or less developed.

Steer—Male of cattle species that has been castrated at an early age.

Stifle—Joint of the hind leg between the femur and the tibia.

Stifled—Having an injured stifle joint. Such injury is common in bulls of breeding age; stifled bulls cannot properly breed females.

Stocker cattle— Cattle of any age or sex, but usually, younger, thinner animals, used for wintering on roughage with little or no grain.

Stocks—Equipment used in restraining cattle, generally consisting of a roller and sling designed to lift animal so that it cannot lie down or throw itself while minor operations are being performed.

Supplement—Feeds, such as many of the high protein feeds, that are used to balance or improve existing rations.

Tattoo—Permanent identification of animals produced by putting indelible ink under the skin; generally performed in the ears of animals while they are young.

Teaser—Animal used to seek out females that are in heat.

Thigh—The outside of the hind leg between the rump and the hock. Should be wide and building, carrying down as close to the hock as possible.

Type—The general desired form of an animal that adapts it for a particular purpose.

Undershot—Having the lower jaw protrude further than the upper jaw.

Vaccination, or *shot*—Introduction of antibiotics into animals to produce an immunity or tolerance to disease.

Vasectomy—Procedure in which some of the sperm cords of a male are cut or removed so that he is sterile yet will go through the act of breeding.

Wasty—Fatty. The term refers to a carcass that has too much fat, requiring excessive trimming; it may also refer to paunchy live animals.

Wean—To remove the calf from its dam so that it can no longer nurse.

Weaner—Calf that has just been weaned.

Yearling—Animal that is approximately one year old.

Index

Page numbers in *italics* indicate illustrations; those followed by t indicate tables.